THE ROLE OF LANGUAGE
IN PROBLEM SOLVING I

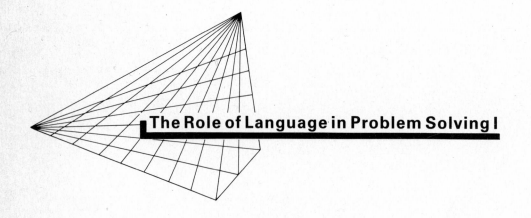

The Role of Language in Problem Solving I

Symposium logo by courtesy of:
Angelika Peck
TIR Group
The Johns Hopkins University
Applied Physics Laboratory

THE ROLE OF LANGUAGE
IN PROBLEM SOLVING I

Edited Proceedings of the Symposium held at
The Johns Hopkins University Applied Physics Laboratory
Laurel, Maryland, 29-31 October, 1984

Editors:

Robert JERNIGAN

Decision Resource Systems
Columbia, Maryland, U.S.A.

Bruce W. HAMILL

The Johns Hopkins University
Applied Physics Laboratory
Laurel, Maryland, U.S.A.

and

David M. WEINTRAUB

The Johns Hopkins University
Applied Physics Laboratory
Laurel, Maryland, U.S.A.

1985

NORTH-HOLLAND
AMSTERDAM · NEW YORK · OXFORD

ISBN: 0 444 87764 9

Published by:
ELSEVIER SCIENCE PUBLISHERS B.V.
P.O. Box 1991
1000 BZ Amsterdam
The Netherlands

Sole distributors for the U.S.A. and Canada:
ELSEVIER SCIENCE PUBLISHING COMPANY, INC.
52 Vanderbilt Avenue
New York, N.Y. 10017
U.S.A.

Printed in The Netherlands

PREFACE

This volume comprises the proceedings of The Johns Hopkins University
Applied Physics Laboratory Symposium on the Role of Language in Problem
Solving, which was held at the Laboratory's Kossiakoff Conference and
Education Center on October 29-31, 1984. The Symposium brought together
some 90 researchers and practitioners in several disciplines to consider
and discuss the influence that problem representation and language conventions have on problem-solving effectiveness and efficiency.

Our original intention was to have a symposium on high level languages.
The theme was to examine language design efforts and the effectiveness of
languages for problem solving (of course, with today's technology, "problem
solving languages" is almost synonymous with "computer languages"). As
planning progressed, the theme was broadened to the language issue itself,
that is, to an examination of fundamental principles underlying language
and its use as a problem-solving tool. We wanted a more general inquiry
into language: an inquiry into those philosophical, scientific, and social
attributes of both natural and artificial languages that assist us in
solving problems.

Some of those attributes take the form of notation systems and representational formalisms, others shape the problem-solving environment. The
approach we chose was to examine the language issues with the intent of
understanding the underlying mechanisms and how those mechanisms assist or
hinder us in problem solving. Papers discussing different problem areas,
programming philosophies, and programming languages were solicited and
selected for presentation and discussion with the intent of examining solution paradigms that are represented in the design of particular languages.

It is generally accepted that the use of the word "language" is appropriate
to describe the various notational conventions for instructing computers.
(It is interesting to note that Russell, Frege, and Iverson all developed
notational systems, but that only one is now referred to as a language,
perhaps because it was implemented on a computer.) This assumption is so
fundamental to the science of problem solving with computers that it is
rarely questioned. The acceptance of the notion of language as the proper
metaphor for the notational conventions of programming activity enmeshes
one in philosophical and scientific problems of language, problems that we
have tried to address in this Symposium.

When we speak of programming languages to support problem solving, we
assume a partitioning of languages at different "levels." Thus one language
can be at a "higher" level than another. "High" in this context has rather
loose-knit associations with ideas like "closer to the knowledge domain of
experts," or "subsumes many lower-level tasks," or "more closely related to
natural language."

One might expect the goal of high level language development to be to
enhance the problem-solving ability of its users. But when is a new language the proper solution to a problem? Is there an ideal or model that we
are striving for - perhaps "natural" language or formal logic - that is
somehow more effective or efficient for problem formulation and solution

than other available notations? And if so, what are the characteristics that make it so?

The process of defining general- and special-purpose languages continues, and there is a proliferation of languages for problem solving rather than a convergence toward some ideal language. The introduction of the computer into the problem-solving environment shatters the universality that we have in mathematical notational conventions.

The chronological sequence of the papers presented at the Symposium reflect a developmental progression from the philosophic issues to the practical aspects of the role of language in problem solving. The organization of this volume parallels the structure of the Symposium Program in order to maintain the continuity that developed through the proceedings:

Introductory Remarks present a brief philosophical view of notation systems for problem solving.

Philosophic Foundations: The Role of Representation in Problem Solving comprises the keynote address and a set of contributed papers focusing on theoretical issues related to how problems are formulated for solution.

How Language Can Affect Actions and Solutions consists of an invited address and contributed papers that concern relationships between problem formulations and approaches to programming language design, with a view to both theoretical and practical issues.

Remarks present an appeal for problem-solving languages to support space age technology.

Overcoming Limitations Imposed by Current Programming Languages comprises contributed papers and an invited address with an orientation toward application issues and a focus on the effectiveness of particular languages in solving problems.

Panel Session: Language Requirements for Effective and Efficient Problem Solving is an open discussion of several theoretically oriented issues in the design of languages, programming environments, and systems.

Panel Session: Comparative Application of Computer Languages to Practical Problems is an open discussion of practical issues concerning implementation of programming languages and systems and their implications for system effectiveness and user acceptance.

Summary Remarks reflect upon issues raised during the Symposium with a discussion and recommendations for future Symposia.

Contributed Papers not presented at the Symposium describe two recent advancements in data languages and architectures.

Banquet Address presents some challenging thoughts about the conceptualization and definition of future programming languages.

We thank The Johns Hopkins University Applied Physics Laboratory for financial and facilities support. We also acknowledge the early interest and continued support of George Weiffenbach and Robert Rich, the counsel of

William Guier, Daniel Brocklebank, and Jack Boudreaux during planning, the publicity efforts of Connie Finney, and the patient and skillful word processing and clerical efforts of Vicky Baker, Debra Coffroad, and Linda Sussman in preparing for and conducting the Symposium and in preparing this volume.

Finally, we express special thanks to Barbara Northrop, who served as Symposium Support Coordinator. The success of the Symposium event in October and the production of this proceedings volume are in no small measure the results of her diligence, patience, and skill.

Laurel, Maryland   Robert Jernigan
February 1985   Bruce W. Hamill
                                                     David M. Weintraub

TABLE OF CONTENTS

PROGRAM COMMITTEE AND REFEREES

PROGRAM COMMITTEE

Robert Jernigan, Chairman
Bruce Blum
Jack C. Boudreaux
Daniel Brocklebank
William H. Guier
Vincent G. Sigillito
Robert L. Stewart
Richard L. Waddell, Jr.

REFEREES

Sudhir K. Arora
Bruce Blum
Jack C. Boudreaux
Daniel Brocklebank
John Carlton-Foss
Ward Ebert
Andrew Goldfinger
William H. Guier
Bruce W. Hamill
Kathi Hogshead-Davis
Roy Rada
Vincent G. Sigillito
Robert L. Stewart
Ralph Wachter
Richard L. Waddell, Jr.
B. P. Weems
Nicholas Zvegintzov

INTRODUCTORY REMARKS

The Role of Language in Problem Solving I
R. Jernigan, B.W. Hamill, and D.M. Weintraub (Editors)
© Elsevier Science Publishers B.V. (North-Holland), 1985

INFLUENCE OF LANGUAGE ON ITS USER

Robert Rich

The Johns Hopkins University
Applied Physics Laboratory
Laurel, Maryland 20707

I am very pleased to see such a select group here (might better be that than more numerous, I guess). I was informed about various boundary conditions on my talk. Since this is actually "remarks", there is supposed to be very little that is original in it (lest I steal the thunder of those who come after me). Since, for that and other reasons, I am not going to say anything original, I thought I might (by and large) use the words as well as the ideas of the people who are providing the substance of this talk. So it is really going to be more an anthology than a set of remarks. The topic, of course, is The Role of Language in Problem Solving. The title of my remarks is "The Influence of Language on the User", because it is still true (and I suspect that it will remain that way for some time) that the most effective solvers of problems still are people, and the thrust (as I take it), of the meeting is to show that choice of a suitable notation has an effect on the way in which we approach and solve problems. In case there are any of you who do not believe that, I have a very simple example, as shown in the figure. Until about the 12th century people used Roman numerals for commercial bookkeeping and any other purposes that might occur to them, and in fact, the books of the Medici were still kept in Roman numerals as a matter (I think) of Law, as late as 1499. So, it took us 500 years in Europe to move from the top panel down to the panel in the lower left hand corner.

What I have done in the top panel is to show the multiplication of 47 x 53 using the method of duplation - not in the form in which the Romans, via the Greeks, originally got it from the Egyptians, but in a form that may be more familiar to the present audience. The top panel shows two columns of Roman numerals. The left column shows the results of successively halving the number 47, throwing away the remainders. The right column shows, above the line, the results of successively doubling 53. Each number in the right column is crossed out if the corresponding number in the left column is even; in this example only one such crossed-out number occurs. Below the line in the right column is the sum of the numbers above the line that did not get crossed out, namely the product 2,491 of the two numbers 47 and 53 we started out with.

All of the operations involved - division by two, doubling, addition and recognition of odd numbers - are easily performed on the abacus. But the whole process is a cumbersome one, especially when compared with the more familiar decimal form of the same computation, as shown in the lower left corner of the figure. Here, we are showing that for certain simple calculations (which are in fact doing the same thing, namely multiplying these two numbers), a choice of notation permits you to do away with the Roman abacus (which was a pretty good sized grooved rock; the ones that have been

MULTIPLICATION

■▪■ ▪ ■▪■ ▪ ■ ▪■▪ ■ ▪ ■▪■ ▪ ■▪■ ▪ ■ ▪■▪ ■

XLVII	LIII
XXIII	CVI
XI	CCXII
V	CDXXIV
II	~~DCCCXLVIII~~
I	MDCXCVI
	—————
	MMCDXCI

```
    53
  x 47
  -----
   371
   212
  -----
  2491
```

$$(X-Y)\ (X+Y) = X^2-Y^2$$
$$(50-3)\ (50+3) = 2500-9$$
$$= 2491$$

recovered weigh anywhere from 50 to 200 pounds), and the paper and pencil that is all that is required for the Arabic numerals was a great advance in portability (and portability, of course, as applied to personal computers, is still very much in the news). But we are still doing the same thing in both of those panels.

As we move over to the panel at the lower right, we are, in principle, doing the same thing, but we are taking a very different approach to our problem; namely, we are recognizing this as a useful special case of the general situation that the first two panels address. We have an algebraic formula which happens to fit the two numbers that I have picked and we get the same answer (2,491) in our heads. Now, I think that during the next two and a half days, we are going to be shown both of these levels and 3 or 4 still higher levels of ways in which language can help us solve our problems.

I would like to start out by saying that natural language and culture (human culture), are obviously very closely related, and human culture is obviously very related to thinking. So we have an influence from language to culture to thinking, and this influence (or at least a specific form of it) is known under the name of the "Whorf-Sapir Hypothesis": In Edward Sapir's words, "Human beings do not live in the objective world alone, nor alone in the world of social activity as ordinarily understood, but are very much at the mercy of the particular language which has become the medium of expression for their society... We see and hear and otherwise experience very largely as we do because the language habits of our community predispose certain choices of interpretation." His student, the insurance adjuster Benjamin Lee Whorf, also has his name attached to this hypothesis that we think the way we do because we speak in the way we do and, in his words, "It was found" (as a result of his study of the Hopi language), "that the background linguistic system (in other words, the grammar) of each language, is not merely a reproducing instrument for voicing ideas, but rather is itself the shaper of ideas, the program and guide for the individual's mental activity, for his analysis of impressions, for his synthesis of his mental stock in trade. Formulation of ideas is not an independent process, strictly rational in the old sense, but is part of a particular grammar, and differs, from slightly to greatly, between different grammars."

The Whorf hypothesis has received a great deal of discussion in the linguistics community, modulated partly by the fact that it is very difficult to demonstrate, partly by the fact that Whorf was in fact an insurance adjuster without formal training in linguistics, and, therefore, anything he said had to be seriously attacked as a matter of professional pride. Incidentally, we have some of that in the programming language community (I might point out); none of you have ever noticed it, I am sure. There is a real question about the way people think and the influence of language in it. Jacques Hadamard, in his delightful little book on the Psychology of Invention in the Mathematical Field devotes a major chapter (the VIth) to the question of whether people in fact do use language in their thinking or do not. I will not try to reproduce that chapter, but I do recommend that book and that chapter particularly to those who are interested in it. He finds a champion for each opinion. Max Muller, the etymologist, felt that there was no possible way for anybody to think except by using words; and Francis Galton, a rather an unfortunate choice, I think, the biostatistician, was the one who pointed out that a lot of people do think without words. I think that the resolution of this is very nicely put by Edward Wilson in the most recent issue of The American Scholar, "The symbols of art, music, and language freight power well beyond their outward and literal meanings. So each one also condenses large quantities of information. Just as mathematical equations allow us to move swiftly across large amounts of knowledge and spring into the unknown, the symbols of art gather human experience into novel forms in order to evoke a more intense perception in others. Human beings live - literally live, if life is equated with the mind - by symbols, particularly words, because the brain is constructed to process information almost exclusively in their terms." He is taking the Max Muller approach to the question, of course, and the other approaches you can very well find spelled out in Hadamard's little chapter. The whole question, I think, is put into perspective by Roman Jakobson, just as he has put so many other questions into proper perspective, "Signs are a necessary support of thought. For socialized thought (stage of communication), and for the thought which is being socialized (stage of formulation), the most usual system of signs is language properly called; but internal thought, especially when creative, willingly uses other systems of signs which are more flexible, less standardized than language and leave more liberty, more dynamism to creative thought." I think probably all of us will go along with that.

William Rowan Hamilton, I think, put it in a nice analogy: if we think of a person digging a tunnel in a sandbank, he finds that after he has dug about a foot, the ceiling starts caving in on him; he has to get out of the way and let a mason build arches (or somebody do some other form of shoring) so that he can come back into the tunnel and dig the next foot. So we have two different people doing very different things, the fellow with the shovel digging the sand out of the way, and the fellow with the trowel and mortar building up the arches behind him. If we think of the dynamism of creative thought as the fellow digging away at the problem and then the transformation and communication of that thought as the mason building the arches, we realize that there can be a single activity carried out by very different means and that the final result (namely a tunnel you can safely walk through in one case or a paper in a journal in the other) really requires both kinds of work. What seemed to be an unsolvable problem really turns out to be no problem at all. I refer once again to the sequence of panels in the figure. This is what we really mean, I think, by problem solving: to the extent that we can find that the problem we have to solve is, in fact, no problem, or a trivial one, we have taken advantage of our intelligence, and perhaps we can teach our machines to take advantage of their intelligence in the same way.

I am reminded of one of H. H. Monroe's little essays (writing as Saki) and I will not attempt to quote that because it is a little bit too long to read every word, but again, I refer you to the full text. He had written a book that had gotten some slight renown, and a friend of the family came and asked where she could get a copy. He pointed out that having recourse to an iron monger or a green grocer would entail delay and disappointment and suggested that she visit a bookshop. She met him at a private view a couple of weeks later and said "It is all right, I bought it from your aunt". This is another example of essentially working around a problem rather than actually addressing it, and I recommend to all of the intelligence people here, artifical and natural, that you look for ways of avoiding your problem before you spend a lot of time solving it.

Let me summarize this phase of my talk, namely the natural language influence on culture and therefore on people, by a little bit of doggerel. (I feel that any talk that does not use a little bit of doggerel is really going to the dogs.)

> The Irish had no word for "no",
> The Romans none for "yes",
> Which language best helped empire grow
> Is easy to assess.

I now move to the second part of my talk, which deals not with natural language and natural people, but programming languages and programmers. (That did not come out exactly the way I meant, but I guess it made the point clearly.) The way I would like to introduce the topic is by a far-out position that we can perhaps swing back toward the center from. This is by David Bolter in "Turing's Man", another recent book that I recommend to your attention, and careful and critical reading, which you do not have to do with Hadamard, of course. "The whole course of linguistic philosophy from Leibnitz to the positivists seems to culminate in the computer, where symbols are drained of connotations and given meanings solely by initial definition and by syntactic relations to other symbols." One of the reasons why I recommend that you read Bolter critically is exemplified here. Those of us who have become fluent in a programming language realize that although what he says is strictly true when we first open the manual, it rapidly becomes untrue as we become fluent in the language and begin thinking in it (as Whorf would say), because the symbols in it pick up their own connotations as well as their denotations in artificial languages just as they do in natural languages.

I think that a way of putting this, which I kind of like, is Richard Connors' statement, "A properly trained programmer thinks primarily in terms of programming, only secondarily in terms of a particular language." That really is going a little bit toward Galton's point of view of the Whorf hypothesis. We can move back to the other side by a sentence from Dijkstra's Discipline of Programming, "A most important, but also a most elusive, aspect of any tool is its influence on the habits of those who trained themselves in its use. If the tool is a programming language, this influence is - whether we like it or not - an influence on our thinking habits." This could be a straight quotation from Benjamin Lee Whorf and subject to the same arguments, pro and con, whether or not, on both sides. I am not going to say very much about specific languages, because there are a number of people here who would violently protest against any specific statement I made about a specific language. I'm quoting Richard Connor in his nice little essay, "Happy 25th Birthday, COBOL" in Computer World last April: "Although we think of COBOL as a language, this discussion" (namely his, not mine) "will treat it as something more - a mentality, if you will." Benjamin Lee Whorf again, you see.

I had one interesting example of the direct influence of culture rather than of language and this is again from Dijkstra. One of his problems, you remember, is to arrange a line of marbles in the order of the colors of the Dutch national flag, which is red, white, and blue, in that order like ours, and you are given a device which can pick up a marble and look at it and determine its color and then move it to one place or another, but it is very expensive to do this so you do not want to have to do it twice to the same marble. The solution to the problem is pretty straightforward: you take the groove in which the marbles are and you say I am going to put all the red ones to the left and then the white ones and then the blue ones, and so as you pick up each marble and determine its color, you have the problem of where in the groove to put it. When he asked his students, either Dutch or American, which pebble to inspect, their first suggestion was always "the leftmost one". He says: "I had the idea that this prefer-ence could be traced to our habit of reading from left to right. Later I encountered students that suggested first the rightmost one, one was an Israeli computing scientist, the other was of Syrian origin." So here we have a secondary influence of language, namely the direction in which you write it, which does have at least this modest effect on our thinking of various kinds of problems.

You will all be pleased to know that I am not going to get a chance to read all of my cards. Let me read one more, about why it is easy for people to continue to disagree on the effectiveness of particular languages. This one is from B. A. Sheil's Psychology of Programming article in the Computer Surveys, in March of 1981, and I wish I had time to read that whole article to you, because I expect it would resolve a number of the discussions we are going to hear in the next two and a half days. Let me read this much. "As practiced by computer science, the study of programming is an unholy mixture of mathematics, literary criticism and folklore. However, despite the stylistic variation, the claims that are made are all basically psychological; that is, that programming done in such and such a manner will be easier, faster, less prone to error or whatever.... Sadly, however, psychological data have been at best a minor factor in these debates." I think that with that very real understatement of the situation, I will terminate my talk. Thank you all.

REFERENCES

Edward Wilson: "The drive to discovery", American Scholar, Autumn 1984.
J. David Bolter: "Turing's Man: Western Culture in the Computer
     Age", University of North Carolina Press, Chapel Hill, 1984.
Jacques Hadamard: An Essay on the Psychology of Invention in the
     Mathematical Field, Princeton University Press, 1945.
B. A. Sheil: "The Psychological Study of Programming", Computing Surveys
     13, 1 (March 1981) 101-20.
Sherry Turkle: "The Second Self: Computers and the Human Spirit", Simon
     and Schuster, New York, 1984.
Benhamin Lee Whorf: "Language, Thought, and Reality", selected writings
     edited by John B. Carroll, MIT Press, 1956.
Richard L. Conner: Computerworld, September 10, 1984.
Edsger W. Dijkstra, "A Discipline of Programming", Prentice-Hall, Inc.
     Englewood Cliffs, N. J., 1976.

QUESTION AND ANSWER PERIOD

HAMILL
Would anybody care to address a question to Bob on the contents, put him on the defensive here for a change?

UNIDENTIFIED MALE
Do you have a favorite programming language?

R. RICH
Yes. Are there any other questions? Thank you, Bruce.

SESSION:
PHILOSOPHIC FOUNDATIONS:
THE ROLE OF REPRESENTATION IN
PROBLEM SOLVING

The Role of Language in Problem Solving I
R. Jernigan, B.W. Hamill, and D.M. Weintraub (Editors)
© Elsevier Science Publishers B.V. (North-Holland), 1985

KEYNOTE ADDRESS:

PROBLEMS OF REPRESENTATION IN HEURISTIC PROBLEM SOLVING

Saul Amarel

Department of Computer Science
Hill Center, Busch Campus
Rutgers University
New Brunswick, New Jersey 08903

Good morning. I am very, very pleased to speak after Dr. Rich. His was a most interesting, scholarly, and delightful presentation, and it sets the background for some of the things that I would like to say. The concern with representations and their effect on thinking about the world and on solving problems in the world has been with us for a very long time. As illustrated by Dr. Rich, problems of mathematical notation, arithmetic notation, algebraic notation, have been with us for centuries. In the sciences there has been a tremendous amount of thinking, by those who are conceptualizers, about means of representing theories. So, the issue of representation has been around for a very long time. Much of the historical background on representation has been presented to us by Dr. Rich. As a matter of fact, I think that Dr. Rich has struck the keynote for our discussion, and this gives me a little more time to get involved in more specific things, kinds of things that a keynote speaker is not supposed to be discussing.

The things I would like to talk about have to do with studies of representation choices within artificial intelligence. AI is a part of computer science, and I am sure that most of you are familiar with the work being done in that field. It has been with us since the mid-'50's, and many of the main concerns of the discipline have to do with ways of using knowledge in order to attain desired ends. So, one of the very important basic problems in artificial intelligence is how to transform knowledge into effective problem-solving action. In many ways we have to confront constantly the problem, "What knowledge, in what form?" The issues of the form of knowledge, and the form of the problem in which it can be used, have been with us from the beginning of the discipline. In what way do we formulate, or should we formulate, a problem for a problem-solving schema so that the schema is going to handle the solution-finding activity in a way which is effective and efficient? In AI we are interested in efficiency of problem solving, not only in effectiveness or completeness that are the domains of the logician and the mathematician - we are interested in those too - but we are mainly interested in the problem of efficiency of problem solving.

From the beginning of work in the discipline we recognized the centrality of the problem of representation in AI. The approaches to the problem have been mainly via probing into specific examples, some of them very simple. The main purpose of these probes was to allow us to think carefully, and concretely, about what is involved in choices of representation and what is involved in shifting from one representation into another. This is a style of work which is characteristic of AI research, where focus on special cases dominates over abstract, general approaches to problems.

In retrospect, looking back at the work done in this area over the past twenty years, I must say that the problems involved in the area are extremely difficult, much more difficult than we thought originally. They include many key problems in AI - that we will have to understand first before we can deal successfully with choices of problem representation and with shifts between problem formulations.

The entire issue of choosing a point of view for solving a problem, and changing a point of view, is usually associated with creativity. Perhaps our ability to have machines that perform tasks that could be considered creative depends on a better understanding of the entire question of shifting points of view and choosing appropriate points of view, where the notion of appropriateness that I am thinking about has to do with tailoring a body of knowledge, a formulation, in such a way that problem solving within that formulation is most efficient. Fundamentally, we are interested in understanding the relationship between various formulations of a problem and the effectiveness with which a computer can handle the process of finding a solution for that problem. This is a basic scientific question. A related, but less general, question is the following: Given a specific problem, how do we choose an appropriate representation for the problem? This also is an extremely difficult question. A third question, which I consider more approachable at present (the one on which some of us have been working), is the following: Suppose that you do have a formulation for a class of problems, and some experience in using that formulation in solving a problems in the class; how can I shift that particular formulation into a new one in such a way that the new formulation is going to be more powerful, more appropriate, so that problems in that class can be solved more efficiently? So, at present most of the thinking in this area is focusing on reformulations, and on processes of shifts between formulations. If we know more about these issues, we will be able to handle better the more general problems of choice of formulation and of relationships between formulations and computational complexity.

In the late '50's, early '60's, the main paradigms of problem solving in AI were already set. We had quite a bit of work on heuristic problem solving at that time. The idea was that if a problem is formulated in an appropriate way to a search procedure, the procedure can take over, and given the problem formulation and some judgmental knowledge in the particular domain of the problem, the procedure can search through a space of solutions in a way which is more or less efficient - it is not exploding exponentially - for a solution to the problem. But just around the early '60's - in '63 - John McCarthy from Stanford and then Allen Newell from Carnegie-Mellon brought up a few concerns that had to do with the following interesting question. We know how to approach the solution-finding task given a formulation of the problem; but where does the formulation come from? How do we go from a verbal formulation of a problem or from a sketch of an initial symbolic formulation to one that could be accepted and assimilated by a given problem-solving schema? This was posed as a problem. This was the opening of the entire issue of how to choose problem representations and how to adapt and tailor them to various problem solving methods.

There was a nice little problem at the time that illustrated the kind of difficulty that we had. (See Figure 1). I am sure that many of you are familiar with this little puzzle - it was called a tough nut - the mutilated array problem. We have an 8 x 8 array which is mutilated in the two corners. We also have dominoes of length two. The problem is: can we

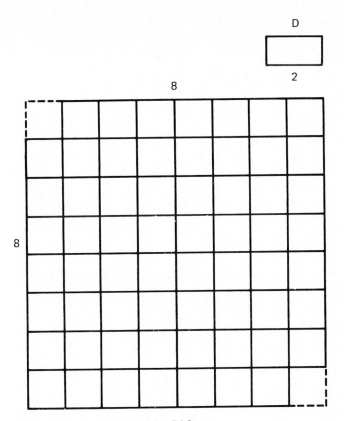

Can the array be covered by D's?

FIGURE 1
MUTILATED ARRAY PROBLEM

completely cover this array by dominoes of length two? Now, one way of approaching this problem is by heuristic search, by having a disciplined search of all the possible covers of the mutilated array; either we do find a cover and the answer is yes, or we exhaust all the possible covers without finding an acceptable cover, and the answer is no. This was the technique that could be implemented with available methods of heuristic problem solving. Now, of course, the way in which this problem could be approached in a much more elegant way involves imposing more structure on the problem itself and reasoning in the following way (see Figure 2). Impose a coloring pattern of black and white in alternate squares of the array, and make out of that array something like a checkerboard; then focus on the process of placing a domino anyplace on that array. Every time a domino comes anyplace on the array, a black and a white element are covered. Then carry out the following argument. A good way of describing any stage of coloring of the array is to give counts of uncovered white cells and of uncovered black cells. This would be a description of the state of the problem. Initially we have 30 white and 32 black cells because two of the white cells have been taken away by the mutilation.

S. Amarel

Every time we place a domino on the array we reduce each of the counts by
one. After placing the first domino we get counts of 29 and 31. After
several subsequent domino placements, we get to a point where we have no
white cells and two black cells. But two black cells cannot be covered by
a domino because a domino can only cover alternate black and white, and
this refutes the possibility of having any cover. This is a very broad
argument, an argument that does not go only into the specifics of any
particular cover, but that handles all the possible covers.

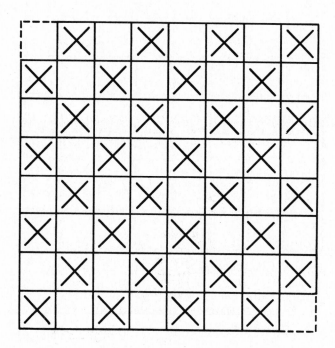

W: Number of white elements not covered
B: Number of black elements not covered

W	B	
30	32	• Initial
29	31	• 1 D covers
28	30	• 2 D covers
.	.	
.	.	
0	2	• ?

FIGURE 2

REASONING ABOUT MUTILATED ARRAY PROBLEM WITHIN
'APPROPRIATE' REPRESENTATION (POINT OF VIEW)

Now what goes on here? Is it possible to capture this kind of reasoning by
a computer, by an artificial intelligence program? More specifically, is
it possible to choose the state description which is essential for the
argument; i.e., to impose on the problem state a structure in the form of
an alternation of black and white cells, and to decide that the description
of the state of affairs should be given not in terms of the coordinates of
particular points on the board, but in terms of numbers of elements of two
specially chosen sets that are derived from the imposed structure? That
particular mode of reasoning is a difficult one. How is it captured?
There is another interesting point here. The choice of state description
is interesting and elegant because it is possible to establish an elegant
argument by using it. So it is a back-and-forth, chicken-and-egg proposition. There is nothing fundamental in the choice of the structure of the
state that is good in the absolute. It is only good in terms of - conditional upon - the successful use in a particular method of reasoning. So
the question is, how do you go about making these choices? Either you have
methods of reasoning in your repertoire, and on the basis of them you look
for certain modes of structuring the state of affairs, or you have certain
ways of looking at the world, such as checkerboards (these are the various
structures in terms of which we perceive the world, and we may have large
numbers of those sitting around), and then you find the best method of
using available descriptions of the world in order find a solution with
relative ease. This is still a difficult question to answer. It could go
one way or the other, or it could be some combination of approaches.

The mutilated array problem, and the issues that it raised, marked the
beginning of concern with problems of choice of representation in AI.
Twenty years after, we still don't have programs that can handle this
problem in a way that captures the elegant approach. As a matter of fact,
I do not think that there are solutions to a related, simpler, problem -
which I tried for some time. The problem is as follows. Suppose that you
know how to handle the mutilated array problem with dominoes of length two,
and you want to solve a similar problem involving dominoes of length three
and some given mutilation of the array; can you use concepts of state
structuring and modes of reasoning that are similar to (transferred from)
the original problem? If you try to transfer the pattern of reasoning from
the 'two-case' to the 'three-case', you will find that it is quite difficult. It takes quite a bit of creativity and thinking by people to carry
out the transition. In general, the ability to transfer pattern and style
of problem solving from one problem to another, similar, problem is far
from trivial.

My own approach to the study of problems of representation, or formulation,
in problem solving has been to look in detail at different kinds of problem
classes and to see how issues of representation and processes of representational shifts appear in them. I have been focusing in particular on
issues of reformulation, and on mechanisms of knowledge acquisition and use
that permit us to move to problem formulations of increased power.

Back in the '60's I looked at theorem proving in the propositional
calculus, at problems of reasoning about actions in some very simple kinds
of transportation tasks (Missionaries and Cannibals), at problems of
parsing, at problems of theory formation, and more recently at another
problem of transportation or scheduling, the Tower of Hanoi problem. What
I would like to do next is to give some examples from problems that I
studied in the area of reasoning about actions, in particular the

Missionaries and Cannibals problem and the Tower of Hanoi problem. My
purpose is to show through these examples some aspects of the key issues
involved.

Before proceeding further, let me mention the names of a few other
researchers who have been working in this area: Newell and Simon in the
'60's (they continue to work to some extent on these problems); Rich Korf
from CMU, now at Columbia, has done quite a bit of work on this in recent
years; and George Ernst, who is now at Case University, has been doing work
in the area together with several of his students.

Let me give you a brief outline of the study that I made several years ago
on the Missionaries and Cannibals problem. (See Figure 3). The problem is
quite well known; it is a puzzle that involves a river, a distribution of
missionaries and cannibals on the two sides of the river, and a boat.
Initially, there are three missionaries and three cannibals on one side of
the river, say the left side, the other side of the river is empty, and
there is a boat with a capacity of two on the left side. The task is to
transfer the entire population of missionaries and cannibals from one side
to the other in such a way that in none of the stages of this process is
there going to be in any one place a number of missionaries which is
smaller than the number of cannibals, for obvious reasons. Furthermore,
the transfer has to be carried out with the least number of trips across
the river. This problem is a little puzzle that has been studied within AI
for some time as a vehicle for developing and trying ideas in the area of
finding plans of action that satisfy a stated goal subject to a variety of
constraints, including a minimality constraint on the solution. Certainly,
the problem can be approached by techniques of heuristic problem solving.
I have a 1969 paper on this in 'Machine Intelligence III'. The basic
emphasis in that paper is on the various possible representations of this
particular problem. The key initial concern is how to formulate the
problem for a heuristic problem solver. Here, the fundamental question is
how to represent situations, or states of the world (i.e., snapshots at one
point in time of the world under consideration), and how to represent
actions in the form of transitions between states of the world. Once we
have these representations, then the problem is to try to find a good way
of moving from some initial state of the world to a desired terminal state
through a sequence of feasible, legal actions. This can be done by using
one of the existing heuristic problem solving procedures. But it is
essential for this entire approach to come up with a good representation of
situations, or world states.

Now, I can imagine a first stage of handling our problem by considering it
as a special case of a more general class for which we have modes of representation and solution procedures. Let us assume, in particular, that the
problem is recognized as a transportation problem. To represent states of
the world in these problems, we define objects of various types, locations,
transportation facilities and their properties, and relations between
objects and locations. States and actions can be described in terms of
this sort of entities. This would lead in our case to state descriptions
where the location of each individual missionary and cannibal and of the
boat are specified. However, if you think a little bit about our specific
problem, you see that there are redundancies in the initial state descriptions that can be eliminated. If you know the situation on one side of the
river, you can infer the situation at the other side, because the total
number of missionaries and cannibals is conserved in the course of the

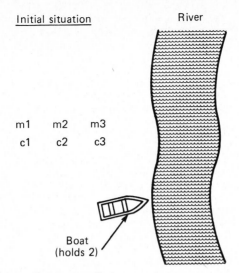

Initial situation   River

m1   m2   m3
c1   c2   c3

Boat
(holds 2)

- Reasoning about actions
- Problem representations
    Situation: (MCB)

- Top part of search tree

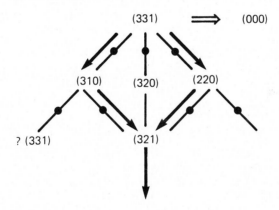

FIGURE 3
MISSIONARIES AND CANNIBALS PROBLEM (M&C)

transfer process. Furthermore, by a certain amount of reasoning, you can
deduce that you don't have to worry about the situation in the river (while
the boat is crossing) if the situations are legal and safe on the two river
banks. In view of these properties of our problem, we can have a complete
description of the state of the world by focusing explicitly only on one
side of the river, say the left side.

We can then start thinking about what is 'really relevant' in state
descriptions. Given the constraints of the problem, it is clear that the

location of individuals is irrelevant. What matters is the size of groups
of individuals: All you want to know is what is the size of the group of
missionaries and the group of cannibals on a river side. This is a very
important transformation in conceptualizing a state - to go from individuals to sets of certain kinds of individuals, and then to numbers of
individuals in these sets. We have seen the same kind of transformation in
the mutilated array problem. The transformation is suggested here by the
constraints of the problem, which are specified in terms of the relative
size of the groups of missionaries and cannibals in any given location.
So, what drives the choice of state description is the requirement to be
able to reason about applicability of actions in the course of the problem
solving process. More generally, the state representation is induced by
the needs of the problem solving scheme; i.e., by the method of reasoning.
Furthermore, the representation is specialized/adapted to the needs of the
specific problem under consideration - even though the initial approach
started with an assimilation of the problem within a broader class.

A situation, or a state of the world, can be represented then as a vector
(M, C, B), where M is the number of missionaries at left in that situation,
C is the number of cannibals at left, and B denotes the place of the boat
(this is a binary variable whose value is 1 if the boat is at left and 0 if
it is at right). An action can be described as a transition between two
situations. Each action describes the transfer of one or two individuals
from one side of the river to the other - because the capacity of the boat
is two, and at least one person is needed to handle the boat transfer.
Using these notions of situations and actions, we can use known techniques
of heuristic search to find a solution in the form of a minimal sequence of
transitions that satisfy the requirements of the problem.

Now, a second stage in problem formulation has to do with recognizing
regularities in the solution itself and in the space of search - in the
space of all possible solutions - and trying to glean out of those
regularities some knowledge that would be useful for a much more effective
way of solving problems in that particular class. A solution can be seen
as a sequence of situations starting with $(3, 3, 0)$ and ending with $(0, 0,$
$0)$ (see Figure 4). There are 12 situations in the minimal solution to the
problem (alternatively, there are 11 transitions, or transfers from one
river-side to another).

By examining the solution, you will find that it has certain interesting
characteristics. One of them is symmetry. If you run the sequence back in
time (from terminal situation to initial) you will find the same structural
features as the regular solution running in normal time. Finding a
symmetry condition is very important for the improvement of a problem
solving process. In our case, it cuts down in half the depth of search.
Once you develop a way of moving forward from the initial situation, you
can simultaneously specify a 'mirror image' sequence that moves backward
from the desired terminal situation; then you try to establish links
between the forward and backward moving sequences - this happens at the
middle of the solution. Finding this type of symmetry in a solution is an
approachable task at present - but by no means easy. Again, identifying
the symmetry, and finding a good way of using it, is very important; it
changes our way of looking at the problem, and it increases the efficiency
with which we can solve it.

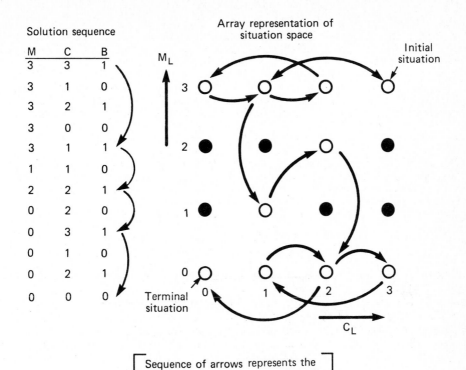

Solution sequence

M	C	B
3	3	1
3	1	0
3	2	1
3	0	0
3	1	1
1	1	0
2	2	1
0	2	0
0	3	1
0	1	0
0	2	1
0	0	0

Array representation of situation space

Initial situation

Terminal situation

[Sequence of arrows represents the solution in array representation of situation space.]

FIGURE 4
SOLUTION REPRESENTATIONS OF M&C PROBLEM – AS A
SEQUENCE AND IN SITUATION SPACE

Another interesting approach, the one that I consider most promising, is to find a way of representing the entire situation space so that it will allow us to find useful regularities about how the problem is solved. This can be done in our case by representing all the possible situations on the left side of the river as a two-dimensional array with four rows and four columns. (See Figure 4.) If M_L and C_L denote respectively the sizes of the group of missionaries and the group of cannibals at left, then the lowest row stands for $M_L=0$, the next higher row for $M_L=1$, and the upper row for $M_L=3$; and the columns from left to right stand for $C_L=0$ to $C_L=3$. Thus, the initial situation is represented by the top right point in the array, where $M_L=3$, $C_L=3$, and the boat is at left. Some of the possible situations in the array are forbidden, because they violate the condition that the group of missionaries should not be smaller than the group of cannibals at left or at right (they are marked as black points in Figure 4). The only legal regions of the space of situations are those defined by the conditions $M_L=3$, $M_L=0$, and $M_L=C_L$. These regions form a Z in the array of situations, with the top of the Z corresponding to $M_L=3$, the bottom to $M_L=0$ and the diagonal to $M_L=C_L$. The transitions between situations can be seen as moves between legal points in the situation array. For example, a boat transfer from left to right can be seen as a move in the array from some point (situation) to another point one or two lattice steps away to the

left and/or down in the array. You can immediately see that some regions
of the array are 'easily traversable'; the top and bottom of the Z are such
regions. The solution can be seen then as starting at the top-right corner
of the Z; moving to the left over the easily traversable region to a point
from which a 'stable jump' can be made down on the diagonal of the Z; and
then jumping down from the diagonal to the bottom of the Z and moving to
the left-bottom corner through easily traversable territory. From the
pattern of this solution, we can see that there are 'critical points', or
'narrows', in the situation space; these are the points on the top region
of the Z and on the diagonal from which jumps to lower regions take place.
The solution path must go through these 'narrows'; so, it is extremely
useful to be able to identify them. Once we have identified them, and we
have recognized the easily traversable regions, we can develop a global
plan of solution that leads to the creation of 'macromoves'. A macromove
embodies an entire routine of transitions that allows us to jump in situation space without thinking about choices at intermediate points. The
development of such macros is very important for increasing problem solving
power. In the present problem, several macromoves can be defined; one for
moving in the easily traversable region at the top of the Z (where $M_L=3$) up
to the critical point (3, 1, 1) from which a jump to the diagonal is
possible; a second for jumping to the diagonal and moving to the critical
point (2, 2, 1) on the diagonal; a third for jumping from that point to the
bottom of the Z (where $M_L=0$) and staying there; and a fourth for moving in
the easily traversable region of the bottom of the Z to the desired
terminal point (0, 0, 0). Note that the development of the macros is based
on the discovery of certain useful characteristics in situation space, and
this is made possible by an 'appropriate' representation of the space. One
of the important preconditions for the representation of the situation
space is the formulation of a good concept of situation where careful
choices must be made to include only relevant information that is necessary
for decisions in the problem solving process under consideration.

To summarize: Here are the kinds of AI problems that were already identified in the sixties as essential for an understanding of problem representation issues. First is the problem of choosing a good representation for
situations, or states of the world. This requires reasoning capabilities
for finding what is essential, relevant and irredundant in state descriptions - through analysis of problem constraints and applicability conditions for moves. Then, we must choose an appropriate representation for
situation space, where appropriateness is judged in terms of the ease with
which it allows us to recognize properties such as symmetries, easily
traversable regions, and critical points. This choice, I think, is very
difficult because it involves trying to find a conceptual framework within
which certain property-finding activities, or theory-formation activities,
are possible. In a sense, it is pretty much like saying I would like to
have a useful theory about goings-on in problem space, but in order to
obtain the theory I would like first to invent a framework in which I can
express the theory. This is hard, but this is also pretty much the kind of
thing that a good theorist is expected to do. A third problem is how to
form macromoves. After you have identified certain regularities in problem
space, you can use the knowledge about these regularities to synthesize
routines of action, or macromoves, that will be more powerful than individual elementary actions. At present, this is an approachable problem; and
it is receiving increasing attention recently.

During the '70's many of us got interested in applications of AI, and basic work on problem representations was sort of left on the back burner. Much effort was devoted to projects that led to the present enormous activity in the area of expert systems. Now, in turn, some of our concerns with the development of expert systems are reviving interest in basic issues of problem formulation and reformulation.

High performance in a specific problem domain is the primary characteristic of expert systems. To obtain high performance, the problem must be formulated in a very appropriate way, and knowledge about the problem must be represented in a way which is well-tailored to the capabilities of the problem solving system. One of the important questions that face us is how does one attain expertise, or how does one develop capabilities for expert behavior starting from novice, or less-expert, behavior? How do you go from a stage in which your problem solving performance is relatively weak to a stage of improved problem solving performance, especially when the improvement is due to changes in problem representation. Here we are back to the basic problems of how to understand and mechanize processes of problem reformulation. I believe that many of the things that experts do in moving from less-expert behavior to more-expert behavior involve rethinking, or reformulating a problem, re-viewing knowledge in a domain and seeking new knowledge that can be used to advantage in handling specific problems in the domain.

In addition to high performance in parts of a domain, an intelligent expert must have also characteristics of robustness and of self-improvement. Usually, human experts who are very good in a particular domain have also the robustness of behavior that permits them to handle, perhaps less effi- ciently, tasks that are somehow removed from their 'domain of specialty'. They have ways of combining the specialized, powerful, techniques used in their domain with a more general way of looking at a broader domain so that if a problem falls outside their specialty, they can still manage to handle it somehow. In addition, human experts have developmental capabilities, that is, the ability of increasing their expertise in their specialty area and/or of building a new peak of expertise in some different area. These developmental capabilities are closely linked to issues of problem reformu- lation. I believe that the problem of reformulation is a significant part of the problem of expertise acquisition - an area of considerable interest for the development of future expert systems with increased capabilities.

A simple graphic way of illustrating my comments on expert systems is pre- sented in Figure 5. Let us consider a graph of performance versus domain size. A contemporary expert system may be represented by the top graph, where peaks of performance may be available in a few specific subdomains, but the performance is zero outside these subdomains. What we would like to attain is a system with a broader domain of operation, where performance may be relatively low (but not zero) outside the few 'specialty peaks'. This is shown in the middle graph. In addition, we would like to have a system with improvement capabilities that can lead to a stage, as shown in the bottom graph, where some of the existing expertise peaks are strength- ened, and new peaks are developed. In order to go from a system represent- ed by the top graph to a more powerful expert system, as represented by the bottom graph, we need basic advances in the area of problem reformulations.

Now, to make progress in this area we must be very clear about notions of problem, problem class, problem solving schema, and about the concept of

S. Amarel

Performance

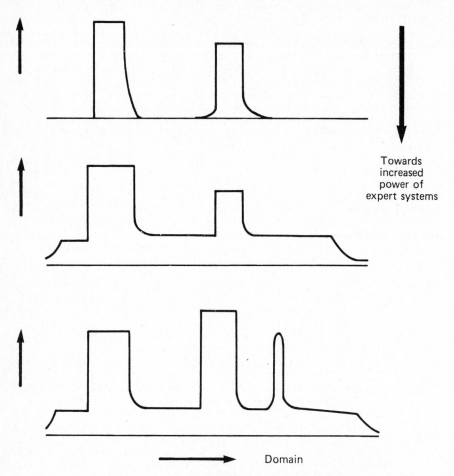

FIGURE 5
PERFORMANCE VERSUS DOMAIN SIZE IN EXPERT SYSTEMS

problem formulation. When we think of transitions between formulations, it
is important to be fairly specific about the objects between which these
transitions take place. I view the notion of a problem statement as
follows: Given a certain domain specification, find a solution such that
it is a member of a given set of legal solutions, and it satisfies certain
given problem conditions; and furthermore, the specification of the set of
legal solutions and the specification of the problem conditions are defined
in terms of the system of concepts given in the domain specification.

So, fundamentally, in order to be able to define or formulate a problem, it
is important to give to the problem solving agent - man or computer - a
specification of a particular domain, and, in addition, knowledge about

what constitute feasible or legal solution objects. Those two things are
very closely related. Sometimes both of them together are conceived as the
knowledge base of the system, but I like to separate them; I find the
separation useful for studies of reformulation.

Another important thing that the problem solving agent needs to know is the
specification of problem conditions. From the multitude of legal solutions
in a domain, the agent must find one or more that satisfy the problem
conditions. These conditions can be in the form of boundary conditions in
a formal problem; they are the initial and terminal situations in a path-
finding problem or in a problem of reasoning about actions; they can stand
for a set of data about a phenomenon for which we would like to build a
theory. The problem conditions are specific things that permit us to test
a candidate solution, and to guide us in the process of finding a solution.

I find it useful for the study of problem formulations to differentiate
between different kinds of problems, and to see them as points in some sort
of a spectrum. I don't think that a theory of problem solving can make
much progress without a recognition of distinctions between problem types.
I am basing the classification of problems on the different ways in which a
solution is structured during problem solving. At opposite ends of my
spectrum of problems I see derivation problems and formulation problems.
Theorem proving and path finding are derivation problems. In theorem
proving you are given a theorem to prove and some initial propositions -
say the axioms - and you try to establish a 'bridge' between the initial
propositions and the theorem. The 'bridge', built by a chain of rules of
inference that constitutes the proof, is the desired solution. The theorem
to prove and the initial propositions specify the problem conditions.
Problem solving in this case is guided quite strongly by the conditions of
the problem. We know much more about problem solving of this kind than
about problem solving in formation problems, such as theory formation and
concept formation. In formation problems, you are given certain problem
conditions, say data about a phenomenon for which you are asked to build a
theory (the desired solution) within a given system of concepts, and you
cannot very well infer the structure of the theory by reasoning from the
data, but in many cases you must generate hypotheses within the given
system of concepts and test them against the data to see if they satisfy
the requirements of the theory. Much of the reasoning in this case goes
from possible solution structures to problem conditions, as opposed to the
case of derivation problems where the bulk of reasoning goes in the
opposite direction. In formation problems, the approach to constructing a
solution is more difficult and less efficient than in derivation problems.

An important type of problem reformulation involves transforming a formation problem into a derivation problem - by finding ways of using problem
conditions to elicit hypotheses rather than to test them. The more you can
move in the direction of using problem data for the elicitation of
hypotheses, or of parts of the desired solution, the more you strengthen
the efficiency of your problem solving activity. There are several
examples of problem reformulation of this kind; they include the DENDRAL
program at Stanford for interpretation of mass spectrograms, and several
programs for medical diagnosis. In all these cases, an improvement in
performance was obtained through developments that brought about an
increased <u>a priori</u> use of knowledge in the structuring of a solution
relative to its use in an <u>a posteriori</u> way for testing a proposed solution.

Now, I would like to get into an example of problem reformulations in a
fairly simple class of derivation problems that I studied recently. The
main goal of this study was to look into the issues that come up in the
process of transitions between formulations. The problems that I explored
are the well known Tower of Hanoi problems. (See Figure 6.) We have three
pegs and we have a distribution of disks over the three pegs, and the problem is to find a sequence of actions that can take us from a given initial
configuration of disks in pegs to another, terminal, configuration. You can
move one disk at a time; you can never take a disk out of a particular peg
if there are other disks in the peg that are smaller; and you can never put
a disk in another peg if in that peg there is another disk which is smaller
than itself. The problem is to find the minimal sequence of moves that can
take you from the initial configuration (situation) to the terminal
situation.

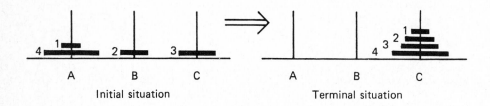

A   B   C   A   B   C
Initial situation   Terminal situation

FIGURE 6
A 4-DISK TOWER OF HANOI PROBLEM

This is the kind of problem that could be considered as reasoning about
actions, or it could be considered as a transportation problem. It could
be assimilated into a general category of problems, such a Missionaries and
Cannibals problem, where you have locations, objects, an assignment of
objects to locations, and constraints on allowable actions. You can
develop a heuristic procedure of the conventional kind that can search for
a trajectory, a sequence of actions, that can take you from the initial
situation to the terminal situation. Now, what are the kinds of things
that you must specify in formulating this simple problem? You have to
define the domain. You have to introduce a domain language for talking
about locations, objects and sets of objects. You have to introduce
predicates in the language: the notions of 'smaller', 'at', and 'in'. All
of these are fairly general things that you might have available in a
system that is familiar with transportation problems. In addition, you
have to introduce a very specific predicate that comes with the statement
of the Tower of Hanoi task: the notion that an object x is moveable at
location 1. This predicate is defined as: there is no object x1 such that
x1 is also at 1 and x1 is smaller than x.

Now you have to specify the notion of situations. One of the ways of
defining situations is to provide the coordinates of each object in the
system, like disk 1 is at 11, disk 2 is at 12 and so on. This is an
object-based description of situations. There is another way of describing
situations. This is a location-based description, in which you specify for
each of the locations the objects that it contains; for example, in location 11 you have the set u1 of objects, in location 12 you have some other
set, and in location 13 you have a third set. So, you have a choice: you

may specify an object-based description or a location-based description. This is an old kind of problem that people have to face in many physics problems. In some situations, for example in hydrodynamics, you find that it is better to think in terms of location-based descriptions rather than object-based descriptions. It depends on the kind of problem that you have in mind. Within the same domain, there may be problems - types of tasks - for which object-based descriptions of situations are preferable, and other problems where location-based descriptions lead to easier approaches to solution.

In addition to the notion of situation, you must specify the notion of action. Actions are to be represented in some way. The representation must show the effect of an action, and the conditions under which the action is applicable from a situation. Consider, for example, the action of transferring the object x from location l_1 to location l_2. The effect of the action can be specified as a pair of situations, the 'before' situation and the 'after' situation, where x is at l_1 in the 'before' situation and it is at l_2 in the 'after' situation. The applicability conditions for the action are that the object x should be moveable at l_1 and it should be also moveable at l_2. We are assuming that the specification of available actions comes to us as part of the domain specification - together with the specification of situations and the definitions of predicates, objects, and locations. On the basis of all this information, we also have an implicit specification of the space of situations, that is of the set of all possible situations and of all possible transitions between situations (these are the possible actions).

Now some of you must be familiar with the situation space (shown in Figure 7), or what is called the state space, of the Tower of Hanoi problem. Each one of the nodes in the triangular graph representation of this space represents a situation, and each one of the branches represents a transition between situations, or an action. This is an interesting representation that is not given to you explicitly with the initial domain specification; it is given to you implicitly, because you are given definitions of situations and definitions of transitions between situations, and you may actually develop the explicit graph representation on basis of these definitions.

A solution is a trajectory in situation space and it can be seen as a path between two nodes in the graph representation of that space. So what we have up to this point is a definition of solutions as trajectories in situation space, and the notion of situation space together with all its underlying concepts. I expect that it is not too difficult to come up with these elements of the problem formulation assuming that we have an initial way of handling, in general, transportation problems and problems of reasoning about actions, and the ability to assimilate a particular problem into the general framework. This is something that I think is doable today.

Fundamentally the definition of an initial formulation of the problem involves the specification of a subject matter domain, with a language for describing things in the domain, and with a specification of a space of situations. The notion of the set of possible solutions can be seen as theset of all the possible trajectories in situation space. The notion of problem class conditions, in this particular case (we are talking about the class of problems, not about a specific problem in the class), can be seen

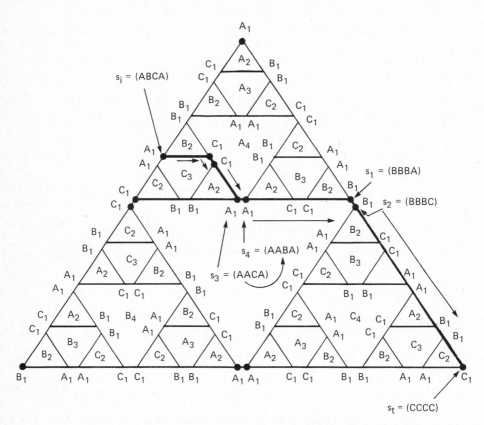

The darkened lines and the arrows show the solution trajectory for
the specific problem which is shown in the previous figure.

FIGURE 7
SITUATION SPACE OF 4-DISK TOWER OF HANOI PROBLEMS

as a condition on all the possible trajectories; namely that a trajectory
should start at a given initial situation, and end at a given terminal
situation.   The class of problems can be parameterized by pairs of situations - the possible initial and terminal situations for all problems in
the class.   The set of situation pairs that are possible end points of
solutions/trajectories in a class of problems defines the data domain of
the class.

Now, given the initial formulation of the problem class, it can be
presented to a problem solving procedure, so that when a specific problem
in the class is given, the procedure can proceed to search for its solution.   There are two major procedure schemas in AI, a production schema and

a reduction schema. The production schema corresponds to the conventional state space approach. You start from the given initial situation and you explore possible legal trajectories until you recognize the given terminal situation. You can then stop and output the solution. This is also known as the forward heuristic search approach.

In a reduction schema, you use a goal-directed approach. In view of the goal and the current situations, you consider a decomposition (reduction) of the problem into parts, you attend to each subproblem in turn, and then you combine the solutions to the subproblems. Such a procedure is powerful, but sometimes it is difficult to apply because of difficulties in decomposing the problem into independent, non-interacting parts. I consider the issue of problem decomposition, and the related representational issues that determine our ability to decompose a problem into independent (or near-independent) parts, at the heart of a theory of problem solving in AI.

In my work with the Tower of Hanoi problem, I found that a transition from the initial problem formulation to some procedural formulation in a production form or reduction form is relatively easy. The important issue is how to move to 'strong' procedural formulations in either of the two schemas. This is equivalent to asking how can we increase problem solving expertise, or skill, in problems of this class.

Quite a bit of work has been done on how to increase skill in solving Tower of Hanoi problems. There has been some interesting work within cognitive psychology in this area, in particular, by Simon and Anzai (in 1978). In their work, a novice starts by using heuristic search of the kind that I described previously, specifically, a forward heuristic search. Then, the problem solver starts using a goal-directed search, which is a reduction approach; i.e., the second type of procedure schema that I discussed. Then he experiments with various problem sizes, of one disk, two, three, four disks. Eventually he invents some important new concepts in the domain. In this particular domain a very important new concept is the notion of a set of blocking disks of a disk (a pyramid sitting on top of a disk). This concept is not given to you to begin with; you extract it from regularities in problem-solving activity. And then eventually the problem solver formulates a recursive strategy for solving problems in this class, which is the end point of this particular developmental process of skill acquisition.

The recursive strategy for solving Tower of Hanoi problems is based on the following argument. (See Figure 8.) If you have, say, a three-disk Tower of Hanoi problem, whose goal is to transfer a pyramid of three disks from peg A to peg C, you first transfer the subpyramid of two disks that sits on top of the largest disk from A to B. This way, the largest disk is free to move from A to C. Then you take the subpyramid that you have left at B and transfer it to C. To move a subpyramid between two pegs, you use the same kind of strategy, recursively, that you used to transfer the entire 3-disk pyramid. Here, a problem is broken down into pieces, and subproblems are solved independently.

I have been interested in how to go (automatically) from an initial specification of the problem as I described it previously to the discovery of the particular regularities that lead to the recursive reduction strategy, and - more specifically - how to use the acquired new knowledge to formulate the recursive strategy. What is involved in these transitions?

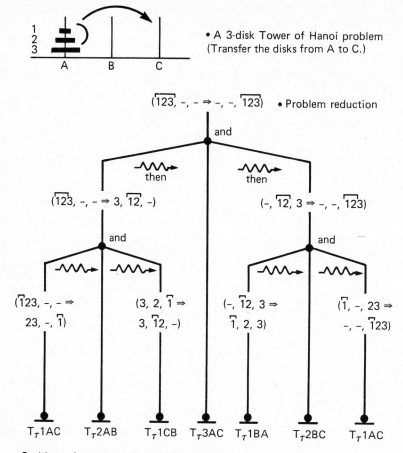

• A 3-disk Tower of Hanoi problem
(Transfer the disks from A to C.)

• Problem reduction

• Problem of representational shift
  [production → reduction]

FIGURE 8
REDUCTION APPROACH TO SOLUTION OF 3-DISK TOWER OF HANOI PROBLEM

I want to show you now a schematic summary of the various kinds of formulations and transitions that I studied, and of developmental paths between
formulations. I am not going to go into any detail, except to make some
comments about what is involved in these transitions. What you have
initially is a set of declarative formulations of the problem. (See Figure
9). The first formulation that I outlined previously, which involves
domain specifications and so on, is a declarative formulation. There are
two kinds of procedural formulations in the course of this development.
One is of the forward-search type. The other is of the goal-directed,
reduction, type. Each transition between formulations gets you to a
formulation which is stronger than the previous one. In order to obtain a
stronger formulation, you need to acquire some knowledge, from either

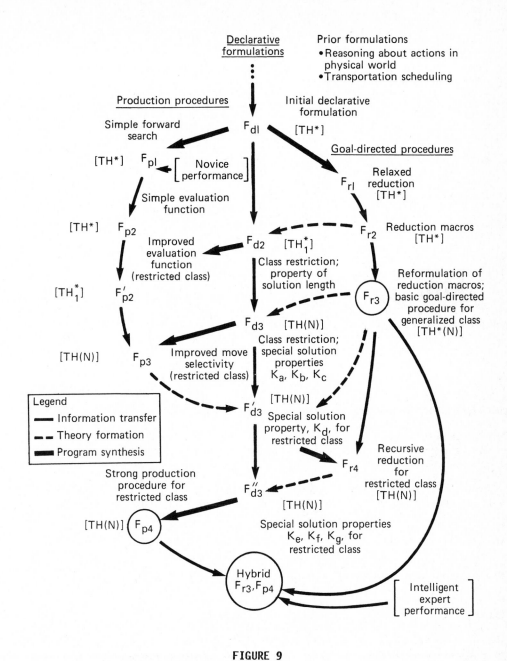

FIGURE 9

SUMMARY OF DEVELOPMENTAL PATHS BETWEEN FORMULATIONS
IN TOWER OF HANOI PROBLEMS

analysis or problem-solving experience, that could be assimilated in the
existing procedure schemas to produce procedures of higher performance.

For instance, starting with a forward-search procedure, you can strengthen
it by obtaining an improved evaluation function on the basis of information
that comes by experience with a goal-directed procedure. One interesting
experience that comes out of that is that when you solve problems, and you
want to improve your skill in solving them, it is a good idea for you to
have several schemas, several methods of solving a problem, at the same
time, because you can use some of the regularities obtained by solving a
problem with one method for the improvement of some other method. That is,
you do not need to be fixed on a particular method to begin with. It is
good to have several methods that you can play with at the same time.

Another interesting thing that I am convinced is very important in problems
of reformulation is the following: You must find ways of inventing
'interesting' restrictions in the domain of a procedure. In order to be
able to use a recursive method of solution in Tower of Hanoi problems, you
must restrict the domain of your procedure to problems that involve
transitions of complete pyramids from one peg to another. This is a
subdomain of the overall domain of Tower of Hanoi problems. Now, if you
start with a very broad method of solving problems in a domain, and you
want to increase problem solving power by developing strong, specialized
methods, you must be able to recognize subdomains of the broad domain for
which these methods are appropriate. Here we are faced with a chicken-and-
egg proposition. Do you start by positing a subdomain that is an interesting one and develop a method for it? Or do you somehow observe that there
are some interesting methods around, and you find that these methods are
associated with certain subdomains of a class of problems, and then you
establish the association? This is exactly the type of problem on which I
am focusing now, not only in the context of Tower of Hanoi problems but
also in problems of theory formation. So, the question of finding
subdomains of a problem domain for which specialized strong methods are
appropriate is an interesting problem in AI, and I think it is very
relevant to issues of reformulation and expertise acquisition.

I do not want to go in much more detail into the specific processes of
reformulation that I studied; but I would like to underline two key
processes that are involved in transitions between problem formulations.
One is the process of theory formation. In other words, it is essential
for us to be able to find useful regularities in a problem domain and to
represent them in some form (I am thinking, in particular, about declarative forms). It is important to acquire this additional information in the
course of problem solving, by analysis or empirically. So, theory
formation is a very important part of expertise acquisition via problem
reformulation. The second process is program synthesis. I think it is
very important to have automatic methods for utilizing knowledge, in
whatever form it appears, in a way that could be applied to, could have
leverage on, a particular procedure schema. We should be able to use it
for changing evaluation functions, changing applicability conditions,
synthesizing macromoves, maybe even for developing a completely new
procedure that would embody in an especially fitting way the new knowledge
that has been acquired during the theory formation process.

So I see that some of the subproblems that come up in the area of problem
representations are basic problems in AI - in such areas as discovering

regularities in a domain, forming concepts and forming theories. Also, given a body of knowledge, how can you assimilate it in a program that already exists, by modifying that program appropriately? These are all open problems in AI at this point, and there is quite a bit of work directed to them. We need to do much more work to understand these problems, and to find solutions for them, in order to be able to make progress in the area of automating processes of shifting formulations. Such progress can bring us closer to the point of understanding how to choose 'appropriate' formulations, and how to relate, in general, problem formulations and computational efficiency.

QUESTION AND ANSWER PERIOD

BRUCE HAMILL
Thank you Professor Amarel. I think we have time for a question or two.

BEN SHNEIDERMAN
I had trouble as I listened to your talk to pick out the points that you were referring to: the efficiency, effectiveness, convenience of a representation. Sometimes you were referring to the human process of solving problems. Other times you were referring to algorithms to be implemented on a machine. It seems to me that you have very different considerations when you are talking about the human's role in trying to create a program and the program's execution or carrying out of some algorithms. Later in your talk, you said "appropriate problem representation", and I would like to ask if you could sharpen these issues by taking the situation, if I arrive before you with two problem representations, how would you choose between them? What would be your criteria or methods of choice?

SAUL AMAREL
I do not know how to answer the second question at this point. The entire orientation of this research is to be able to answer this question later on. The orientation is specifically to representations that could be accepted by a problem-solving schema so that the schema can be more efficient. I am really not talking about questions of effectiveness in terms of expressiveness. For instance, sometimes you may not have a language which is powerful enough, or a conceptual framework which is powerful enough, to be able to express a regularity, let's say, in a domain. I am assuming that the languages and the available conceptual frameworks are powerful enough, but I am concerned about how to transform the knowledge which is available so that it can be best tailored to specific problem-solving actions within a procedural schema. So when I am talking about appropriateness, I mean appropriateness relative to a particular schema of problem solving. My sense throughout is that using knowledge in problem solving is a very relative thing. There is nothing absolute to that. A body of knowledge could be used very well if it exists in a certain form for a problem such as finding a path between two points in some domain, and it can be disastrous from the point of view of solving some other problem in the domain. So I think the issue is extremely relative: for different classes of problems and tasks, certain kinds of forms of knowledge and representations are better than others, in the sense

of problem solving efficiency. This is exactly the objective of my
research, to be able to put your finger on these relationships. What I
have at this point is: for the specific classes of problems that I studied,
observations about what representations are more powerful than others.

JAIME CARBONELL
I just want to underscore your necessity for being able to reformulate a
problem representation in order to bring additional knowledge to bear. Let
us go back to your dominoes. The answer, it turns out, for the dominoes
that consist of 3 squares, is trivially to reformulate to a simple counting
argument. There are 62 squares in the mutilated chessboard, and that is
not divisible by 3. In fact, that argument goes to any number of dominoes,
any number of sizes, any number of squares in the domino, even allowing for
irregular shapes. Except, perhaps, 31. Sixty-two is divisible by 31. So
let me change the problem and make it come in a different shape of 31. Can
you have two of them in order to cover the mutilated chessboard? That one
can be reformulated, too, by the way.

SAUL AMAREL
I am thinking that even if you pass some obvious global tests, like the
number of squares is divisible by the length of domino, 2 or 3 or whatever,
under these conditions also you are uncertain as to whether a cover is
possible, because you are uncertain about what kind of imposed structure
can really lead to the refutation argument. Now, there are global things
that you can do, like divisibility arguments. But that is not enough,
because in addition you have to find a cover, a coloring, like 3 colors in
the case of a domino of length 3, that is distributed in a way that can
allow a refutation argument to take place.

The Role of Language in Problem Solving I
R. Jernigan, B.W. Hamill, and D.M. Weintraub (Editors)
© Elsevier Science Publishers B.V. (North-Holland), 1985

PROGRAMS AS SYMBOLIC REPRESENTATIONS OF SOLUTIONS TO PROBLEMS

Stuart H. Hirshfield

Hamilton College
Department of Math and Computer Science
Clinton, New York 13323

At a suitably high level, a computer program can be
described as a symbolic representation of a solution to a
problem. At that same level, the activities associated
with computer programming are a potential source of
insight into the relationship between problem-solving and
linguistic capabilities. Common methods for writing programs are described and found to shed light on how such
skills interact to produce a program. Methods for teaching programming are analyzed in terms of which skills
they address and their relative effectiveness. It is
speculated that the skills one needs to write programs
are equally linguistic and logical in nature and that,
contrary to current practice, the ultimate programming
strategy is one which "plays both ends against the
middle" - i.e., uses available linguistic constructs to
aid in the organization of a solution.

1. INTRODUCTION

At its most general level, this paper addresses the relationship between
problem-solving and symbolic (linguistic) capabilities as manifested in
computer programs and the activities associated with programming. Programs
are described as symbolic representations of solutions to problems, and in
this light, the behavior of novice programming students is seen to be rife
with examples of how such skills interact, as described in Section 2.
Section 3 presents a variety of traditional approaches to teaching and
learning about programs, and discusses briefly what some call the "I can't
compute syndrome" as it relates to these approaches. The theoretical basis
for and the practical implications of a new approach, dubbed "Symbolic
Problem Solving," are presented in Section 4. This approach stresses the
interaction of problem-solving and language skills and amounts to a "both
ends against the middle" strategy. Section 5 is a discussion of both the
theoretical implications of this strategy - i.e., the insights it affords
us into human behavior - and future plans with respect to expanding this
work.

2. PROGRAMS AND PROGRAMMING

The fundamental assumptions underlying this study are that computer programs are linguistic entities created for the purpose of solving problems
and, as such, they provide us with a potentially fruitful domain for
exploring the relationship between what have traditionally been distinguished as problem-solving and linguistic skills. One need only look at a

program written in a modern algorithmic high-level programming language to
appreciate this potential. To an experienced programmer, a program conveys
information about both the logical organization that has been imposed on a
problem, and the linguistic conventions adhered to in expressing its solution. The fact that such programming languages are formal and severely
restricted in expressive (power when compared to natural languages) lends
credence to the hypothesis that programs are manifestations of a process
that involves (equally) planning, structuring and transliteration. Examining this process - i.e., the activities associated with programming -
supports these notions further still.

The programming process has been analyzed and decomposed according to a
variety of criteria. At a minimum, there is a consensus that a variety of
interrelated tasks are involved in what we call "programming" and that
different combinations of skills are required to accomplish different
tasks. Consider, for example, the tasks of designing a program, coding a
program, and debugging a program.

The design task involves the formulation of an abstract plan of attack -
i.e., a language-independent algorithm. A designer must, among other
things, be able to assess a stated problem, recognize constituent subproblems, define dependencies between sub-problems, and coordinate and
integrate activities to produce a solution to the whole problem. As
Weinberg states, "The job of system design calls for an eye which never
loses sight of the forest ...". Such skills are rightfully classified as
being predominantly of a problem solving nature.

The activities associated with coding a program are decidedly more linguistic in nature. Historically, a coder was one whose sole job it was to
perform the translation step from algorithm to program in a particular
computer language. This translation was considered to be mechanical, to
the extent that the coder could be expected to accomplish his/her work as
long as the algorithm was expressed according to some agreed-upon notational conventions. Knowledge of either the problem being solved or the
global method of solution were considered extraneous to the task. The
coder needed only to know the mappings from algorithmic notation to
programming language, and be thoroughly conversant in their respective
syntaxes. While pure coders are today a dying breed, top-down approaches
to programming have perpetuated the notion that , if an algorithm conforms
to some programming language independent standard, the coding task should
amount to one of straightforward translation.

In contrast to both design and coding, the debugging task is seen to
involve nearly equal amounts of problem solving and linguistic acumen.
Problem solving skills help the debugger to analyze error messages and
program output, formulate hypotheses as to the cause of error, devise plans
to test such hypotheses, and, hopefully, to develop and incorporate program
"fixes" so as not to upset the logical structure of the program. All fixes
must ultimately be expressed in terms of a programming language and, thus,
have a linguistic aspect. In fact, many errors can be attributed to
improper use of the programming language itself (e.g., syntax errors
detected by a compiler), in which case the process of debugging that error
involves still more linguistic expertise.

By continuing along these lines to associate particular kinds of skills
with particular phases of the programming process, one is led to construct

the following model of program composition as a skill [Note: For the time
being, I will restrict my attention to the subset of programming activities
involved in program composition, on the following grounds: while the
debugging and testing tasks appear to involve both problem solving and
linguistic skills, they do so because they can be thought of as sequences
of (re)analysis- (re)design-(re)code operations. Also, the discussion to
follow focusses on approaches to teaching and learning. A vast majority of
programming texts and courses do not explicitly teach debugging and testing, but instead concentrate solely on issues of composition.]: the early,
abstract phases of problem analysis and program design relate most directly
to general problem solving skills; whereas, the acts of formalizing the
design to produce an algorithm, and translating that algorithm into syntactically correct code, are more linguistic in nature. This view of program
composition has fostered the two most prevalent approaches to teaching and
learning.

3. TRADITIONAL APPROACHES TO TEACHING AND LEARNING

Historically, the first course in computing was, in reality, an introduction to the syntactic details of a particular programming language.
Language manuals were used as texts, and teaching involved explaining the
notational intricacies of one or more still-evolving high level languages.
Hindsight being what it is, this approach to teaching programming is
explainable in terms of both our initial fascination with the idea of a
high level language, and our faith that such languages would in general
render computing machines more accessible. Instead, our fascination began
to fade as we pushed these new languages to (and beyond) their expressive
limits. As is often the case, a new tool did not solve all problems; it
simply freed us to concentrate on other ones.

As the domain of programming expanded and problems became increasingly
complex, educators and practitioners began to recognize the problem solving
component of the programming process. To be sure, the words "problem solving" began to appear more frequently in course titles and descriptions.
Still, it was not until structured programming and top-down development
became practical goals, that our collective attention turned appreciably
toward developing general problem solving skills. More recently, courses
have addressed these skills independent of a particular programming language, on the assumption that coding (while a necessary evil) is mechanical
and trivial.

While both of these approaches to teaching have a more or less clear
perspective on what they are trying to convey, the role of the student-
learner in these scenarios has never been clear. Questions of what skills
are most valuable to a student of programming, or how a student is expected
to develop these skills (if the student indeed is to), are still subjects
of great debate. Early on, mathematical aptitude, itself a sticky issue,
was considered prerequisite to (or, at least consistent with) programming
aptitude. General problem-solving aptitude as reflected by a variety of
ill-defined measures has been used as a paradigm for computing skills, with
only minor success. The Computer Programmer Aptitude Battery, a series of
tests intended to isolate and measure specifically those skills critical to
programming, has been offered in many forms since its development in 1974.
A recent evaluation of it, conducted by Hostetler, has shown that two of
these tests appear to correlate more highly with success in an introductory

programming course than do others. The two tests, as defined by their
author, are:

Reasoning   - a test of ability to translate ideas and operations
              from word problems into mathematical notations, and

Diagramming - a test of ability to analyze a problem and order the
              steps for solution in a logical sequence.

It is no coincidence that the abilities these tests address have been
identified by students as being among those that are the most problematic
in their first course [Ulloa].

Student problems are nothing new to anyone who has ever taught such a
course. What is continually amazing (even to experienced teachers) is the
profound and uniform nature of the more disabling ones. Every teacher has
been confronted with students (who sometimes make up a disproportionally
large part of the class and seem to be randomly distributed with respect to
background, etc.) who simply "cannot compute." These are students of
otherwise normal intelligence for whom no amount of explanation or instruction helps. In talking with them, one senses an insurmountable barrier
blocking forever their progress in a programming course. Hostetler's
analysis of the CPAB supports my contention that this barrier is encountered somewhere in the middle of the program development process.

In most cases, the average "cannot compute" student is fully capable of
understanding a program as presented. That is, they can recognize a
program as addressing a given problem and can distinguish programs whose
logical structure reflect the problem at hand from those that don't.
Frequently, they can express in English an informal algorithm which demonstrates their understanding of the problem. Similarly, they seem to have
little trouble adapting to the syntactic details of the language. What
they find much more difficult (if not impossible) is creating a program
from scratch. This difficulty appears to stem from an inability to perform
what the CPAB refers to as "diagramming tasks". More generally, these
students seem incapable of bridging the gap between generic problem solving
skills and facility with a particular language.

At the outset of this paper, programming was described as being equal parts
of planning, structuring, and transliteration. I define "structuring" as
that process which formalizes a problem solution in terms that lend themselves to translation. That is, structuring bridges the gap between
activities that are essentially problem solving and those that are purely
linguistic. In these terms, "cannot compute" students lack the skills
associated with structuring.

What, then, must a student of programming know in order to structure a
problem? Simply stated, a student must have an appreciation for what the
machine is capable of. That is, he/she must know, at a level of detail
consistent with the ultimate target language, what actions the machine can
perform and how he/she as a programmer can combine these actions into a
formal algorithm. Shneiderman refers to such information as "semantic
knowledge" and suggests that "once the internal semantics have been worked
out in the mind of the programmer, the construction of a program is a
relatively straightforward task." Examples of semantic knowledge include
knowledge of what an assignment operation accomplishes, what an array is,

how conditions and loops are expressed, etc. The relative ease with which students pick up a second programming language once they have mastered these concepts, supports Shneiderman's contention. The problem facing novice programmers, though, is the development of this semantic base.

More generally, semantic knowledge can be thought of as specifying the building blocks available to a programmer for expressing a program. These building blocks, which ultimately may themselves be complex structures, constitute the programmer's repertoire for attacking a problem. By asking students to write programs without having had a chance to develop such a repertoire for themselves, we are, in Piagetian terms, distinguishing the learning process from what is being learned. That is, we are focussing on the composition process at the expense of appreciating what a program is. As Papert would say, to understand how one learns programming, we should instead study programs.

I am suggesting a slightly more liberal interpretation of the noun "program", based on the view expressed earlier that programs are symbolic solutions to problems. In this context, the main purpose of a program is to present what can be thought of as the "deep structure" of a solution. The deep structure is derived by imposing semantic constraints on the solutions, and these constraints insure that the solution has a linguistic representation. That is, a program expresses both the logical and linguistic structures of a solution to a problem. Thus, in order to develop a program, one must understand how the logical components of a solution relate to its linguistic components - i.e., one must know the structuring information. It is this information that at once allows programmers to (1) assimilate new program structures and thus expand their programming repertoire, (2) learn new programming languages, and (3) address problems in a "top-down" fashion. Perhaps it is this information that we have failed to impart effectively to our "cannot compute" students.

4. SYMBOLIC PROBLEM SOLVING: COORDINATED APPROACHES

In an attempt to address the notions of "structure" and "semantic information," many courses and texts have presented solving methods and programming language syntax in parallel. That is, they have addressed both ends of the spectrum in the hope that the middle ground - i.e., the formal logic of programming with an algorithmic language - will become visible. This hope appears to be unfounded for "cannot compute" students, many of whom have difficulty seeing this middle ground (particularly at that point in the course when they are most likely to be dealing with their fear of computing machines). It is interesting to note that such students, in addition to having trouble writing programs, also tend to have difficulties performing "step throughs", a task that requires of them that they assume the role of computer in processing a program. Whereas students seem to understand programs both at a linguistic and problem-oriented level, without an appreciation for the machine's basic operations - i.e., its set of potential deep structures - they still have trouble "playing computer." Only recently, and in limiting contexts, have we recognized the need to address the development of semantic information explicitly.

One method which has been suggested for developing in students an appreciation for a computer's capabilities, is the "user-oriented" approach. According to this approach, students are first exposed to programming by using programs and documentation provided to them. They are shown how to invoke the programs, and are asked to evaluate the programs based on their ease of use, quality of documentation, etc. Students are introduced to a particular programming language gradually by providing them with code representing the programs they are already familiar with. Ultimately, students modify and extend these programs on their own.

The "reading intensive" approach further embraces this notion of an intro- duction to writing programs. Students in reading intensive courses spend most of their early time reading and predicting how a computer would inter- pret a variety of programs. Again, the intention (I think) is that this type of exposure to programs will help students to see (and to develop for themselves) rules of structure. The development of original programs is considered only after students have had experience modifying and correcting programs provided to them.

A third approach is what I refer to as the "active approach," and is based on the premise that one develops a structural "vocabulary" by means of experimentation and hands-on programming experience. Students of this approach are encouraged from the outset to develop simple but syntactically well-formed programs, in the hope that they will simultaneously overcome their fears of a computer and become conversant in the machine's vocabu- lary. Problem solving skills are left unaddressed until students have developed their structural vocabularies, at which point it becomes more realistic to consider issues of program design.

A number of courses and texts advocate the "metaphorical" approach for presenting semantic information. That is, they make use of a model to describe for students the machine's primitive operations. Prior to writing or seeing any programs, students become familiar with the workings of the machine, from the machine's point of view. The underlying supposition is that by developing this familiarity, students will learn to structure their solutions to problems so as to be consistent with the computer. This approach has been used extensively in teaching assembly languages - i.e., in cases where the analogy between programming language instructions and primitive machine operation is more or less direct. Researchers are just now exploring the potential of this approach to high level languages.

While each of these approaches goes about it in a slightly different way, each is seen to focus on the development of structural, semantic informa- tion as a means interrelating problem solving and linguistic skills. Each has to be shown to be useful in limited situations. Practically speaking, it may be that some combination of these approaches will prove to be most effective at helping students to overcome the "cannot compute" syndrome. For the moment, it is of substantial interest that such a focus is justifi- able on theoretical grounds.

5. LANGUAGE AND PROBLEM SOLVING

Piaget describes the first manifestations of clearly intentional behavior as being wholly dependent on familiar actions. That is, children attempt to solve problems in terms of actions that are already part of their sensory-motor repertoire. As children develop and come to address more complex problems, they resort to active experimentation with patterns of familiar actions. Ultimately, children are seen to perform such experimen- tations internally, and new coordinated forms of directed behavior seem to develop spontaneously. These coordinated forms of behavior then become part of the child's ever-expanding repertoire of actions. My contention is that an infant's actions can be thought of as a program - i.e., a physical realization of a plan for solving a problem - and that doing so encourages us to speculate about the role of language in problem solving activities.

Sensory-motor infants are restricted in that they can express their solutions to problems only overtly. Novice programmers are required to represent their solutions symbolically. In both cases, expressing the solution appears to be less of a problem than does the formulation of the solution. In both cases, the solution is expressed in terms of a "vocabulary", the development of which is prerequisite for solving the problem. The infant's vocabulary is a repertoire of familiar actions. The programmer's vocabulary is a collection of generic programming constructs. Awareness of this vocabulary is what I have referred to as "structural information."

There is evidence to support the hypothesis that language has its origins in the primitive symbolic behavior of the sensory-motor period, a la Piaget. One can also make a case, a la Whorf, that a developing capacity for language influences profoundly the abstract logical and conceptual skills we attribute to adults. Our experience with teaching programming seems tentatively to suggest that the "linguistic" structures one has at his/her disposal - i.e., the vocabulary available for expressing oneself - influences how one attacks certain problems, and that so-called linguistic and problem-solving skills develop coincidentally. Furthermore, these skills are seen to nurture each other in a cyclic fashion. That is, the development of problem solving skills may depend at least initially on the development of an attendant vocabulary which, in turn, serves to structure solutions.

More important than these specific observations are the prospects for future work to explore further the potential of programs and programming as sources of insight into the relationship between problem solving and language. Both the product and the process seem intuitively to provide a wealth of behavioral data useful for addressing this relationship. In particular, the effectiveness of the aforementioned teaching techniques on a large scale would seem to warrant investigation. Also, some researchers have already begun to investigate language and problem solving skills as they manifest themselves in programming activities involving, e.g., non-algorithmic languages. Finally, program synthesis (or, automated programming) efforts may shed some light on the specific nature of structuring information (e.g., how much and what kind of information is required to formalize an algorithm, and how does such information reflect the target language).

BIBLIOGRAPHY

Cherniak, B., "Introductory programming reconsidered - A user-oriented approach," Proceedings AMC SIGCSE-SIGCUE Joint Symposium, Feb. 1976, pp. 65-68.

Flavell, J. H., The Developmental Psychology of Jean Piaget, Van Nostrand Co., New York, 1963.

Hill, J. C., "A Model of Language Acquisition in the Two-Year Old," COINS Technical Report 83-08, Univ. Massachusetts at Amherst, 1983.

Hirshfield, S. H., "A Computer Model for Sensory-Motor Intelligence and its Relation to Natural Language Acquisition," Diss. Syracuse University, 1978.

Hirshfield, S. H., "Program Synthesis as a Tool for Teaching Programming," SIGCSE Bulletin, Vol. 16, No. 2, June 1984, pp. 4-7.

Karmiloff-Smith, A., A Functional Approach to Child Language: A Study of Determiners and Reference, Cambridge University Press, Cambridge, 1979.

Motil, J., Programming Principles: An Introduction, Allyn and Bacon, Inc., Boston, 1984.

Palormo, T. M., Computer Programmer Aptitude Battery: Examiner's Manual, 2nd ed., Science Research Associates, Chicago, 1974.

Papert, S., MINDSTORMS: Children, Computers, and Powerful Ideas, Basic Books, Inc., New York, 1980.

Pirsig, R., Zen and the Art of Motorcycle Maintenance, Morrow and Co., New York, 1974.

Polya, G., How To Solve It, Doubleday Anchor, Garden City, N. Y., 1957.

Schneider, G. N., Weingart, S. W. and Perlman, D. M., An Introduction to Programming and Problem Solving With Pascal, Wiley & Sons, Inc., New York, 1978.

Shneiderman, B., Software Psychology: Human Factors in Computer and Information Systems, Winthroop, Cambridge, Mass., 1980.

Ulloa, M., "Teaching and Learning Computer Programming: A Survey of Student Problems, Teaching Methods and Automated Instructional Tools," SIGCSE Bulletin, Vol. 12, No. 2, July 1980, pp. 48-64.

Weinberg, G., The Psychology of Computer Programming, Van Nostrand Reinhold Co., New York, 1971.

QUESTION AND ANSWER PERIOD

SAMMET
I have a question and it is sort of sarcastic. You started out by saying that there was a high correlation between people who could do word problems and translate those into math notations and their success in an introductory programming course. Did I get that right?

HIRSHFIELD
Yes.

SAMMET
I don't know whether you are aware of it, but there was a PhD thesis done in the mid 1960's which did exactly that, namely, take the classical high school algebra problems, translate those into the necessary formalism and

execute them. Do you think that program would do very well in the introductory programming course? But more seriously, given that that thesis did
exist, and it was a very nice piece of work, do you care to comment at all
in this context?

HIRSHFIELD
Are you talking about Bobrow's thing? Yes. I am still not sure what the
question is. Whether that program would do well in an introductory
programming course, it probably would do better than a lot of the students
with question marks on their face.

SAMMET
Do you really believe that? My serious question ....

HIRSHFIELD
That is a tough nut. I have no idea, I mean, I would like to believe that
there is a lot more than just that involved in programming and I'm not
saying, certainly, that that is the only thing. All those correlations
showed, was that aspect of problem solving seemed to correlate with its
success in the course, and I agree, there is a machine to do it.
Congratulations.

GUIER
My comment is that I wonder if you are worrying a lot about people who are
just naturally not going to be good programmers. I am reminded of an
experience that happened long before computers were invented. I don't
know, there may be very few people who are as old as I am here, but it was
during a music club class, and we started playing games - and all of these
were very accomplished musicians - we were playing games of get up and
suddenly, out the top of your head, play a new piece of music that you made
up. And we were astonished to find that some people, who were not necessarily the best musicians, were extremely good at invention of that sort,
and others were not, and there was clearly a very innate talent that simply
had nothing to do with learning how to play music.

HIRSHFIELD
I agree, and it is not as if I am overly worried about these people - and
if I was I would be in class right now - (laughter) I do not mean that
sarcastically, I mean I am not interested in them in an attempt to convert
them into programmers. I do not regard this as a religious experience.
But I am interested in it from a behavioral standpoint because I think they
provide us with, at least, human kinds of evidence about what kinds of
skills it does take and does not take to communicate in these languages.

McALLISTER
I would like to just make a comment. I particular enjoyed your list of
things it cannot do, and I think I know by experience because of experience
what you are talking about. I would like to add though, as a computer
user, a problem which I found extremely harmful to the understanding of
many, many topics. In your list, may I add after documentation what I
call, validation or variation on the theme. Students, for example, would
they be able besides catching the parenthesis that you showed on your slide
as missing, to catch the additional set of parenthesis which were missing
when computing the average? Normally they would look at the output and be
happy with it and I think that is one item that is very very hard to teach,
so I propose it for an addition for your list.

HIRSHFIELD
Thank you, I agree.

The Role of Language in Problem Solving I
R. Jernigan, B.W. Hamill, and D.M. Weintraub (Editors)
© Elsevier Science Publishers B.V. (North-Holland), 1985

43

PHYSICS AND TECHNOLOGY, COGNITIVE PSYCHOLOGY, AND COMPUTER LANGUAGES: TOWARD AN EXPERIMENTAL R&D EPISTEMOLOGY USING LANGUAGES AS A TOOL

John A. Carlton-Foss

Human-Technical Systems, Inc.
Box 151, Lincoln Center, Massachusetts 01773

ORIENTATION

The fundamental purpose of computer software is to provide successful connections between the hardware of the computer and the structure and nature of the real world, either directly through analog-to-digital conversion devices or indirectly by translating professionals' implicit and/or explicit ways of cognizing the real world into electromechanical information:

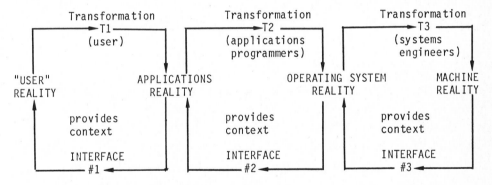

To a certain extent the software/hardware engineer can rely on the efforts and good will of users to assist in this translation. But as technologies become more capable and complicated, and as operators must track an increasingly large universe, requiring learning and higher cognitive processing, it becomes increasingly important to provide people with proper assistance so they do not suffer from cognitive overloads. As more is known about the translations between environment and individual for a particular region of human experience, it becomes increasingly possible and, indeed, desirable to create algorithms for the translation of informa- tion from environment and machine to person, and to make explicit the knowledge, the models, the ways of forming constructs associated with that translation. To reduce the effort required for development of individual applications, and to increase operating efficiency, it is advantageous to represent these directly as languages. In this way one language may be built up within itself (e.g., as with LISP) and/or one may serve as the ground for future developments (e.g., as LISP on a UNIX-based system or FORTRAN on a VMS operating system).

Though spreadsheets and other very high level languages for business applications have been in vogue for several years now, only recently has the more general opportunity to match professional requirements with computer languages been recognized. Some of the high level matches already developed and/or in development are:

LANGUAGES	PROFESSIONS
CAD (Computer-Assisted Design)	Engineering and Architecture
TK!Solver	Physics and other disciplines requiring algebraic problem solving.
Robotics (still very primitive)	Repetitive but flexible manufacturing.
Spreadsheets	Office tasks
Decision support systems	Executive support

These are important areas for product development, and the products produced can be very profitable. However, the related issue I wish to bring forth here is the quality of the match between what is developed and the the requirements of the users. The best product will provide technical value, will match with task and environmental demands, and will match with the cognitive demands and styles of the user. These dimensions (and profitability of the product) are all interlocked. Wilson, noting this, has begun a project to design software to match "normal" physicists' problem solving styles. TK!Solver (Software Arts, 1984) may already have provided a tool for solving algebraic equations that will be difficult to improve upon. Greenberg, too, has made a very strong statement regarding architectural CAD systems which are touted as "user-friendly" by the computer industry:

> "Perhaps of even more concern than ease of use are the
> limitations of the software systems offered to archi-
> tects. They are not really design systems. They can
> help in the production process, but available tools are
> simply not flexible enough to use for preliminary design
> and testing of alternative strategies. Nor is the output
> realistic enough to permit esthetic or design evalua-
> tions" (Architectural Record, September 1984, p. 150).

As part of improving the fit of computer systems (hardware and software) to the patterns of working people in the workplace, we must address how to formulate languages which will help us.

Ultimately the language of computer hardware reduces to one of switches-on and switches-off, corresponding to a binary arithmetic and logic which can be used to directly represent numbers, letters, and other symbols. For our purposes, the transformation between this digital reformulation of the noncomputer world and the binary logic of the hardware world is complete once assembly language has been written. This is done on a case-by-case basis by a skilled programmer, or through use of compilers and interpre- ters, or when very high level languages are used to attack a problem. At

issue, then, is how to structure these transformations, how to engineer
their "front ends" (i.e., their interfaces with practitioners), and to what
ends computer "languages" should be redesigned from new principles rather
than left as they are. Related to this is the need to make decisions about
which parts of knowledge, which patterns of thought, should be represented
directly in the computer language, and which should be represented in other
ways (e.g., direct input by such means as A-to-D conversion and keying;
database approaches including descriptive representation, analogical representation including icons, or other languages based on patterns which are
difficult to represent directly and must therefore be approximated using
non-ideal languages). A key job of software engineers is to interface
higher level languages (e.g., FORTRAN and others) to the assembly language
of the microprocessor used in a particular computer. The flexibilities and
constraints of the assembler language and higher level languages, and how
they implicitly structure and represent knowledge is a key issue. Software
engineers can profit from professionals who truly understand the issues of
users and transformations.

PROBLEMS AND SOLUTIONS

Implicit structuring underlies both problems and solutions. At the risk of
overstating the case, I will say that without an appropriate language there
is no problem and no solution. Appropriate language and an immense amount
of skill permit us to bring real but implicit problems to conscious awareness, to formulate them, and to develop solutions. As a Spanish interpreter
pointed out to me some years back, it is difficult to discuss anything
about apple cores in Spanish, because there is no word in Spanish for the
core of an apple. Problems are also often not recognized or acknowledged
because some of their aspects may be repressed or denied by key groups of
people. People who identify "problems" can find themselves accused of
"creating" problems. Thus, I learned from Carroll Wilson the importance of
emphasizing the positive nature of the "problem" and the "problem solver."
I often like to speak in terms of "challenges," and the opportunities
presented through problem solving.

Applying this to computer languages, an excellent example is provided by
selection menus in "user friendly" software design. The problem is: How do
you make it easier for users to know what their options are, and to select
the one they want? If you are a physicist or engineer and like to use
FORTRAN, the question is likely to be a "bother" rather than a "problem."
One does not require menus, for one knows what the options are, and there
are very few of them. It is much easier to remember than to read through
many menus, is it not? One of my colleagues operating from this viewpoint
recently made such statements as: "I derive the formulas, and I do the real
programming. Then I give it over to Joe Human Factor and he makes it look
nice. Boy can he make it look nice." Of course, the beautification is
relevant and nice but trivial to him, and consistent with this, FORTRAN
does not permit one (without going into Assembler) to create the highlights
and icons that would make prompts and menus easier to read and understand.
Thus, neither the problem nor the opportunity for a solution is brought
forth using FORTRAN as a language.

The opposite is true with C, and by now we know the tremendous opportunities that have been opened as a result. This language permits, and perhaps
implicitly encourages, such things as screen design to increase legibility,
highlighting to accent certain key information, multiple screen colors,

user-defined functions, and very high level languages such as TK!Solver that permit users to begin to think in their own terms rather than being burdened by the cognitive demands of the computer system. It makes it easy for programmers to provide roll-down menus and icons which provide quick access to useful information without cluttering the screen (as would happen with FORTRAN). Thus, the problem of intelligibility comes up and is readily answered when C is used.

With this in mind, I will take a cut at defining the terms "problem" and "solution." A problem is a perplexing question put forward for discussion, consideration, or solution. This implies that the answer is not known, and that problem solvers will have to invest some degree of effort to "work through" to a solution or resolution. It also implies that there is no straightforward, established procedure or algorithm which can be used to arrive at the "solution." There is a real "unknown" to be found. For our purposes there will always be a strong intellectual component to the work, as well as components involving economic and business realities, the need to shed preconceptions and personal biases in order to consider various frames of reference, and a personal emotional commitment tied to any real solution (e.g., see Polanyi 1962 for a discussion of personal commitments to knowledge). Polya and others have written extensively on the topic of problem solving itself, and it need not be discussed further here.

The result of problem solving is one or more "solutions. "During undergrad- uate work, many of us were taught to believe in a unique, perfect solution, perhaps one derived mathematically. But that is not the only kind of solu- tion. For example, Christiansen at Harvard Business School has written about "satisficing" solutions - solutions that are good enough even though they may not be optimal. I will define a solution as a defensible, accept- able answer to a problem. The kind of solutions we are considering will include a strong intellectual component, a considerable economic opportun- ity, and an emotional resolution and acceptance of the finding as a solution.

A research collaborator's comment many years ago provides a concrete example of this definition. After we had spent months on an elementary particle experiment, he quipped that, of course, you stop the experimental procedure when you believe you have proved or disproved the hypothesis. Logic, statistics, experimental and survey results help, but in the final analysis there is a crucial gut sense that we rely on to decide whether we have a solution. One might well ask whether there can be a computer lang- uage at whatever level to capture, or (to aid and abet), this sense as a part of problem formulation and solution.

DIFFERENT LANGUAGES FOR DIFFERENT PURPOSES

Assembly languages are for the most part extremely adaptable, so it is possible in principle to create higher level languages tailored to the requirements of any given profession or application. However, proper determination of requirements and high quality design for an application requires considerable investment by software engineers to do the program- ming, human factors experts to clarify the nature of the interaction between people and computer hardware/software and the problem to be solved, and professional groups such as physicists to identify ever more clearly and in adequate generality how they think or want their thinking reflected in the applications for which the computer is to be used. For example,

FORTRAN represents an initial formulation of a language for the physics community. It is quite efficient for operations involving numbers, and has been the computer language of choice for physicists for more than two decades. Meanwhile, even though LISP was chronologically the second computer language developed, and could conceivably be very useful because its flexibility would permit physicists to create and work with a wide variety of different structural representations of the relations among, say, elementary particles, it has not been attractive to physicists, but rather to the artificial intelligence community: it offers great flexibility and generality in treating both numbers and alphabetical (or even nonverbal) symbols. The tradeoffs between these very different languages designed for different purposes, include:

o FORTRAN makes life difficult for programmers who are trying to generate code to manipulate or "understand" letters and words. LISP makes life easy for them.

o While it is possible to set up sophisticated data bases with FORTRAN IV on large mainframes, it is a very cumbersome matter to do so. With LISP one can easily set up many different types of data structures.

o FORTRAN is computationally simple and elegant, but it is not structured to reflect the way that a physicist is trained to solve problems or perform calculations. Instead, a physicist must conceptualize a problem, derive analytical formulas that can be used to compute a numerical solution (or to compute values that the physicist can correlate and/or interpret to derive a solution) and then translate necessary parts to FORTRAN code. LISP as such is comparatively inelegant and inefficient with numbers, and reflects physicists' ways of thinking even less than does FORTRAN -- although it may be easier to adapt LISP to physicists' thinking patterns than to so adapt FORTRAN: LISP is non-procedural while FORTRAN is procedural.

o FORTRAN is a procedural representation of one particular set of knowledge about the physical world. It is made for modeling on n-dimensional manifolds, and when one wishes to model physical phenomena on other types of algebraic or topological spaces (e.g., SU(3) or other unitary groups) one must translate, or computerize only the portion of a model that can be programmed.

o The seeming complexities of LISP functions and algorithms may appear unnecessary if one is dealing only with numbers.

o Concepts such as recursive functions required in LISP are inconsistent with the ways that physicists and many other professionals are taught to think. Although recursion is a very powerful tool, it seems at first glance that most professionals need not be required to understand or to use it.

o Some versions of LISP have only fixed point numbers, and others are limited to three-dimensional vectors. This, of course, makes them of limited use in the physical sciences.

This illustrates a proposition which I believe to have general truth: there is no perfect computer language, there are only languages which are more or less well adapted to particular applications. All computer languages have their special benefits, as well as their downside tradeoffs with respect to work patterns for which they were not designed. This applies to very high level languages such as spreadsheets and TK!Solver, as well as lower level languages such as FORTRAN and C, and finally to assembler and machine language.

LANGUAGES FOR DIFFERENT PARTS OF PHYSICS

Heretofore I have addressed the matter of a language for physics, as
contrasted with languages for other purposes. The same arguments can be
made for languages reflecting different fundamental ways of looking at the
physical world. Eleven years ago I developed the notion of root metaphors,
which I have called "images." This notion came from Fermi's conclusion that
(as of the 1940s) there were only six fundamental problems in physics, and
Holton's notion of "themata."

Computer languages and languages in general, physical theories, ways people
perceive and interpret and model, are based on certain "images." Different
images are held dear by different groups of collaborating colleagues.
Sample images, interactions, and languages relevant to physics include:

IMAGES
wave propagating freely in a medium

object, or its complement, a hole of the same size

force field

uncertainty (e.g., Uncertainty Principle)

orderings (hierarchical, relational, etc.)

creation, and its complement, destruction
(e.g., the creation and destruction operators of quantum electrodynamics)

IMAGES OF INTERACTIONS
diffraction and/or refraction of wave from a small object
(e.g., a slit, or its complement, a knife edge)

resonance
(e.g., a wave is trapped briefly in a central force field)

collision
(e.g., of two hard bodies)

damping
(e.g., the quenching of light as it enters a light-absorbing material,
the decrease in amplitude of a pendulum swing as it passes through water)

system
(e.g., galaxy = objects held together by mutual attraction)

time evolution
(e.g., the "big bang" theory of the origin of the universe
leading to known matter, stars, galaxies)

```
┌─────────────────────────────────────────────────────────────────────────┐
│                        MODELING LANGUAGES                                 │
│                        the phenomenon itself                              │
│                                                                           │
│                     phenomenal description                                │
│                                                                           │
│                          arithmetic                                       │
│                                                                           │
│                   algebra and trigonometry                                │
│                                                                           │
│                    elementary mathematics                                 │
│                                                                           │
│               differential and integral calculus                         │
│                                                                           │
│      difference equations and other discontinuous mathematical forms      │
│                                                                           │
│    functions of a real variable on Hilbert and other subtract spaces       │
│                                                                           │
│                    groups, rings, and algebras                            │
└─────────────────────────────────────────────────────────────────────────┘
```

What makes the solution to a problem "satisfactory" or "unacceptable",
"elegant" or "inelegant" is heavily influenced by the congruence of the
solution's inherent images and languages with those preferred by a particular group. (Note: This helps explain why one group's elegant solution may
be unacceptable to another group.) We see this most clearly in the debate
which raged between Einstein and Bohr. Einstein refused to accept that God
might "play dice with the universe" using statistical chance, while Bohr,
Heisenberg, and the new quantum mechanics embraced such statistical treatment. We further see the role of images in the contradictory experimental
results which led to the formulation of the wave-particle duality. One set
of physicists, following upon and embracing the work grounded in the
particle "image," arrived at objective experimental results which seemed to
contradict the likewise objective experimental results of the other set who
embraced experimental procedures that elicited the wave nature of matter.
There were even different forms of mathematics used for each: particle
mechanics for one, trigonometric functions and Fourier analysis for the
other. Bohr put forth a further "image" (NB, not necesarily in and of
itself "deeper," although in the context it may have represented a "deeper"
cut into reality) which resolved the matter -- the "image" of contradictory
opposites forming a whole. Fortunately, he and his colleagues were able to
formulate a mathematical treatment which, although not completely rigorous,
was good enough to be accepted by practical scientists.

We are soon to run into a similar tension between a variety of computerized
problem solving techniques using very high level languages (e.g., TK!Solver
developed upon the work of Milos Konopasek and recently released by Software Arts; see also Kenneth Wilson's ideas as discussed in Kolata, 1984).
With TK!Solver one can set up a problem by keying in the equations defining
the relations among them (see "Rule Sheet" in Figure 1). The computer
recognizes any new variable typed in and adds it to a separate and complete
listing of all variables (independent and dependent) used in the equations
(see "Variable Sheet" in Figure 1). The user can then define the units in
which each variable is measured, and can provide an English description of
the variable for future reference. To solve the problem, the operator
defines the numerical values for a critical number of the independent

J.A. Carlton-Foss

variables, and the computer calculates the numerical values for the rest of the variables (see "Input" and "Output" in Figure 1; see also discussion in Konopasek and Jayaraman, 1984).

```
═══════════════════ TK!Solver ® by Software Arts ═══════════════════
(1) Status:                                                    14 /!
┌──────────────────────── Variable Sheet ────────────────────────┐ ⇧
  St  Input       Name    Output    Unit     Comment
  |   16          m                 g
      80000       v1                cm/s
                  v3      46188.022 cm/s
                  v4      46188.022 cm/s
                  th4     30        deg
                                                                   ⇩
┌────────────────────────── Rule Sheet ──────────────────────────┐ ⇧
  S Rule
    m*v1 = 2*m*v4*cos(th4)
    0 = 2*m*v4*sin(th4) - m*v3
    m*v1^2 = m*v3^2 + 2*m*v4^2
```

FIGURE 1

Such non-procedural problem solving makes it more feasible to develop a variety of models of varying complexity. In part, this is because the equation solving is done by the computer, and conceptually desirable approaches one learns to avoid because of their mathematical difficulties now become attractive and feasible. For example, one might wonder to what extent various "billiard ball" models would provide the same answer. In Figure 2 the same problem is diagrammed and in Figure 3 it is solved using such a purely geometrical model of the collision dynamics. The computer software not only calculates individual solution sets, but also plots them in graphical form (Figure 4). (Even with such aids, one may note, the user must still understand the physics. The unexpected discontinuity in the velocity of the impinging particle is a result of the definition of the coordinate system, not of a computer error. The user must know enough to be able to clarify such ambiguities.)

This approach should (and will ultimately) be accepted as providing a complete solution to an analytical (or in some cases a non-analytical) problem, though I expect considerable resistance. That is, many physical scientists will "feel" that the computer approach yields numbers but not solutions. The basis for this has to do with the "images" held by physicists. Though physicists use computers to calculate numerical values for models, these models all have analytic expressions as their base. Closed analytic expressions are programmed in FORTRAN to compute a

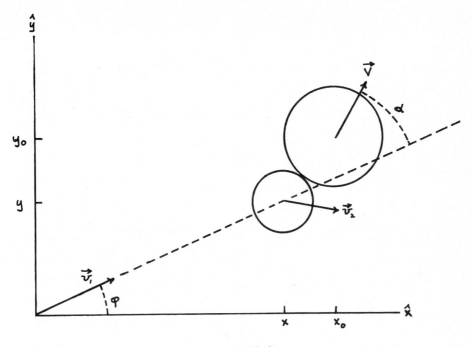

FIGURE 2

numerical value procedurally. The "solution" was derived by the scientist;
the numerical value was calculated by the computer or hand calculator in
the same way that a slide rule might have been used 20 years ago. The
similarity is that TK!Solver represents solutions as numbers. The
difference is that the (sometimes non-procedural) solution algorithm is
represented internally to the computer. Although this makes the "solution"
invisible to most users, the solution does exist, has been found, and it
permits the user to go backward and forward from various critical sets of
variables to the corresponding dependent variables. Thus, I would argue
that TK!Solver provides very high level language "solutions" to problems
expressible algebraically, even though some people will feel uneasy about
it.

I believe that we should develop many computer languages (e.g., very high
level languages, or also low level languages, complete with any necessary
compilers, interpreters, etc.) like (or improvements upon) TK!Solver,
spreadsheets, and word processing packages that are the best possible
representations of each of the key root concepts or "images" of physics,
engineering, business, or other fields. In the same way that we have
extended the application of specialized mathematical expressions developed
as an expression of one image to other problems characterized by unknown
images (e.g., Lagrangian and Hamiltonian formulation for particles and
sinusoidals for waves, as applied to the photoelectric effect), we can
create and apply specialized computer languages. An example is the way
that LISP is being used to create (perhaps limited but also intriguing)
models of human cognitive processes. Ideally, these languages would be

```
RULE SHEET
S Rule
- ----
* m*v1x = m*v2x + M*Vx              " conservation of momentum, x-projection
* m*v1y = m*v2y + M*Vy              " conservation of momentum, y-projection
* m*v1^2 = m*v2^2 + M*V^2           " conservation of energy

* v1x = v1*cos(phi)
* v1y = v1*sin(phi)
* v2^2 = v2x^2 + v2y^2
* Vx = V*cos(alpha+phi)
* Vy = V*sin(alpha+phi)

* phi = atan2(y,x)
* x0 - x = (r+R)*cos(alpha+phi)
* y0 - y = (r+R)*sin(alpha+phi)

* thV = alpha+phi
* thv2 = atan2(v2y,v2x)

* phi0 = atan(y0/x0)
* phimax = phi0 + asin((R+r)/sqrt(x0^2+y0^2))
* phimin = phi0 - asin((R+r)/sqrt(x0^2+y0^2))

* place('alpha,element()+1) = extrapol('alpha)
* place('v2x,element()+1) = extrapol('v2x)
```

```
VARIABLE SHEET
St Input      Name     Output     Unit     Comment
-- -----      ----     ------     ----     -------
                                           *** COLLISION OF TWO BALLS ***
      1       M                            mass of target ball
      1       m                            mass of incident ball
      1       R                            radius of target ball
      1       r                            radius of incident ball
      8       x0                           x-coordinate    /initial position
      1       y0                           y-coordinate    \of target ball
              x        7.403816            x-coordinate    /position of incident
              y        -.9090743           y-coordinate    \ball on impact
L    -7       phi                 deg      angle of departure of incident ball
L             alpha    79.656961  deg      angle of impetus
      1       v1                           incident ball initial velocity, total
L             v1x      .99254615           "       "        "       "    x-comp.
L             v1y      -.1218693           "       "        "       "    y-comp.
L             v2       .98375045           incident ball final velocity, total
L             v2x      .93902635           "       "        "       "    x-comp.
L             v2y      -.2932481           "       "        "       "    y-comp.
L             V        .17954123           target ball final velocity, total
L             Vx       .0535198            "       "        "       "    x-comp.
L             Vy       .17137877           "       "        "       "    y-comp.

              phi0     7.1250163  deg      departure angle for head-on colision
              phimax   21.488319  deg      maximum departure angle
              phimin   -7.238286  deg      minimum departure angle

L             thV      72.656961  deg
L             thv2     -17.34304  deg
```

FIGURE 3

adapted to one set of problems characterized by one image, and applied on
an exploratory basis to new problems characterized by as yet unknown
images. Since the development of physical theory and models seems to
involve comparisons with known images and models before new ones can be
identified and used, such "languages" would help make research and practice
conceptually clearer, functionally more efficient, and less subject to the
error which now arises when the limits of validity of languages and models
are reached. For example, the wave image and its form of mathematics are

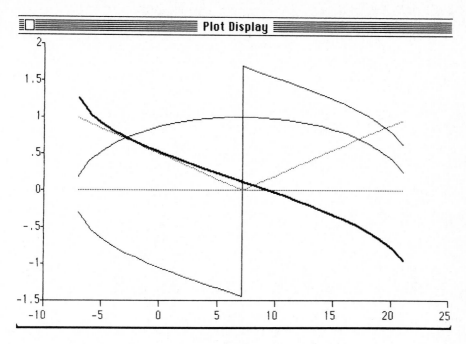

FIGURE 4

often used in treatments of the Heisenberg discussion of the Uncertainty
Principle (e.g., see Messiah, 1961, for a wave-picture discussion of
meaning of the Uncertainty Principle). While they may be useful for
students who are first learning to understand a new concept, the image of
the Uncertainty Principle is not a wave-image, and errors can result if
professionals do not draw a careful line and address the principle in its
own right. Carefully constructed alternative computer languages would
permit practitioners to reconceptualize quickly. They would not have to
succumb unknowingly to the implicit assumptions of languages such as
FORTRAN which encourage (or require) use of trigonometric functions (wave-
picture).

It would be splendid if we would carefully augment our computers' current
series of "languages," languages which transform the user's personal and
environmental "realities" to the "reality" of the computer. These
"languages" include the machine language, assembler, the operating system
with the included compilers and interpreters and so forth, higher level
languages (e.g., FORTRAN), and in some cases very high level languages.
With the addition of the new languages, the languages themselves would
reflect the enormous variety and depth of human knowledge. They would
parallel the analogous but less accessible multitude of mathematical fields
which represent some of the ways that the human mind is capable of
structuring the world. However, we must also recognize that there may be
intrinsic limits. There are phenomena in the user "reality" that cannot be
transformed accurately enough to the machine "reality". Phenomena which
currently defy computerization include those which cannot be transformed
into digital format, but which must be kept in analog format (e.g.,

emotions, values, gestures, meaning), and phenomena which include ambiguity or paradox or significant cultural or interpersonal differences. I have found these to present a particularly knotty challenge when working with LISP to generate programs modeling psychological areas which included people's subjective side.

ISSUES OF VALIDITY AND RELIABILITY

Of particular interest is the "perception" and "representation" of "reality," about which we can ask a series of questions:

1.   What tools can we build to help us to perceive and represent reality more efficiently?

2.   What perceptions do the use of those tools make possible? What biases do they project into our work? For example, if we use field theories which model fields in terms of particles we can focus very clearly on the edge between what is known and what is not known, but we are also forced into a mental set in which bumps in plots of scattering cross sections as a function of angle, for example, are interpreted as new particles. Such tenuous assumptions must almost necessarily be regarded as biases, and we must be extremely careful about how we use them.

3.   What biases do these tools help us avoid? For example, computerization of standard procedures (including calculations) can reduce oversights and make researchers and practitioners less inclined to avoid the difficult or onerous parts of their jobs.

4.   What new opportunities do these tools make available?

PERSONAL EXPERIENCES WITH LANGUAGE AND PROBLEM SOLVING

While I first thought about these issues as a physicist, I also had the privilege of learning several new computer languages not generally associated with work in physics, to complete a doctorate in psychology, to practice psychology in relation to science and technology, and to do engineering. The different needs of the different professions are blatantly obvious. HVAC engineers have traditionally used look-up tables for typical solar radiation on the earth's surface, and no statistical data is provided on hourly or daily variation due to atmospheric and other effects (e.g., see ASHRAE, 1981, pp. 27.19 - 27.36). Such formulation would be unacceptable to the physicist, who would be almost driven to model solar radiation analytically on a computer using data about the sun, the solar system, the earth's atmosphere, and the relative orientations of surfaces. A scientist would also be concerned with statistical fluctuations, while an engineer would be concerned with the "safety factor" (the multiplicative margin of error provided in a design for possible error) -- each being uneasy about the other's method for gauging possible error and for keeping it within acceptable tolerances. An engineer might well consider the physicists' analytical modeling and statistical analysis unnecessary, inefficient, or even undesirable (e.g., very inefficient, unless you happen to carry a computer with you out into the field). A psychologist, on the other hand, might try to use either or both of the above modeling procedures, and also some even more sophisticated data analyses (e.g., factor analysis and non-parametric statistics) to determine

possible trends, the degree of statistical significance of possible obser-
vational error by researchers, possible observer bias, and the impacts on
people for whom thermal comfort and other solar-radiation-dependent
phenomena would be important. An important part of the psychologist's
model would also include the meaning of the "numbers" for real people.
Though I believe that much of people's emotions and actions can be modeled,
they won't be until someone is ready to invest a major effort in the
area. In general, what a particular group of professionals require of a
computer language and applications program has very much to do with how
they perceive and understand a situation, and where they place the accent
of their reality.

The problem of language is deep and the limitations of a language for
solving a given problem are often difficult to identify. The problem is
found in computer languages and representations, and is also more general.
There are a number of examples in mathematical physics, in which a major
activity is the discovery of appropriate mathematical "languages" to model
phenomena, though the more conventional is the modeling itself which is
done once the language has been (wisely or unwisely) decided upon. In
1968, I was working with Regge models for particle collisions, scattering
and production of elementary particles. We used "functions of a complex
variable" to model the scattering amplitudes for various multipoles and
channels (and from these were calculated the scattering cross sections).
Singularities in these scattering amplitudes corresponded to "elementary
particles" and "scattering resonances." At the time there were several
crucial theorems which were believed to be true but which were unwieldly to
prove using the formalisms of mathematical physics. In one case, the
Mandelstam hypothesis, this was because we were using functions of one
complex variable. The hypothesis held that a single analytic function
described the amplitudes for scattering and creation of particles in high
energy physics phenomena. A weaker form posited the existence of a simple
relation among the amplitudes for the various "channels" in which a
scattering event could take place. This simple relation was present if the
amplitude for one channel could be analytically continued in s-t-u-space
(s,t, and u being the complex momenta of combined particles in different
relativistic frames) to the amplitudes for the other channels. In 1963
Omnes and Froissart wrote that such an analytic continuation could not be
proved, and as late as 1969 to my knowledge, no one knew whether such
analytic continuation was possible. But the barrier to the solution was
that the mathematical language used was "functions of one complex
variable." In fact the problem had already been solved by mathematicians
using "functions of several complex variables." In 1913 William Osgood had
delivered lectures on this topic, and on p. 65 of the Dover printing of
these lectures, he presented theorem 1:

> Let f(x,y) be analytic at every point of the boundary of a regular
> region, T, of the 4-dimensional space of (x,y). Then f(x,y) admits an
> analytic continuation throughout T, and the resulting function will be
> analytic in T.

One can apply this theorem to the scattering amplitudes as functions of
paired relativistic invariants s,t, and u to conclude that the required
analytic continuation exists. Thus, the difficulties encountered with a
problem framed in the language of a single complex variable was reduced to
a matter of developing the correct (and perhaps new and different) corres-
pondences between mathematics and physics in the more sophisticated

mathematical language of several complex variables. The downside was that "functions of several complex variables" was a very complicated, highly abstract, non-computational field, one in which there were very few theorems.

We are all victims as well as beneficiaries of our cognitive commitments at a given place and time, and being heavily committed to the analytic cognitive style (no less concrete than abstract operators on n-manifolds in Hilbert spaces) and images of the physics at MIT, I didn't proceed further. I wasn't interested in abstract mathematics -- nor were the physicists I was associated with. As I came to understand later, this had to do with our allegiance to certain "images" and associated languages which characterized our field. Even if we could "prove" a theory easily with a "deviant" language, we wouldn't "believe" it, because it had not been done in our language grounded in our images. This kind of mind-set poses a serious challenge for those who would develop computer languages at any level to address and solve different problems:

> How can one match the language to the structural demands of the problem and at the same time match the cognitive demands of the problem solver so that a "solution" can not only be defined and computed but also accepted as a solution?

RECOMMENDATIONS

I conclude by proposing that programming languages, subroutines, very high level languages, and data base management systems be developed to assist the physicist who seeks to work with varying images theoretically and experimentally. The development should progress through a series of stages. The early ones should be of immediate applicability to work done by physicists performing "normal science" (see Holton), and in fact have already been partly implemented. The later stages, built upon the earlier ones, have not been implemented.

Development 1. Subroutines permitting standard mathematical functions (e.g., Bessel and Legendre functions, integrals) and physical calculations (e.g., relativistic mechanics) to be called the same way a programmer can call a function from the transcendental library in the usual FORTRAN or C compiler. This would facilitate the usual calculations which for decades have been programmed.

Development 2. A very flexible data base management system with extensive, perhaps domain-specific, user-matched (i.e., not necessarily "user-friendly" which focuses on ease-of-use for novices, but "matched" to the requirements of physicists who may be rather sophisticated and may not have the patience to put up with the usual wait times and menus of "user-friendly" interfaces) data manipulation capabilities. Such DBMSs would help the professional cope with the increasingly challenging task of keeping track of relevant data without premature commitment to an analytical model. They would permit configuration of data in varying patterns so that multiple alternative theories could be generated and tested.

Development 3. Specific calculation, data analysis, or theoretical-predictive subroutines can be provided in a readily accessible library. This would be a precursor to an "expert system" for physicists: it would permit professional physicists to call on routines to perform normal calculations and analyses without extensive intervention on their parts.

Development 4. Non-numeric routines are needed. Physicists are not alone in needing multi-sensory feedback loops at their fingertips, but because of the professional's strong commitment to analytical results, such non-analytical loops offer special value to physicists. A single keystroke could translate a numerical result into a graph, sound train, or other sensory or pictorial mode to facilitate reasoning. There are machines which provide facilities ranging from graphics packages, to computer assisted design/engineering packages, to those which permit the scientist to plot data points in N-dimensional space, and then to perform projections, rotations, and transformations of the points to achieve a readily recognizable visual pattern. The goal here is to translate data whose meaning and pattern is opaque to data whose meaning and pattern is transparent.

Development 5. Indigenous representations of root phenomena can be translated into computer languages which, in turn, can be used to generate models which are at their roots different, and which will therefore provide increasingly different results as the limits of applicability of a given root metaphor are reached and exceeded as applied to a given phenomenon. Some of the metaphors have been presented earlier in this paper: central field, spinless particle, wave, particle with angular momentum, etc. Some of the dimensions characterizing differences include: continuous versus discrete variables, space-time versus frequency-energy representations, and varying representations of data (e.g., analytical, numerical, pictorial, verbal.) While a single computer language may permit representation of more than one root metaphor as applied to a selected new phenomenon (e.g., FORTRAN can be used to represent both the space-time and the frequency-energy pictures), some require different computer languages (e.g., FORTRAN and TK!Solver are appropriate for calculating certain aspects of symmetry breaking in SU(3) but inappropriate for representing and calculating the algebraic properties of SU(3) symmetry models for elementary particles; it may be possible to specialize LISP or other symbol manipulation languages based on more abstract mathematical forms to represent selected applications domains). LISP-based artificial intelligence models of human cognition (which is projected into the measurement technologies of physics and generally regarded as yielding an "objective" measure of physical reality) would also help to quantify the subjective dimensions of physical science, and thereby to reduce problems which arise when cultural as well as paradigmatic biases appear in work performed.

Development 6. An overarching language will eventually have to be created to integrate all the components of the above extensions. At present it is impossible to be certain of an appropriate language (though LISP may suffice), for the experimental data of modern physics have gone far beyond the calculus of Newtonian physics, and the appropriate mathematical forms for models are not yet apparent.

Development 7. The bridge can be established between physics research/application and artificial intelligence. This effort is connected with those for developing extensions 5 and 6. To accomplish this it will be necessary to identify frames in which experimental and theoretical material is developed. This presumes that the physics of Western culture (and the corresponding "physicses" of other fundamentally different cultures) is an activity of human thought and action, and therefore can be modeled using artificial intelligence techniques. Preliminary work on identifying cultural frames of "indigenous science" has been done (Carlton-Foss, 1975) and the results suggest that the framing task, though very challenging, can be accomplished more readily than the modeling of the emotional patterns of the "common man" in the street.

REFERENCES

Aitchison, P., Concurrent windows on Unix enhance productivity, Systems & Software, April 1984, 112-115.

American Society of Heating Refrigerating and Air-conditioning Engineers, ASHRAE Handbook: 1981 Fundamentals, Atlanta: ASHRAE, 1981.

Card, S. K., Moran, T. P. and Newell, A., The psychology of human-computer interaction, Hillsdale, N.J.: Erlbaum, 1983.

Carlton-Foss, J. A., Indigenous science and the psychosocial domain, [Unpublished manuscript.] Cambridge, MA: Human Technical Systems Inc., 1975.

Casson, R. W., Language, culture, and cognition, New York: Macmillan, 1981.

Cohen, P. R. and Feigenbaum, E. A., The handbook of artifical intelligence, (3 Vols.), Los Altos, CA: William Kaufman Inc., 1982.

Evans, R., The Atomic Nucleus, New York: McGraw-Hill, 1955.

Frautschi, S. C., Regge poles and S-matrix theory, New York: Benjamin, 1963.

Greenberg, D., Architectural Record, September 1984, 150-159.

Holton, G., Thematic origins of scientific thought: Kepler to Einstein, Cambridge: Harvard University Press, 1973.

Jacobsen, S., Operating system creates portable environment for window-based program, In Systems & Software, April 1984, 116-122.

Kolata, G., Computing in the language of science, Science, 13 April 1984, 140-141.

Konopasek, M., Personal communications.

Konopasek, M., TK!Solver, Wellesley, MA: Software Arts, 1984.

Larribeau, R., Integrated applications get a friendly face with windows and a mouse, In Systems & Software, April 1984, 125-130.

Messiah, A., Quantum Mechanics, (Vol. I), Amsterdam: North-Holland, 1961.

Omnes, R. and Froissart, M., Mandelstam theory and Regge poles, New York: Benjamin, 1963.

Osgood, W. F., <u>Topics in the theory of functions of several complex variables</u>, New York: Dover, 1966.

Polanyi, M., Chicago: University of Chicago Press, 1962.

Whorf, B. L., <u>Language, thought and reality</u>, Cambridge: MIT Press, 1956.

QUESTION AND ANSWER PERIOD

JERNIGAN
Do we have any questions from the audience

PIETRZYKOWSKI
It is more of a remark than a question. I have a feeling that it really needs a little bit of clarification. We talked already quite a bit about computer languages and sort of throw them, more or less, in one basket; but there exist very distinct boundaries between different types of languages. For example, languages like FORTRAN or PASCAL or C or ADA, all of them are procedural languages which believe it or not are now called Von Neumann type of languages which reflects the structure of the Von Neumann type of architecture. Language like LISP, its only problem was MacCarthy thought ten years too early about it. It was not properly appreciated because of the low level of computer technology and their general software awareness; however, we are dealing with a language of a completely different structure. Functional language which recently, finally is brought to its proper dimension by development of data flow architecture and stuff like that. So indeed it is very interesting, it is a pity that in the previous talk the discussion was concentrated on FORTRAN, was no comparison how students, for example, would react on something like LISP. From my very limited experience, I never had any statistical, I don't like it, but I thought that they reacted very, very, differently. Simply differently, I am not saying better or worse, but we are dealing with quite a different conceptual framework. We will go even a step further and to go to the logic programming languages like PROLOG or its various now parallel extensions, again it is a completely different thing. The criticism which was before brought and rightly so, that is, a teaching structure, top-down programming, in languages like FORTRAN or PASCAL or whatever is a joke, because you do all this, I am not only teaching, also developing, it applied to both. It is simply possible we are just kidding ourselves, you can't do it. You just write a little statement and then you jump and make a code. However, it is true that in such a language as LISP, to some extent it is possible. In such a language as PROLOG it is really possible. So I am not proving any point; however, to the last part which is quite interesting in the talk, the issue of images in and relating in language, it is a very refreshing thing I heard, because I have heard about specialized languages for let us say Physics and whatever already 20 years; nothing, of course, has happened except various bunches of subroutines in FORTRAN, that is what has happened. But here I heard something really refreshingly new, which may find, like seeing work by a close type of analytic equations, etc. That is a very interesting view, which I first heard here, and hopefully maybe this direction will be pursued.

CARLTON-FOSS
I would just say that I like your comments very much about LISP versus
procedural languages, and I would say that it is extremely important to
have scientists and perhaps engineers, but especially scientists, people
who are doing basic research and applied research, to almost be forced to
have several different frames of reference that they are working with, to
learn both LISP and FORTRAN, so that LISP can provide a different template
with which to work with a problem.

The Role of Language in Problem Solving I
R. Jernigan, B.W. Hamill, and D.M. Weintraub (Editors)
© Elsevier Science Publishers B.V. (North-Holland), 1985

AI LANGUAGES SHOULD EXPRESS GRADUALNESS

Roy Rada

LHNCBC, Computer Science Branch
National Library of Medicine
Bethesda, MD 20209

Artificial Intelligence (AI) deals with complex problems
whose solution requires substantial "problem-specific
knowledge". This knowledge can be embedded into the
problem description but requires a language with special
properties. The hypothesis of this paper is that the
language must express gradualness. Gradualness here
means that small changes in the structural presentation
of a string of symbols in the language correspond to
small changes in the function or meaning of that string.
This hypothesis is supported by mathematical characteri-
zations and studies of natural and AI systems. The author
also presents a sequence of experiments in "knowledge
acquisition" for expert systems. These experiments show
the role of gradualness in problem description.
KEYWORDS: Artificial intelligence, cognitive science,
knowledge acquisition, languages, machine learning,
problem solving.

1. INTRODUCTION

What is the role of language in problem solving? In formal language theory
[Hopc79] a language is simply a set of strings of symbols. In more common
usage, a language is a means of expression. Problem solving involves a
problem solver and a problem. The problem solver and the problem are both
represented in a language. The point of this paper is that the language
appropriate for problem solving takes advantage of regularity in the
problem.

To illustrate the multiple levels at which representation or language must
coincide with structure in the problem and problem solver, knowledge
acquisition for expert systems is examined. The knowledge acquisition
problem is different from many of the problems that have attracted computer
scientists in the past, because it relies heavily on complex, non-mathemat-
ical domains of human experience. A language for knowledge acquisition
must take advantage of the structure in knowledge.

The goal of Artificial Intelligence (AI) is to create Problem Solving
machines. The Problem Solving activity requires a Problem and a Problem
Solver. We assume that a fundamental and necessary mechanism of a Problem
Solver is the universal, weak method of "generate and test". The challenging and largely unanswered question remains: "what additional properties
are necessary and sufficient for Problem Solving". One necessary property

of successful problem solving clearly concerns the relationship between
Problem and Problem Solver. Given that the Problem Solver is natural, we
hypothesize that the Problem should appear natural. Since the appearance
of the problem depends in part on the language for representing it, and
since we are interested in AI problems, the first hypothesis of this paper
is that AI languages should be natural.

The second hypothesis of this paper (see Table 1) concerns the attributes
of a language that allow it to naturally represent a problem. Problems may
be viewed as state-spaces. A key property of naturalness is gradualness,
where gradualness means that similar states tend to have similar values.
More-natural AI languages should be able to more easily represent information about which states have similar values. One key to acquiring this
gradualness is to decompose the state-space of the problem. The development of "sub-state-spaces" allows credit assignment to more precisely say
which operators will give new states whose value is closer to the goal.
Finally, the decomposition of states may make human knowledge about states
and similarity of states more accessible and helpful. Each state in the
initial formulation of a problem may be so complex that no one can tell
which other states are similar to it. But once the states are decomposed
so that, for instance, a network of medical terms is viewed no longer as a
single state but each medical term becomes a state, then medical knowledge
about terms and their relations can be applied to the problem. The knowledge that certain terms are similar can be the key to a graceful move
through the state space.

2. LITERATURE REVIEW

Since the weather is now mild, we anticipate that it will remain mild over
the next hour-changes should be gradual. Typically, for living systems
success is founded on gradualness. [Conr79] The crossover operator in
genetics takes advantage of good leading to good. [Holl75] Work in
cognitive psychology points to the importance of regularity. The more a
person rehearses a phrase the more likely the person is to remember the
phrase. [Kint82] Operant conditioning shows how behaviors are gradually
acquired or lost. Philosopy also provides evidence of the role of gradualness in problem solving. New theories of science evidently arise from
combinations of good parts of earlier theories. [Dard82] Likewise,
reasoning by analogy requires that two experiences be represented in the
same language and that similar units in the representation of each
experience be compared. [Dard82]

The optimum of a smooth, unimodal function can be easily found with a
simple gradient search. Interest during the Second World War on strategies
for finding enemy submarines led to the development of search strategies
which show that effort should be proportioned to the expected gain.
[Ston75] In the area of artificial intelligence, the importance of
gradualness can be seen repeatedly. Lenat [Lena83] talks about the
importance of "continuity" (which is identical to gradualness). Lebowitz
[Lebo83] uses the term "predictability"; Carbonell [Carb83] the term
"similarity"; and Amarel [Amar85] the term "regularity"-all of which terms
have much in common with "gradualness".

Ernst and Newell's General Problem Solver [Erns69] solved problems by
creating subgoals when the step from initial state to goal state was too
big. Subgoaling is a way to achieve gradualness in a search space.

HYPOTHESES	
Hypothesis 1	AI languages should capture gradualness
Hypothesis 2	Decomposability facilitates gradualness

ROLES OF GRADUALNESS	
Domain	*Role*
Biology	Small changes in amino acid sequence of enzyme tend to result in small changes in its pattern recognition ability
Psychology	Proximity of reward to behavior allows conditioning.
Philosophy	Similarity between 2 scientific theories facilitates discovery of new theories
Computer Science	Smooth, unimodal functions are easily optimized

AI Terms for Gradualness	
term	*author*
continuity	Lenat
predictability	Lebowitz
similarity	Carbonell
regularity	Amarel

TABLE 1
A SUMMARY IN TABULAR FORM OF HYPOTHESES AND LITERATURE REVIEW

Rosenbloom [Rose84] has been emphasizing the role that knowledge can play in guiding the choice of subgoals. Pearl [Pear83] argues that problems are solved by consulting simplified models of the problem. The key to the applicability of these simplified models is the decomposability of the problem. Decomposability means in part that all subgoals can be solved independently. Pearl tries to discover the decomposability or gradualness of a problem by constructing models of the problem and manipulating parameters or constraints.

One of the ways that nature helps problem solving is by providing states that can be readily changed without drastically affecting their value. This kind of gradualness is on the surface different from decomposability. An enzyme is a folded transcription of a genome and has an active site. Enzyme$_1$ in Figure 1 has 4 components; **enzyme**$_2$ in Figure 2 has two "buffer" components that **enzyme**$_1$ lacks (these Figures come from Conrad [Conr79]). Changing of a component in **enzyme**$_2$ leads to a less drastic

R. Rada

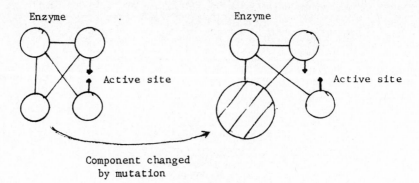

FIGURE 1
ENZYME MUTATION WITHOUT BUFFER

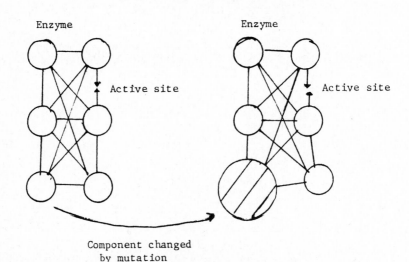

FIGURE 2
ENZYME MUTATION WITH BUFFER

change in active site than a similar change to the structure of $enzyme_1$.
Gradualness is different from reliability. Reliability can be achieved
with redundancy. A reliable system is immune to small changes in its
structure. A gradual system responds in small ways to structural changes.

3. FORMAL ASSESSMENTS

One way to argue for the importance of gradualness involves probabilities.
Given that the proper operator to apply at each decision point is unclear,
the problem solving system takes a chance. The probability of reaching the
goal within a certain number of steps depends on the extent to which the
problem allows mistakes to be gracefully corrected. Suppose the probability of a successful step is x and that the goal can be reached after a

certain set of steps have been taken. If any intervening steps are
allowed, then the time to reach the goal scales as x. If, however, 2 steps
must occur in sequence, then the time to reach the goal scales as x-
squared. When x is 0.001, x squared is 0.000001, and reaching the goal
takes 1000 times longer when a pair of steps must occur together. [Conr79]

Problem solving has been described as a process involving a Problem Solver
and a Problem. A problem can be defined as a Search Space SS where SS
itself is a quadruple of: [Barr81]

    (1)   States S,
    (2)   Operators O for going from state to state,
    (3)   Initial state I, and
    (4)   Goal state G.

Problem Solver PSer searches through **SS**. PSer typically utilizes a
heuristic evaluation function f to assess the value of any given state in
S. [Nils80] The representation of **SS** should allow the PSer to predict the
value of some states in **S** based on the value of other states in **S**. An
efficient search must be able to examine some states in **S** and make reliable
predictions about the value or nearness to the goal G of other states in
S. No PSer can effeciently search all possible **SS**. Some problems are
exponentially hard, [Gare79] and an ideal learning machine can not exist.
[Oshe82]

The "continuity" of a **SS** can be formalized. The continuity C of **SS** depends
on the extent to which similar states in **S** have similar f values. A
measure of **SS** continuity might build on:

    (1)   similarity between states **s** and s' is proportional to the number
          of operators required to go from **s** to s' and

    (2)   the similarity of f(s) and f(s') is proportional to f(s) - f(s').

A measure of gradualness or continuity then needs to assess the extent to
which similar states have similar values.

One way to measure gradualness first defines equivalence classes of states
S9i8 and heuristic-predictions h. S_i is defined to be [s | s ES and s can
be reached from I by a sequence of i operations]. The heuristic-predic-
tion function h on s E S_i is h(s) = best [f(w) | w E S_{i+1} and 1 move
connects s to w]. To determine the continuity C the following 3 steps are
performed:

    (i)   Order all s E S_i so that $f(s_1) \leq f(s_2) \leq o\ o\ o \leq f(s_m)$.

    (ii)   Compute h for each s and create the sequence $h(s_1)$,
               $h(s_2), \ldots, h(s_m)$.

    (iii) $C_k(S_i, f)$ = number of inversions in the sequence $h(s_1), \ldots,$
               $h(s_k)$.

C has been applied to the Minimal Spanning Tree Problem and the Traveling
Salesman Problem (see Table 2). The results suggest that computationally
less tractable problems have less gradualness. [Rada83]

R. Rada

Definition of TSP	
S	power set over weighted edges
O	adds an edge to a set of edges
I	empty set of edges
G	complete tour of minimal cost

TABLE 2
DEFINITION OF TRAVELING SALESMAN PROBLEM

Often the first definition of **SS** does not produce enough continuity to facilitate problem solving. PSer might then transform **SS** into an **SS'** that better reflects the continuity in **SS**. One example of such a transformation occurs in the literature on the Traveling Salesman Problem (TSP). The intuitive, greedy PSer attack on TSP involves starting at the shortest edge and always adding the shortest available edge. [Rein77] When the edge weights of TSP are constrained to satisfy the triangle inequality, an approximation algorithm is able to guarantee a tour whose cost is no more than 1.5 times the cost of the optimal tour. [Chri81] This best-known approximation algorithm first finds the Minimal-Cost Spanning Tree (MST) in the graph. The original TSP **SS** is now transformed into a **SS'** where S' has fewer states than **S** and **O'** must maintain the skeleton created by the MST. This transformation creates a more continuous search space-in **SS'** the application of **O'** is more likely to lead from valuable state to valuable state than in **SS**.

4. WEIGHT REFINEMENT

In acquiring knowledge for an expert system, a human usually begins by studying textbooks and talking with experts. The knowledge of the domain is partly structured for computer usage and the process of refining the knowledge continues. In trying to further understand how this building process occurs, one has the opportunity to examine any phase of it. In machine learning the favored approach is to take a working system, destroy some small part of it, apply a learning or refinement method to the system, and test whether the system returns to its previous high level of performance.

4.1 FRAMEWORK FOR WEIGHT REFINEMENT

In the weight refinement problem, an expert initially supplies a set of weighted rules which perform suboptimally. The form of these rules may be restricted to

$$p_1 \; G \; ...G \; p_j \; \sim \; w \; x \; p_{j+1},$$

where p_i is a predicate and $w \; E \; [-1,1]$. The syntax and semantics of the rule system discussed here are similar to that of EMYCIN. [Mell79]

The problem of finding the correct weights for the rules can be viewed as a search through a search space. For the weight refinement experiments, the initial state **I** is some given set of weighted rules RULES. The set **S** of states = [s | s has the same form as **I** but may differ in the weight assignments]. An operator in **O** may add or subtract from weights in **s**. The goal states are defined in terms of a Learning Set **LS**. **LS** = [ls$_1$,...,ls$_m$], where ls$_i$ = (Givens, Diagnoses). Both Givens and Diagnoses are sets of (predicate, certainty) pairs. A state **s** in **S** performs correctly on **ls**, if the rules in **s** operate on Givens so as to produce Diagnoses. The goal states **G** in **S** are those which perform perfectly on **LS**.

A control strategy must choose which weight change operators to apply. The strategy may involve testing rules against the learning set and adjusting weights that are responsible for mistakes. Although the weighted, rule-based system resembles in some ways the perceptron, a set of rules is more powerful than a single perceptron. This distinction is pertinent because the perceptron learning algorithm works on a single perceptron and has not, to this author's knowledge, been shown to work on a system like the weighted, rule base.

Greedy search [Aho83] fails for weight adjustment. Consider the simple case where weights are either 0.5 or 1.0 and the Initial State and Learning Set are as in Table 3. A greedy search would notice that by changing the weight on a → 0.5 → b from 0.5 to 1.0 that 3 diagnoses, d, e, and f, would be correctly concluded and a score of +3 attained. On the other hand, only by keeping a → 0.5 → b and increasing weights with b → 1 → d, b → 1 → e, and b → 1 → f can all 4 diagnoses be correctly reached. A greedy search would not be willing to postpone the achievement of the +3 score, and would thus make a mistake.

Initial State		Learning Set	
		Givens	Diagnoses
	b→.5→c	(a,1)	(c,.25)
a→.5→b	b→.5→d		(d,.5)
	b→.5→e		(e,.5)
	b→.5→f		(f,.5)

TABLE 3
WHERE GREEDY SEARCH FAILS FOR WEIGHT REFINEMENT

4.2 A PROGRAM FOR WEIGHT REFINEMENT

An algorithm, called LEARNER, for the systematic correction of the weights is based on the strategy of raising the attenuations when a proposition is concluded without enough certainty and lowering the attenuations when a proposition is concluded with too much certainty. The amount by which a rule's weight is to be changed depends on the past performance of that rule. LEARNER has a Credit Assignment (CA) and Weight Adjustment (WA) phase. Each phase reads an element **ls** from **LS**. RULES are then applied to the Givens of **ls**. The conclusions of the rules are placed in a group called Conclusions. The predicates in Conclusions and Diagnoses are divided into 3 categories:

(i)   **Perfect** are those that have the same certainty in Conclusions and Diagnoses,

(ii)   **TooMuch** are those that have more certainty in Conclusions than in Diagnoses, and

(iii)   **TooLittle** are those that have less certainty in Conclusions
             than in Diagnoses.

Each rule has a property list with the following identifiers: Perfect,
TooMuch, and TooLittle. For a given **ls**, CA determines the sequences of
rules that lead to each predicate in Conclusions. For cases (i), (ii), and
(iii), each of those rules has its Perfect, TooMuch, or TooLittle property,
respectively, augmented with **ls**.

WA changes the weights to improve the performance of RULES on **LS**. For case
(i) no adjustments are necessary. For case (ii), the attenuations of those
rules that lead from Givens to Diagnoses are lowered, if the the number of
ls on TooMuch - TooLittle - Perfect is greater than 0. Each attenuation on
the path is lowered in proportion to the difference between the certainty
in Conclusions and Diagnoses. Case (iii) is handled similarly to case
(ii).

LEARNER, with a number of embellishments not described above, was
implemented in a 500-line LISP program. By the time the design and
implementation was complete, however, 117 rules for interpretation of
computed tomograms had been hand-tuned on 25 cases from the radiology files
of Henry Ford Hospital. We decided to test LEARNER on these 117 rules by
randomly altering weights and applying LEARNER. About 40 experiments
covered small and major weight changes to anywhere from 1 rule to all
rules. In all but 1 of the trials, LEARNER moved step-by-step closer to
correct weights and reached the goal state in 1 to 33 iterations of the
CA+WA cycle.

The kind of initial state and learning set on which LEARNER can fail is
shown in Table 4. The rule $b \rightarrow 0.5 \rightarrow c$ is marked by CA as both Perfect for
ls_1 and TooMuch for ls_2. Since the rule's history shows TooMuch and
Perfect, WA makes no changes to the rule. The Goal state requires that
$a \rightarrow 0.5 \rightarrow c$ and $b \rightarrow 0.25 \rightarrow c$.

Initial State= { $a \rightarrow .25 \rightarrow c$, $b \rightarrow .5 \rightarrow c$ }

Case	Learning Set	
	Givens	Diagnoses
ls_1	(a,1),(b,1)	(c,.625)
ls_2	(b,1)	(c,.25)

TABLE 4
DIFFICULT SITUATION FOR LEARNER

Two experiments were performed on alternative strategies for weight refinement. In experiment (1) each time an incorrect conclusion was reached for
a learning case, every rule on the path to the conclusions of that learning
case was amended. In experiment (2), at least, one rule was changed each
time a conclusion was wrong, but the strategy for choosing that rule was
based on a comparative assessment of the histories of the rules.

In experiment (1), the amendments were scaled so that the "worst" rules were the most changed. This scaling was achieved for the case of a TooMuch conclusion by first evaluating for each rule the number of TooMuch - Perfect - TooLittle and then mapping this quantity into the positive numbers by changing a negative number to a tiny positive number. A similar calculation was done for the TooLittle case. Tests of this method on the computer showed a partial resolution to the earlier-discovered difficulty. The cases that were easily handled before were still easily handled and the previously impossible cases were now being slowly corrected. The problem was that changes for the difficult cases occurred very slowly, due to the tiny scaling factor.

Method (2) was designed to improve efficiency, and computer experiments with it have thus far confirmed our expectations. If there are one or more rules on a TooMuch path whose history contains no Perfect and no TooLittle, then those rule(s) are changed and other rules are unaltered. If there are no such rules, then a rule that has the lowest value of Perfect + TooLittle is found and changed. With method (2), the cases thus far tested have been quickly solved.

An exhaustive search for the correct weight assignments would examine a number of states that was an exponential function of the number of rules. If there are r rules and each weight may assume one of w values, then the number of states in the search space is w^r. If the number of elements in LS is on the order of r, and if each path for an ls has no more than one weight incorrect, then WA would test $O(r)$ states.

In the weight refinement experiments the accuracy of the greedy search is increased through decomposition of the search space SS. A rule-based system that is to adapt through experience should have properties of each rule that record the performance history of that particular rule. The operators in SS must, in turn, respond to these performance histories. Several levels of granularity are helpful, as the operators must note which rules have a history of interacting in confounding ways.

5. PREDICATE GROUPING

5.1 ISSUES

The weight refinement problem is partially solved. The next knowledge refinement problem in the descent from easier to harder problems is the correct grouping of predicates in rules given that all the correct predicates have been determined for the domain. Holland's [Holl75] genetic algorithm includes both a language for problem representation which is sets of strings and a search strategy which hinges on reproduction with change and selection of the fit. If the predicate grouping problem is attacked with the genetic algorithm, serious limitations are discovered. In a search space of strings where each string corresponds to a sequence of predicates from an if-then rule, the genetic algorithm finds insufficient continuity to allow it to progress smoothly. Bethke [Beth80] has shown the importance of smoothness in the search space to success of the genetic algorithm. Perhaps a representation of the rules can be found such that that smoothness for the genetic algorithm exists. A more straightforward approach, however, is to use a more powerful language for the search space.

The principle of the genetic algorithm is that in natural systems good
leads to good. One can take advantage of this fundamentally sound
principle for the predicate grouping problem, by adding more sophisticated
credit assignment and more heuristics and knowledge for guiding "reproduction with change". The credit assignment must attach more than a scalar of
fitness to a rule. The history of the rule should testify to whether the
rule made conclusions which were too specific or too general and if so,
exactly which conclusions were too specific or too general. The importance
of generalization and specialization hierarchies for machine learning has
been stressed by Mitchell. [Mitc82]

When changing a rule R by exchanging predicates between R and some other
rule or body of predicates, additional knowledge about the relationships
among predicates seems necessary. The genetic algorithm does not
call for such extra information, but experience with knowledge refinement
has indicated repeatedly the relative weakness of diagnostic rules alone.
Reggia's [Regg80] work on coma localization showed the need for geometric
languages. Clancey's [Clan83] explanation systems needed languages for
expressing physiological relationships and not just the rules of diagnosis
and treatment in MYCIN.

5.2 EXPERIMENTS

Twenty-five cases were collected from the neuroradiology computed
tomography files of Henry Ford Hospital. [Acke84] The cases were randomly
drawn so as to have 5 instances of meningioma, 5 of subdural hematoma, 5 of
infarct, 5 of metastasis, and 5 of normal. A hierarchy of terms was
created for classification of the radiological images, and the same terms
were used in the construction of rules for diagnosis.

To experiment with predicate refinement a Search Space SS was formalized.
From a small sample of the 25 cases, rules were extracted that corresponded
directly to the descriptions of the cases. This set of rules was the
Initial state of SS. Weights were ignored in this experiment so that rules
were simply of the form

$$p_1 \; \lambda \ldots \lambda \; p_j \; \rightarrow \; p_{j+1}.$$

The Operators in SS replaced a predicate p_j with a predicate p_k. The Goal
in SS was that a state perform well on the LS cases. This resembled the SS
of weight refinement except now predicates were altered rather than
weights.

Two sets of experiments were performed: the first showed weaknesses in the
genetic+heuristic strategy, and the second showed the importance of extensive heuristics and domain-specific knowledge. The experiments were done
by hand and the results must be considered more suggestive than definitive.

The genetic algorithm is an iterative 3-phase process. Strings (which
could be interpreted as rules or as chromosomes) are the basic data
structure of the genetic algorithm. In phase 1, each rule is assigned a
value based on its performance on the learning set. In phase 2, the rules
are copied in proportion to their value. In phase 3, predicates of one
rule are exchanged with predicates of another rule (see Figure 3). The
fundamental unit of a rule is a predicate, like "location lesion vascular".
This structuring of the rules provides considerable assistance to the

genetic algorithm. Nevertheless, the genetic algorithm without further assistance from heuristics did not perform well in the experiments.

RECOMBINATION	
OLD RULES	NEW RULES
rule ▭▭▭▭▭▭	▭▭▭◯◯◯◯
rule ◯◯◯◯◯◯◯	◯◯◯◯▭▭▭

FIGURE 3
ILLUSTRATION OF RECOMBINATION

To guide the changing of predicates, a semantic net was constructed. This net added relationships to the hierarchy of predicates. For instance, since infarct and metastasis have similar findings on a Computed Tomogram (CT), and since an enhancement study on CT is recommended for suspected metastasis, the knowledge base was augmented by the addition of labeled arrows connecting infarct to metastasis and metastasis to enhancement study (see Figure 4). When a change was to be invoked in the rules, the semantic net was consulted to guide the choice of which predicates in which rules to exchange. The principle was that predicates which were one link apart in the net were similar predicates and should be given high priority for swapping.

infarct—*similar*—metastasis—*do*—enhance

FIGURE 4
PART OF SEMANTIC NET

The experiments with semantic-net-plus-genetic-algorithm led to better rule performance than the experiments with only the genetic algorithm, but the performance was still not impressive. The genetic algorithm with a few heuristics did not successfully refine CT rules. The semantic net gave an extra dimension of structure-function relationship to the domain and added some smoothness to the search space, but more heuristics were needed.

Additional heuristics were installed for the final set of experiments. These heuristics affected the choice of the initial set of rules. They also substantially augmented the role of credit assignment. Whereas in the early experiments the initial rules were arbitrarily selected, in the final experiments each rule in the Initial State had location, density, and shape predicates. The "weak" rule in Figure 5 correctly diagnosed one of the metastasis cases, but also wrongly responded to an infarct case. This credit assigment further guided the choice of rule alterations. For instance, since the "weak" rule was too general, a heuristic added a predicate to the rule that was close to metastasis in the semantic net. In this instance, enhancement positive was added to the "weak" rule. The resulting "strong" rule (see Figure 5) caused the rule to continue to fire correctly for metastasis and to no longer conclude infarct. With this heuristically enriched approach, the initial base of rules was changed so that performance of the new rules was optimal. The credit assignment phase can

tell about the need for generalization or specialization in the rule,
[Mitc82] and the semantic-net then guides the choice of the predicates to
be tested in the rule.

Weak Rule	
Antecedents	*Consequent*
location intracerebral density hypodense shape irregular	diagnosis metastasis

Strong Rule	
Antecedents	*Consequent*
location intracerebral density hypodense shape irregular **enhancement positive**	diagnosis metastasis

FIGURE 5
WEAK AND STRONG RULE

6. TEXT PROCESSING

To the extent that techniques are available for handling predicate grouping, the more fundamental question arises of how can a computer read text
and build an expert system. A different and perhaps simpler problem is
addressed routinely by researchers in the field of bibliographic retrieval.
A popular methodology requires that human indexers read text and determine
from a thesauri of key words which words characterize the text. [Salt83]
An inverted file then stores pointers from these keywords to the text.
Retrieval with computer is then easily performed, given that the library
patron can pose a query in terms of the predefined vocabulary of keywords.
Existing automated systems identify certain words in a text based principly
on frequency of occurrence of those words. [Salt83] Again, an inverted
file is created and retrieval of text is straightforward. In the parlance
of this paper, the keyword plus inverted-file approaches transform the
retrieval problem into a SS with marked continuity. As each new key word
is added to the library patron's query, the set of retrieval articles
changes in a small way. Unfortunately, this transformation of the problem
has denuded the text of much of the vital classification information that
library users would like. The structure of the text as manifested in the
complex relationships of the natural language has been largely lost in the
final disconnected set of keywords that are chosen to characterize the
text. Accordingly, much effort is ongoing to parse and represent text more
robustly.

Extracting knowledge from text requires a substantial knowledge base to
start. Carbonell [Carb84] uses a hierarchical, case-frame representation
of knowledge and text built on the language SRL. Baskin [Bask78] proposed
a knowledge acquisition experiment in which text was processed by a frame-
based knowledge system. The role of sophisticated models of the domain in

trying to build rules from more fundamental text-like knowledge is also
discussed in Swartout. [Swar83]

Which approach to extracting knowledge from text is most appropriate can be
answered in part by referring to work in cognitive psychology. Kintsch
[Kint82] has done extensive experimentation on the ways that people read,
store, and retrieve text. His models which have fared well in tests
against people and machines, should serve as inspiration to computer scientists who are designing systems that are to take advantage of knowledge in
text. Waltz [Walt82] has suggested that for robot-type systems, a language
component must rest on similar primitives as the vision component, and this
has motivated research into neuroscience-motivated schemes for language
processing. Which models of language processing would help development of
expert systems depends on the performance expectations of the expert
system-a robot may need to connect to perceptual primitives, while a
medical diagnostician may best connect to some other set of primitives.

The National Library of Medicine manages one of the world's largest online
bibliographic retrieval services. Articles are classified by employees of
the library. The classification scheme depends on a graph where each node
is aterm from the Medical Subject Headings (MeSH). [Medi84] Typically, a
user at one of many locations around the world interacts with a librarian
who translates the user's request into a set of MeSH keywords. [Hump84]
This translated request is then typed onto a computer terminal, and a set
of citations is returned. The author has devised an algorithm to reference
the MeSH graph and, in response to a user query, to order the articles in
the textual database according to their relevance to the query.

To improve the relevance ranking the MeSH graph can be augmented. Dominiak
[Domi84] adds weights to the edges in the MeSH graph, so that the relevance
algorithm can return better relevance rankings. His algorithm also adjusts
weights through experience in much the way that LEARNER, discussed earlier
in this paper, does. This MeSH weight-adjuster depends on data from users
that shows how they rank articles.

Another approach to the augmentation of MeSH is to link nodes not otherwise
linked. MeSH is relatively edge-poor. Additional, meaningful edges can
guide the relevance algorithm in its matching of articles to queries. We
have developed a strategy for augmenting MeSH that depends in part on other
"knowledge-bases". [Cocc84] One such knowledge-base is Current Medical
Information and Terminology (CMIT), which gives many important relationship
among terms. [Fink81] Guided by various credit-assignment schemes, our
edge adjustment algorithm refers to CMIT, as the earlier described
predicate-adjustment algorithm referred to the semantic net of Computed
Tomography primitives. A kind of reasoning by analogy occurs between CMIT
and MeSH.

7. DISCUSSION

In Artificial Intelligence the issue often arises too early of what
computer language should be used. The first question should be what
concepts and relations best capture the regularity in the problem space.
In knowledge acquisition for expert systems via text, the first questions
should concern the nature of the text and operation of the expert. Only
later might one ask whether the knowledge would be better represented in an
object-oriented language like LOOPS [Stef83] or a frame-based language like

KRL. [Bobr77] Likewise, before one considers an ATN parser like RUS
[Wood73] or a case-frame parser like DYPAR [Carb84], one should determine
what are the basic features of a parser that would facilitate gradual,
predictable motion in the search space of the parser.

The main issues in languages for problem solving have to do with the
conceptualization and formalization of the problem solving. The actual
computer languages which are used are of secondary importance. The AI
literature abounds with descriptions of high-level languages built for
special purposes, but clear arguments for why one language is better or
worse than another are often lacking. [Broo79] The best language is the
one that is easiest to use, which at the deepest level means that the
language has the same kind of regularity that the search space has.

If a problem is defined as a state space, then certain relationships which
exist among states must be recognizable by the problem solver. For many
complex tasks, such as knowledge acquisition for expert systems, the proper
representational scheme should have terms that correspond to the knowledge
which people use in the tasks. The popular AI computer languages like LISP
or PROLOG usually do not directly correspond to the natural terms of the
problem solving task, but can be used to build higher-level computer languages that are then appropriate for certain complex problems. The strategy
followed in this paper has been to take a greedy-like search and to test
how well that search performs on a simple representation of a problem.
Additional knowledge is then added to the problem, and the efficiency of
the new search is assessed. To add knowledge, the state space is decomposed. Credit assignment on the sub-space is made more detailed, and
knowledge about similarity of substates is obtained from various sources.

The theme guiding the addition of knowledge is that "knowledge gives
gradualness" and "gradualness facilitates learning". Any problem solving
system invokes some kind of generate-and-test mechanism. Given generate-
and-test, one also needs for successful problem solving to have gradualness. To achieve gradualness, the program should know which states are
likely to have similar values. The importance of assigning values by a
critic is well-recognized in the machine learning literature. [Mitc83]
One of the key challenges is to properly percolate reward among those
components of the program which most deserve the reward.

One way to circumvent some of the difficulty of credit assignment is to
take advantage of whatever information exists about "similarity" of states
and hope that similar states will have similar values in tractable problems. The attractiveness of working with numbers (as in weighted, rule-
base systems) is that the numbers have a natural metric of similarity (they
have a total ordering). A state which includes numbers may be slightly
changed by slightly changing a number, and the resultant effect on the
value or performance of the state may well show small improvement. In more
complex situations, such as changing edges on a graph (which corresponds to
grouping predicates in rule refinement), knowlege about how the nodes are
related in other, similar graphs, may be a useful guide. A reasonable
heuristic in human problem solving is that similar states have similar
values. The vast stores of human knowledge which already exist on
computers often speak to the similarity of states. A language for
knowledge acquisition should be able to represent a search space at several
levels of granularity, and to detect similarity between a substate in the
search space and a concept in another knowledge-base.

ACKNOWLEDGMENTS
Much of this work was supported by NSF Grant ECS-84-06683 while the author
was at Wayne State University. C. Coccia, S. Humphrey, N. Miller, Y.
Rhine, and J. Smallwood, have contributed significantly to the experimentation. Parts of this paper have appeared elsewhere. [Rada84]

REFERENCES

Acke84. Lauren Ackerman, Matthew Burke, and Roy Rada, "Knowledge Represen-
    tation of CT Scan of the Head," Proc SPIE, Vol 454: Applic Optical
    Instrument Medicine XII, pp. 443-447, 1984.

Aho83. A Aho, J Hopcroft, and J Ullman, Data Structures and Algorithms,
    Addison-Wesley, Reading, MA,1983.

Amar85. Saul Amarel, "Problems of Representation in Heuristic Problem
    Solving," Proc. Conf. Role of Language in Problem Solving, p. to
    appear, North-Holland, Amsterdam, 1985.

Barr81. A Barr and E Feigenbaum, Handbook of Artificial Intelligence, Vol.
    1, William Kaufman, Inc, Los Altos, CA, 1981.

Bask78. A B Baskin, "MEDIKAS-An Interactive Knowledge Acquisition System,"
    Proc. Second Annual Symp. Computer Applic. Medical Care, IEEE Computer
    Soc Press, 1978.

Beth80. Albert Bethke, "Genetic Algorithms as Function Optimizers," PhD
    Thesis, Dept Computer and Communication Sciences, Univ Michigan, Ann
    Arbor, Michigan., 1980.

Bobr77. D Bobrow and T Winograd, "Overview of KRL, a Knowledge
    Representation Language,"Cognitive Science, 1,   pp. 3-46, 1977.

Broo79. R Brooks and J Heiser, "Transferability of a Rule-Based Control
    Structure to a New Knowledge Domain," Proc. Third Annual Computer
    Applic. Medical Care, pp. 56-63 ,IEEE Computer Soc Press, 1979.

Carb83. Jaime Carbonell, "Learning by Analogy," in Machine Learning, ed. T
    Mitchell, pp. 137-161, Tioga Publishing, Palo Alto, CA, 1983.

Carb84. Jaime Carbonell and R Frederking, "Natural Language Interfaces to
    Knowledge-Based Systems," in The Factory of the Future, ed. D Reddy,
    Digital Press, 1984.

Chri81. N Christofides and A Ningozzi, "Exact Algorithms for Vehicle
    Routing Problem Based on Spanning Tree and Shortest Path Relaxations,"
    Math Program, 20, 3, pp.255-282, 1981.

Clan83. William Clancey, "The Epistemology of a Rule-Based Expert System-a
    Framework for Explanation," Artificial Intellengence, 20, 3, pp. 215-
    251, 1983.

Cocc84. Craig Coccia, Susanne Humphrey, Nancy Miller, and Roy Rada,
    "Knowledge Base Augmentation and the Improvement of a Bibliographic
    Retrieval System," unpublished document, National Library of Medicine,
    November 1984.

Conr79. Michael Conrad, "Bootstrapping on the Adaptive Landscape," Bio
    Systems, 11, pp. 167-182, 1979.

Dard82. Lindley Darden, "Reasoning by Analogy in Theory Construction,"
    Proc. Philosophy Science Assoc., pp. 147-165, 1982.

Domi84. Michael Dominiak, "Weight Refinement for Information Retrieval,"
    unpublished, Department of Computer Science, George Washington
    University, December 1984.

Erns69. G Ernst and A Newell, GPS: A Case Study in Generality and Problem
    Solving, Academic, New York, 1969.

Fink81. A Finkel, B Gordon, M Baker, and C Fanta, Current Medical
    Information and Technology, American Medical Association, Chicago,
    1981.

Gare79. M Garey and D Johnson, Computers and Interactability, Freeman, San
    Francisco, 1979.

Holl75. John Holland, Adaptation in Natural and Artificial, Systems Univ
    Michigan Press, Ann Arbor, Michigan, 1975.

Hopc79. John Hopcroft and J Ullman, Introduction to Automatic Theory
    Languages, and Computation, Addison-Wesley, Reading, Mass, 1979.

Hump84. Susanne Humphrey, "Biomedical Computing Awareness via MEDLINE,"
    SIGBIO Newsletter, 6, 4, pp. 21-32, ACM, March 1984.

Kint82. Walter Kintsch, "Memory for Text," in Discourse Processing, ed. W
    Kintsch, pp. 186-204, North-Holland, Amsterdam, 1982.

Lebo83. Michael Lebowitz, "Concept Learning in a Rich Input Domain," Proc.
    Internat'l Machine Learning Workshop, pp.177-182, 1983.

Lena83. Douglas Lenat, "The Role of Heuristics in Learning by Discovery,"
    in Machine Learning, ed. T Mitchell, pp. 243-306, Tioga Publishing,
    Palo Alto, CA, 1983.

Medi84. National Library of Medicine, "Medical Subject Headings, Tree
    Structures, 1985," NLM-MED-84-4, Bethesda, Maryland, July 1984.

Mell79. William van Melle, "Domain-Independent Production-Rule System for
    Consultation Programs," Proc. Intern'l Joint Conf. Art. Intell., pp.
    923-925, 1979

Mitc82. Thomas Mitchell, "Generalization as Search ," Artificial
    Intelligence, 18, 2, pp. 203-226, 1982.

Mitc83. Tom Mitchell, Paul Utgoff, and Ranan Banerji, "Learning by
    Experimentation: Acquiring and Refining Problem-Solving Heuristics," in
    Machine Learning, ed. T Mitchell, pp. 163-190, Tioga Publishing, Palo
    Alto, CA, 1983.

Nils80. Nils Nilsson, Principles of Artificial Intellengence, Tioga, Palo
    Alto, 1980.

Oshe82. D Osherson, M Stob, and S Weinstein, "Ideal Learning Machines," Cognitive Science, 6, 3, pp. 277-290, 1982.

Pear83. Judea Pearl, "On the Discovery and Generation of Certain Heuristics," AI Magazine, 4, 1, pp. 23-34, 1983.

Rada83. Roy Rada, "Characterizing Search Spaces," Proc. Intern'l Joint Conf. Art. Intell., pp. 780-782, 1983.

Rada84. Roy Rada, Yvonne Rhine, and Janice Smallwood, "Rule Refinement," Proc. Soc. Computer Applic Medical Care, pp. 62-65, IEEE Press, 1984.

Regg80. J Reggia, T Pula, T Price, and B Perricone, "Towards an Intelligent Textbook of Neurology," Proc. Fourth Ann. Symp. Comp. Applic Med. Care, pp. 190-199, IEEE Press, 1980.

Rein77. E Reingold, J Nievergelt, and N Deo, Combinatorial Algorithms, Prentice-Hall, Englewood Cliffs, NJ, 1977.

Rose84. P Rosenbloom, J Laird, J McDermott, A Newell, and E Orciuch, "R1-Soar: AN Experiment in Knoweldge-Intensive Programming in a Problem-Solving Architecture," Proc. IEEE Workshop on Principles of Knowledge-Based Systems, pp. 65-72, IEEE Computer Soc Press, Silver Spring, MD, 1984.

Salt83. Gerard Salton and Michael McGill, Introduction to Modern Information Retrieval, McGraw-Hill, New York, 1983.

Stef83. M Stefik, D Bobrow, S Mittal, and L Conway, "Knowledge Programming in Loops," The AI Magazine, 4, 3, pp. 3-14, 1983.

Ston75. L D Stone, Theory of Optimal Search, Academic, New York, 1975.

Swar83. William Swartout, "XPLAIN: A System for Creating and Explaining Expert Consulting Systems," Artificial Intelligence, 21, 3, pp. 285-325, 1983.

Walt82. David Waltz, "Event Shape Diagrams," Proc. National Conf. Artificial Intelligence, pp. 84-87, 1982.

Wood73. William Woods, "An Experimental Parsing System for Transition Network Grammars," in Natural Language Processing, ed. R Rustin, pp. 111-154, Algorithmics Press, New York, 1973.

QUESTION AND ANSWER PERIOD

GUIER
Thank you, I found that fascinating. I sense that, in a broad overview, you were saying that by gradualness, you mean that the person is going to iterate on the problem. He is going to go around and around and he is going to try a way of looking at it, both from the start of the problem to the finish, find out what is wrong with it, go back and iterate again. Is that essentially a reasonable paraphrase?

RADA
I would not have used that characterization to try to capture what was my
main point, although perhaps my main point might be properly transformed
into that.

GUIER
That was not a rhetorical question in the sense that that, to me, is
another way of describing a gradualness in approach to a solution and in
particular in getting into computer languages and the PROLOG-like ones -
this business of iterating through it and adding a little bit at a time or
a little complexity at a time or another dimension of the knowledge in
order to get to the conclusion - I find the PROLOG type of language a very
natural way to do that.

RADA
You mean in the sense that you could add a rule easily?

GUIER
Yes. In a very small, gradual way.

RADA
I have another slide where I mention modularity. One of the big arguments
in favor of rule-based systems for representing knowledge is that it is
supposedly easy to add or delete a rule - you move gradually towards better
and better rule-bases. Or at least the change in performance is small when
you add or subtract a rule - that is the claim of some people. I would
argue that that kind of gradualness is important in problem solving and
that for certain kinds of problems, rules are natural and they are a
convenient way to represent knowledge because they lend to modifiability.

BLUM
Just a couple of philosophical comments I would ask you to comment on. One
is, I understand what you are doing in terms of looking for a smooth search
space. That essentially implies continuity, which means that you probably
won't do things which are counter-intuitive, such as bringing a cannibal
and a missionary back across the river. If you find yourself wrapped in
some sort of paradigm, where you are trying to preserve smoothness, you may
end up losing intuition or counter-intuitive creativity. Secondly, the
algorithms that you use are still relatively mathematically sloppy, in the
sense that it may have been a sort of Monte Carlo thing. As you know the
Bayesian pattern matching, and training sets, approaches define things
quite precisely even though we don't understand why they happen. On the
other end are the people working in expert systems saying, we want to use
subjective measures to help us in our decisions because we're really using
the knowledge from our experts. And yet I sense in the smoothing
philosophy you seem to be in part coming between the two ends, that is, we
are starting with expert knowledge and then we are using some of these
other techniques to refine what the experts are doing, and in the process
you may lose some of the expertise. Do you want to comment on those?

RADA
The first of the two, I have earlier addressed. On Friday, when I was
working on this talk I sensed myself to be making a leap relative to what I
had been doing and was very excited. I was talking with somebody on the
phone about this leap, and I thought, "Here I am arguing for gradualness,
but in life one often points to big catastrophic steps. I don't think

leaps necessarily relate to intuition. I think, on the other hand, one can argue that there is a role for gradualness, that in fact, the bigger jumps, in a sense, build on the opportunity to make small steps along the way. If you have a system where you can't make a small change in it and get some small change in its performance, you are inhibited from experimentation. Imagine looking for a solution, and either getting punished, having no effect or causing a drastic change. I don't think that is the way nature works, although what we mark historically are the big changes and leave everything else in the background.

As to the second question, I wasn't certain, at first, whether you were saying that I should make my algorithms more precise, or you were saying, that if these were probabilistic-type algorithms, then in a given field of expertise, such as medicine, people might feel uncomfortable dealing with these methods of changing information. That is a good point. I can say, that in terms of library retrieval we have more flexibility than we have in interpreting radiographic images.

CARLTON-FOSS
I have a couple of associations. The basic question is what are the tradeoffs with the assumption of gradualness? The first thing that popped into my head when I heard you talking about that was the problem of, I have a number between 1 and 100, how do you go about guessing what that is? It is a slightly different problem, but it might be related. Second one is, I have an HP calculator and it has something that solves equations a little bit, or evaluates polynomial expressions, actually, and the final solution depends upon the initial number that you put in. The starting point is a key issue and I presume the calculator uses gradualness as its approach. So there are likely to be some tradeoffs to the gradualness assumption somewhere. Related to that I noticed that you used a metric for your definition of distance in your space, and I'm a little bit more accustomed to using metrics in topological spaces that are continuous rather than discontinuous. Your metric involves operators which are discrete, it looks to me, rather than continuous, and that may have a tremendous impact on what you really mean by gradualness.

RADA
That is the reason I didn't use the word continuity. As to your earlier comments, I agree that the notion of a tradeoff is applicable almost any time. You gave some examples of trying to guess a number and also about the initial values. How you choose the initial value or the initial state makes a big difference in how you get the answer, but I'm not sure how that is a question about tradeoffs. A tradeoff would mean that if you have gradualness you lose something else.

CARLTON-FOSS
What are the down-side tradeoffs, assuming gradualness. For example, maybe a bunch of experts agree on this being a certain constellation of givens implying something over here. But maybe there is another constellation over here that they missed. So if you assume gradualness, you close off possibilities.

RADA
There are a number of other comments about gradualness. You could, for instance, put an enormous amount of redundancy into a system. You could make it such that with a small change it didn't fall apart or you could

make it so redundant that with a small change nothing happened. But that
is not what you want. What you want is that you can make a small change
and get some small meaningful change in behavior, where meaningful would
include that you aren't going to get stuck in a local minimum. This is the
part of gradualness which makes it difficult to create.

GUIER
That is coming close to what I wanted to say. I think that the question
about, can you find the counter-intuitive solutions or get onto counter-
intuitive paths is a very important one. But I would say that those kinds
of things are the sorts of jumps that humans make very well and computers
may not make very well. I have a personal bias that says that we are not
really trying to teach computers how to think like us, we are trying to
teach computers to really help us. I like the thought of your gradualness
in the sense that the computer can take over and give you a lot of support
there. The human may still be needed always to do the counter-intuitive
jumps and to think up brand new experiments to try just in case that is the
better way.

WEINTRAUB
I think part of what we are losing here is that when one thinks of a person
intervening, the question should arise of where the person intervenes.
Computers are tools, and we have end-users, and the problem is between the
end-users and the tool. What has happened in the past has been that we
have a whole cadre of people in this country called analysts and "underneath" them (sometimes they are the same person) programmers who take
applications and map them into things which the machine can handle, and
sometimes we have data analysts who take the output of the computers and
map back into things that people can understand. What some say we are trying to do is move into an area where a novice with computers but an expert
in an application can walk up to the machine and say this is what I want to
do. When Bill Guier talked just now about removing people, I think it is
very important for us to make sure we say where we want to eliminate them.
For example, there was a series of guffaws before when people were talking
about FORTRAN application packages. Having been a particle physicist for
something like seven years and having worked with a program called
Ashmedai, which is a series of FORTRAN routines, which is a devil of a
thing to write up (Ashmedai means devil in Hebrew), but anyway, it was a
series of FORTRAN subroutines where one could feed it a Feinman diagram.
This diagram embodies the ultimate high-level question in quantum mechanics. The program produces the exact probabilities. You didn't need to
know anything about FORTRAN to use it. It was a beautiful package, nice,
user friendly, worked like an HP calculator, and it was written totally in
FORTRAN, and all one needed for it was FORTRAN. For us to have written a
computer language to do exactly the same thing would have involved man
years, whereas Ashmedai was written by two people over one summer. And
when we talk about eliminating people we have to be careful. There we were
eliminating the programmers. We want to be careful not to eliminate the
people on the top end who make the non-intuitive jumps (which are required)
which the computers can't do. I rambled a little bit there, but I think I
made the point.

SESSION:
HOW LANGUAGE CAN AFFECT
ACTIONS AND SOLUTIONS

The Role of Language in Problem Solving I
R. Jernigan, B.W. Hamill, and D.M. Weintraub (Editors)
© Elsevier Science Publishers B.V. (North-Holland), 1985

INVITED ADDRESS:

AI LANGUAGES FOR PROBLEM SOLVING

Jaime Carbonell

Carnegie-Mellon University
Department of Computer Science
Schenley Park
Pittsburgh, Pennsylvania 15213

I.   INTRODUCTION

The topic of "AI Languages for Problem Solving" can be divided into three
parts: First, a very brief assessment of where we are with respect to the
artificial intelligence languages. Second, a very lengthy discussion of
where we want to be, focusing primarily on the problem solving and knowledge acquisition processes themselves to determine the right criteria for
languages and systems in which problem solving methods are encoded. Here
we focus on reasoning strategies and knowledge acquisition as two primary
aspects. And finally, touch upon the shape of tomorrow's problem solving
languages.

II.   EXISTING AI LANGUAGES, A COMPARATIVE ANALYSIS

Let me start now with some criteria that one can use to compare some of the
existing (or past, in the case of Microplanner) languages for artificial
intelligence. I selected the languages in Figure 1 as representatives, not
as an exhaustive set, and each one of the criteria that I will talk about I
am evaluating as to how well each language addresses these particular
desiderata. This is not meant to be a Consumer Reports article from which
one selects which language to buy next week. Each criterion can have a
very different impact for different kinds of applications and for different
kinds of reasoning processes. (I never did like the way the Consumer
Reports rates from 0 to 5 the safety record of an automobile and from 0 to
5 how much wind noise it generates, and both criteria are equally weighted
in the final recommendation.)

First of all, let us look at the criteria for expressive power. Expressive
power means the ability to encode in that language almost any structure or
process you wish to encode in it. Does the language itself limit you in
some way toward certain kinds of structures and away from others? For
example, you can encode almost anything as an s-expression, so LISP rates
high in expressive power. In Microplanner, if you drop down to LISP you
can encode everything that you want to, but directly in Microplanner you
can encode very few things. Some languages, such as OPS, the production
system language in which expert systems like R1 were developed, and Prolog
happen to have very limited expressive power.

Now you may notice that rating low on some of these criteria may actually
be a bonus rather than a detriment. It really does depend upon what you
are trying to do with it. In OPS, for example, one can have only rules and
working memory elements. Working memory elements are one level deep attribute-value lists. The rules are condition-action pairs that, upon matching

CRITERIA	LISP	µPLANNER	OPS	PROLOG	FRAME
Expressive Power	+	=	−	−	+
Built-in K-rep.	−	−	=	=	+
Control Flexibility	+	−	−	−	+
Built-in Infer.	−	+	+	+	=
Speed/Toy	+	−	+	+	=
Scale-up Char	+	———	+	−	=
Dec./Proc. Mesh	=	−	=	+	+
Knowledge-Acq. Support	−	−	=	+	+
Development Environment	+	=	−	−	−

FIGURE 1
COMPARISON OF AI LANGUAGES

these attribute-value lists, perform the action of adding new ones or deleting them from working memory. But that is the range of its expressive power. And perhaps that is also a reason for its effectiveness in solving large classes of problems. Conceptual simplicity is a powerful aid if problems can be cast into such simple structures. Prolog has a particular Horn clause form for expressing information, and this is also fairly restricted.

There are many frame-oriented languages - and here I use KRL, FRL and SRL as instances of a larger generic set because they differ mostly in syntactic sugar (or maybe in the syntactic vinegar, depends upon whether you like the conventions they use). In their underlying structure, they are all sufficiently similar that I am lumping them into one group. The expressive power of frame-based languages is fairly large. Frames can be embedded within each other; one can represent rules as frames; one can represent graphs, trees, procedures, or any other well structured objects as frames - via a fairly direct mapping.

Let us consider some of the other attributes. Some languages have a built-in knowledge representation with well-defined semantics, and others do not. LISP and Microplanner do not have any kind of well-defined semantics. What is the semantics of an s-expression? Anything you want it to be. OPS and Prolog do quite a bit better at that. There is usually some interpretation that one can give to the clauses in Prolog or the working memory attribute-value lists in OPS. Frames have an even better defined semantics, or will allow for, at least, a better defined semantics in that the roles of the respective frames are well defined with respect to each other and with respect to some inheritance criteria as well.

Now what about the flexibility and range of control structures? Can you write any kinds of programs, or are you confined to a particular given control structure? LISP, being the closest to a standard programming language, allows you to do whatever you want but does not provide you much help in doing it. Microplanner, except when ignored by dropping down to the underlying LISP, has only one control structure, and that is depth-first theorem proving with complete backtracking. When a solution attempt fails, it always backtracks chronologically until the search space (or the programmer) is exhausted. OPS and Prolog also have very confining control structures. Now again, having a confining control structure may clarify the ways you think about a problem. If you can cast your problem as an instance of this control structure, you already have it solved. And if it works, it probably solves the problem quite fast because the method has been optimized in its implementation because it was built in by the people who designed the language.

On the other hand, if the problem and control structure do not match, you are in trouble. Let me give you an example of a control structure that does not match a class of problems. Many of the diagnostic expert systems, such as MYCIN, are done with a back-chaining control structure. You start out with a set of hypotheses that you are trying to prove or disprove, a finite set, and you try to back-chain from those hypotheses by considering all the rules that could have given rise to them. MYCIN tries to match the right hand side of each rule to the hypotheses. Then, by going backwards one or fifteen or as many rules as necessary, it attempts to connect the hypotheses to the data that would predict them. Thus, back-chaining determines which conclusions are supported by observables. The presence of supporting data adds evidence in favor of each conclusion over the others. Its absence detracts from the plausibility judgment of that conclusion. Well, that kind of control structure works great when you can enumerate all the possible answers and you must only select among them. In MYCIN's case, it has to select among 50 or 60 possible infecting organisms to be able to diagnose which one is present, and therefore, what drug to prescribe.

Let us now take the R1 task. R1 is the expert system that configures computers, developed by John McDermott at CMU and later on extended and put into real everyday use at Digital. There are somewhere between 10^{52} and 10^{54} possible configurations of a VAX, i.e. possible ways of wiring together any of the thousands of components, cabling them, laying them out on the floor plan, connecting them to the unibus structures and the back planes, and so forth. So it does not make any sense to try to enumerate all of the 10^{52} possibilities in trying to amass evidence in favor of one over the others. The only way you can configure a VAX is by starting from the individual components, building subassemblies that will work, and then assembling these subassemblies into greater, larger ones, backtracking only if and when necessary. So you need a forward-chaining constructive strategy rather than a back-chaining diagnostic strategy.

Now I am going to use that as an example of a mismatch between a language and a problem. One could not use Prolog directly, for example, because it also employs a back-chaining control structure. It would require going through all kinds of machinations to essentially bypass a default control strategy. On the other hand, OPS has a kind of forward-chaining control strategy that happens to match the problem well.

Now you may notice that a built-in inference procedure is almost in opposition to flexibility in control structure. LISP does not have any built-in inference procedure. Microplanner, OPS, and Prolog all do. In some of the frame systems, you notice that these two measures are not quite in opposition, and the reason is because we do not have a zero-sum game here. A particular language that allows you to represent declarative concepts could have more than one way of using these concepts and thereby provide you with a range of possible control strategies. Maybe all the ones you want - maybe not. It is just that, typically, when a control strategy has been built into a language, it has been the only one.

Another consideration one can use is the efficiency, or running time, of a prototype system or any small pilot system. Most AI languages do pretty well except Microplanner. How about trying to scale up the problem to a real system? Microplanner is even worse there (this is an example of a language whose day has come and gone). LISP, on the other hand, at least the new implementations of it, are pretty good for building large systems. OPS is very good too. Prolog, unless we get to some of the parallel decompositions of Prolog that are starting to be investigated now, does not scale up. What large-scale successful expert system has ever been written in Prolog? The answer is none. What large scale AI task has ever been implemented in Prolog? The answer is again none. So Prolog is a little bit of a craze which may not become a mainline, useful language. The jury is still out. Too many people speak as though it were the "replacement for LISP". I do not believe that one language ever replaces another one. COBOL is still with us. But also, Prolog's utility in terms of solving large-scale problems is open to question.

Another important consideration is whether one can mesh declarative and procedural information into the same language. One of the greatest non-issues in AI, one that wasted the time of a large number of good researchers, was something called the declarative-procedural controversy. This raged in the '70's, especially the early '70's. (I must plead guilty to having taken sides on that although I will not tell you which side I took). At that time, the dichotomy was that a language had to encode knowledge in a procedural form, such as the body of a function or the code of a program to be executed but not introspected upon, nor reasoned from, nor modified, etc., or it had to represent it in a declarative form, it could not be run as a program; it was a different kind of knowledge altogether. At best, declarative knowledge would only serve to select which one of a set of external procedures to run. Nowadays, we recognize that it is better to have a knowledge representation that enables interpretation and introspection. The best knowledge representation serves both for reasoning and for execution. That has been one of the most major breakthroughs of Prolog. Information and how that information is applied is represented in the same form. Clause forms can tell you what to do when, and they can also encode declarative, factual information. In fact, the same information that can be used declaratively in one sense can then be used procedurally in another. You can use it to build up a control structure and then you can apply that control structure in solving a problem. Therein Prolog is a huge win, by promulgating that idea and making it concrete. Frame systems are also of this nature, especially the newer frame systems, such as SRL. A frame can represent a rule or can represent a procedure. Through the knowledge acquisition phase, a frame can be augmented, refined, modified, and then executed as a procedure.

Let us now turn to knowledge acquisition support. Is it possible to augment the knowledge base of a system in a natural way? Well, it may be possible to augment the more conventional languages like LISP or Microplanner, but the languages themselves do not help you to do it. Prolog, because of its modularity and uniformity of encoding, sure helps, and many of the frame systems also help. This is tied to being able to mesh declarative and procedural information. Needed information tends to come in a declarative form and then must be used and tested in a more procedural form.

The last criterion has more to do with traditional programming concerns. What about the development environment? Are these languages interactive? Do they provide run-time and support? Can you examine internally what is going on? Can you edit? Can you modularize your data base? What kind of programming environment does there exist? And that has to do more with the maturity of these languages. LISP, both Common LISP and INTERLISP, have very useful supporting environments that allow one to build systems well and to examine how they work internally. The others do not; probably as a "not-yet" type of do not rather than as a "never-will" type of do not.

III. PROBLEM SOLVING STRATEGIES IN AI

At the risk of oversimplification, let me say that what one wants to do with a problem is to find a sequence of actions that map the initial state onto the desired solution. Whereas the bottom line of a solution is just simply a sequence of actions, there are reasons why these actions were taken, i.e., the subgoals that comprised the intermediate steps in solving the problem, and so forth.

We will not go into a detailed description of the state space approach because Saul Amarel has already presented that quite eloquently; instead, let me try to characterize the control flow for making decisions when one searches a state space. A state space is a reduced description of the world, focusing only on those aspects we are interested in to solve a problem, and the transitions from states occur by means of application of

particular operators. An operator corresponds to an action performed in
the world that changes one state to another state. Each time an operator
is applied, the new state is compared with the goal state. If you did not
achieve it, you want to check whether or not you want to quit anyway; for
example, you run out of resources, or, you have explored the whole space.
If you did not achieve the goal and you have not reached termination, you
save the state, typically, and then you select another state to expand.
That other state can be any previously explored one. Depth first means
that you continue to persevere in the last state that you have generated.
Breadth first means that you generate all descendants to each given state
before proceeding on to the next one. Heuristic search tends to be a
combination of the above, one where you expand the state that you believe
is closest to the answer at the moment.

There is a backplane that has four slots — 1
and 4 are of width 4, 2 and 3 are of width 6

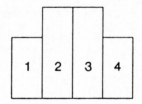

There are four boards to be placed into the
backplane — A of width 6, B of width 4, C of
width 4 but must precede A, and D of width 6
but must go in slot 2

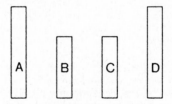

FIGURE 2
AN R1 COMPUTER CONFIGURATION TASK

Let me use a concrete example, a simplified version of a small subtask of
R1. The problem is that you have a back plane in a computer that has four
slots - slots 1, 2, 3 and 4 - two of which are of width 4 and two of which
are of width 6. The four boards have to be placed into those four slots.
But there are certain conditions they must satisfy. You know that board A
is of width 6 and board B is of width 4, and board C is of width 4 also,
but it must precede A - namely, it must occur to the left of A in the set
of boards. And D is of width 6, but it has to go into slot number 2. That
is a description of the problem, and we left out the implicit knowledge
that a board can only go into a slot that is big enough to hold it. So
board C could go into any one of them, but board A could only go into slot
2 or 3.

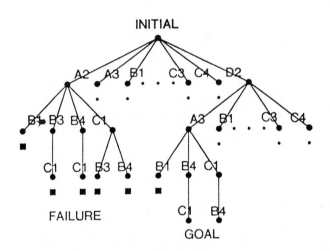

FIGURE 3
AN R1 HEURISTIC SEARCH TREE

That is a description of a problem, and I am sure that if any of you think
about 11 seconds you can probably arrive at an answer. One way of doing it
is through heuristic search (I have not bothered to go through the breadth
first and depth first expansion of this space). You start in the initial
state - the initial state corresponds to all the boards being outside the
slots - and then you look at all the possible moves that you could take.
You could put board A into slot 2, or board A into slot 3, or board B into
slot 1, etc., down the line. Two of these, according to the heuristics,
look like feasible things to do, putting A into 2 or D into 2, and then
after that you could place board B into slot 1 or board B into slot 3, or
board B into slot 4, or board C into slot 1. The heuristics help you
eliminate many choices, but you still have a tree, an expanding tree, to
search. Notice that for this example the tree will never be more than four
links deep because there are only four boards to fit into the slots, and
each action that you take consumes a board.

That is a standard way of solving a problem in the original problem space. But let me suggest to you now that this is an inferior way to solve this problem, because we are not using our knowledge effectively. It is not a bad way because it is impossible; it certainly will work. There are always algorithms that also will work, but that one does not necessarily want to do, such as reading through the entire white pages to find the name of a particular acquaintance for a phone number.

I am going to cast this problem into a constraint satisfaction problem space. Remember, the problem space consists of states and operators that transition from one state to another. The specific problem space that we talked about was the object level problem space. The problem space consisted precisely of the objects and actions that were described in the initial problem without any transformations. In order to map it into a constraint space, we will have to apply some transformations. So let us first identify the constraints; they are conditions that must be kept true at each stage of the problem solving. For instance, the boards can fit into slots only big enough for them. The way we are going to represent constraint states is by possible assignments of boards to slots. A possible assignment for board A is either 2 or 3. A possible assignment for board B is any slot. We know that board C has to occur to the left of A, but as A has not been placed yet, we cannot make that constraint operational. So we will hold it in abeyance for the time being. Board D must go into slot 2. That is a given constraint - that is an easy assignment. So, if we represent the problem in the constraint propagation problem space, the initial state in this space is the set of all possible assignments. All possible solutions are contained within that state.

We are going to add a heuristic to this space called the <u>least commitment strategy</u>. That is, we reduce the amount of guessing, reduce the indeterminacy. So if we have one of these boards which is confined to a single choice, we make that move. If we did not have one that was confined to a single choice, we would select most constrained assignment. The selection of one assignment constrains the remaining ones, as illustrated above, since each board can only be assigned to one slot. The process continues making the most constrained assignment at each step and propagating any new constraints that become operational. For example, after board A is placed, the "left-of" constraint comes into play and "C" has only one legal assignment left. In this example (illustrated in Figure 4), no search was required in the constraint space. In general, assignments may have to be reconsidered - and the least commitment strategy minimizes the need for reconsideration.

Another problem solving method is means-ends analysis. One starts with the "ends" - i.e., a description of initial and goal states, and one seeks a "means", i.e., a method of transforming the former into the latter state. Search is driven by a comparison of the initial state with the goal state, and a selection of an operation that can reduce that difference. It is important to note that a complete description of the goal state is required for comparison.

The R1 task cannot be cast into this form, because if we had a description of the goal with all the slots filled by boards, then we would know which ones went where already. Means-ends analysis works by letting the current state be the initial state, finding the most important differences between the current state and the goal state, comparing the two states to figure

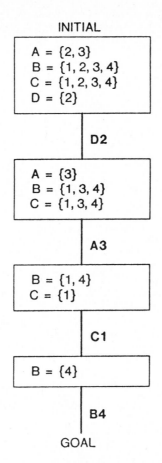

INITIAL

A = {2, 3}
B = {1, 2, 3, 4}
C = {1, 2, 3, 4}
D = {2}

D2

A = {3}
B = {1, 3, 4}
C = {1, 3, 4}

A3

B = {1, 4}
C = {1}

C1

B = {4}

B4

GOAL

FIGURE 4
AN EXAMPLE OF CONSTRAINT PROPAGATION IN R1

out how they differ, and now selecting an operator that can reduce that difference. Operators are indexed by the set of differences they can reduce. If the preconditions of an operator are satisfied, you apply that operation, and the current state is updated. If the preconditions are not satisfied, you generate the subgoal of establishing the set of preconditions for this operator; that subgoal is then recursively subjected to means-ends analysis. So, for example, if my problem was to get to the West Coast from here, the operator that would be most applicable would be to take an airplane, but a precondition of taking an airplane states that I must have a reservation and a ticket and so forth. So now I have to solve those subgoals: how do I get a reservation, how do I buy a ticket, how do I get to the airport?

An example of a difference table that indexes operators to the differences they reduce in a chemical expert system done by a means-ends analysis is illustrated in Figure 5. For example, if you wanted to change one material to another, i.e., if the difference is that you have the wrong kind of material, or the wrong amount of material, or the wrong temperature, or the wrong pressure, or it is in the wrong phase (it is a liquid when you want a gas, etc.), then you apply any operator relevant to that specific difference. You must have this knowledge in order to be able to apply means-ends analysis; otherwise, you have to resort to heuristic search or other weaker techniques.

DIFFERENCES \ OPERATOR	PUMP	TURBINE	DISTILLATION COLUMN	EXTRACTOR	REACTOR	MIXER	SPLITTER	HEAT EXCHANGER COOLER	HEAT EXCHANGER HEATER	FURNACE	CONDENSER	REBOILER
Type materials			−	−	+	+						
Amounts of materials						+	−					
Temperature	S^+	S^-						−	+	+	S^-	S^+
Pressure	+	−										
Phase											−	+

+ = increases, adds

− = decreases, removes

S = side effect

FIGURE 5
A MEANS-ENDS ANALYSIS DIFFERENCE TABLE

Thus far, we have looked at heuristic search, means-ends analysis, and constraint propagation - i.e., problem solving in the constraint space. Now let us talk about another method, transformational analogy. Transformational analogy, stated in a few words, works as follows: You encounter a problem similar to one you have a previously solved, but the two are not identical. Maybe the old solution would work, changing some parameters or adding a new operator to the solution sequence or otherwise editing the old solution. The way it works is that you try to match a new problem to previously solved problems and you get a partial mapping: pieces of the old problem map identically onto the new one, and other pieces differ. Then you use the partial mapping to aid in the transformation process of the old solution into a solution to the new problem. You want to edit the old solution such that you account for all the differences between the new problem and the previously solved problem. This is a three phase process:

(1) You match the old problem to a similar one, (2) you recall the solution of the similar problem (so you have to have the problems and solutions indexed), and (3) you transform the recalled solution to match the requirements of the new problem.

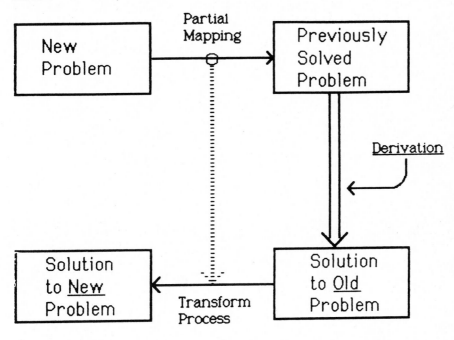

1. <u>MATCH</u> old problem similar to new one

2. <u>RECALL</u> final solution to the old problem

3. <u>TRANSFORM</u> recalled solution to satisfy the constraints of the new problem

　　　– use match to guide transformation

　　　– ignore the solution procedure (derivation)

FIGURE 6
TRANSFORMATIONAL ANALOGY

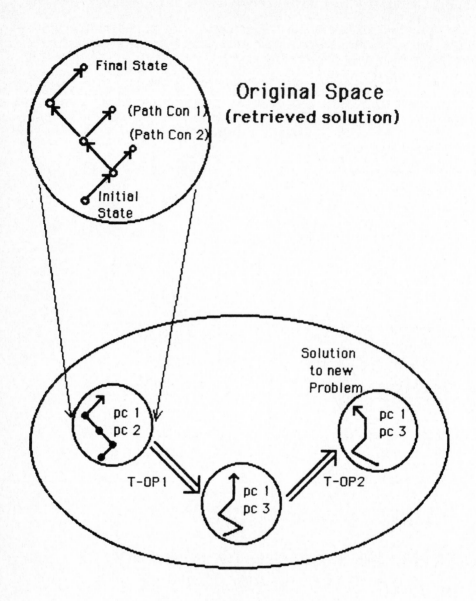

FIGURE 7
THE ANALOGICAL TRANSFORM SPACE

Initially in the original space, the untransformed space, we had a set of
states that described states of the world and a set of operators that tran-
sitioned from one to another, and a solution was a particular path through
the state space. Now, in the transformed space, each state corresponds to
an entire solution. The initial state corresponds to the retrieved solu-

tion of the similar problem, and the goal state corresponds to a solution that satisfies the description and the constraints of the new problem. A transform operator is something which takes a solution and maps it into another solution that is very closely related to it by making small directed alterations. If the distance between the initial state and the goal state in the transform space is much less than the distance between the initial and goal state in the object space, the answers can be arrived at far more effectively and efficiently by transformational analogy. In other words, we can take advantage of past experience to solve new problems by virtue of the fact that they were related to old ones.

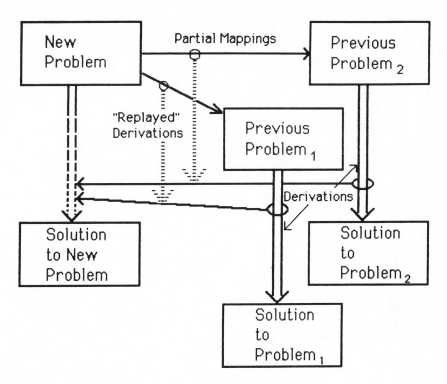

• The <u>derivation</u> is mapped & replayed, not just the solution

• <u>Multiple</u> past problem solving episodes can be integrated

• Recall of previous problems entails partial match or
  (partial or exact) match of <u>initial segment of derivation</u>

• <u>Learn strategies,</u> not just generalized plans

FIGURE 8
DERIVATIONAL ANALOGY

Now, let us consider another way of solving a problem by analogy. This
method is called derivational analogy, and it works as follows: Start with
the partial mapping between new and previously solved problem descriptions.
Before, we tried to map the solution to that problem into the solution to
the new problem by means-ends analysis in the transform space. But now, we
transform the derivation - in other words, the set of decisions made to
arrive at that solution, such as the way that you decomposed the problem
into its subgoals, the reasons why you preferred a particular solution to
another. This process requires you to keep an audit trail of your reasoning process, so you can try to re-apply that audit trail, perform that same
inference process on the new problem and rederive a solution to the new
problem by exactly the same means that you found successful for solving the
first problem. To the extent that old and new problem do not differ, all
you do is copy the corresponding steps. In the places where they do
differ, you have to fill in those gaps by reasoning in the original problem
space, but with a much smaller, reduced subproblem. In other words, if
there is some glitch in the new problem that was not addressed in the solution path before, it now needs to be addressed before continuing to replay
some other subproblem.

Another issue in the derivation process is that sometimes you can draw
information from more than one previously related problem. If they both
say the same thing, everything is fine, the same kind of derivation process
works. If they say different things, you just simply select from one of
them as your main one and from the other one only to fill in the gaps, if
possible. It turns out that integrating knowledge from multiple previous
subproblems has proved effective in our trial implementation for some very
simple automatic programming. The reason that I was talking about a very
simple automatic programming environment is because this is a class of
problems that we could not solve very effectively by transformational
analogy. It is hard to transform a completely written program in PASCAL
into a completely written program in LISP. What you really transform is
the high level structures, the algorithm, your decision as to why you
selected one data structure over another one, etc. A research group at
Information Sciences Institute - Jack Mostow and David Wilde - are trying
to extend this idea to some real automatic programming in design decisions,
not the toy one on which the technique was initially developed.

A derivation consists of the audit trail of the solution process, as illustrated in Figure 9. For instance, if you pursue a bad alternative for a
long while and then realize that it is bad, you need only save the reason
for failure, and the next time around, you do not have to bother going down
this particular dead-end path. Creation of subgoals, plan instantiations
that were found useful, etc., are also recorded in the derivation. There
are two kinds of justifications for each step taken by the problem solver:
internal ones and external ones. External justifications are based on your
entire knowledge; internal justifications are based on interactions among
the steps in the problem solving. For instance, a step to achieve the
requisite preconditions of an operator is an internal justification. The
reason for that particular step is that it is needed for something else in
the problem-solving process.

When a new problem comes up and you are trying to follow the same derivation, you try first the main path; if it works, everything is fine, and if
it fails, at some point - usually towards the end when you add the more
domain-specific knowledge, rather than at the beginning when you are trying

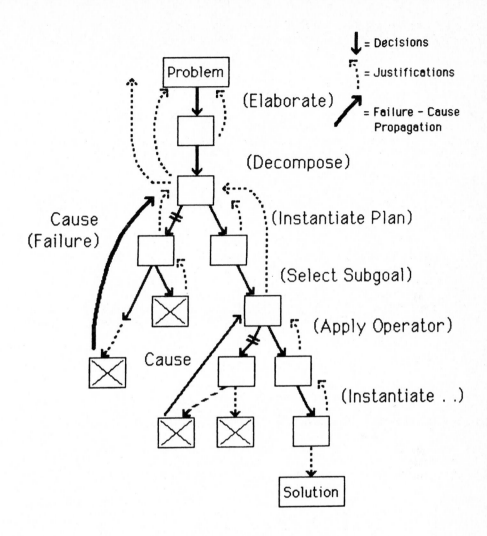

- A sequence of decisions is kept as an audit trail of the reasoning process:
    - Justifications (RMS)
    - Alternatives
    - Failure Causes

FIGURE 9
THE ARIES METHOD

to decompose and analyze what the problem is to begin with – you may have to diverge away from the known solution. When you diverge, you can very

often come back to the solution if you can maintain the justifications for
the subsequent steps in the process intact.

The only other thing that I should mention here is that as a function of
how much effort you want to put into problem solving, you may want to
explore false paths that you considered before but did not work in the old
situation. You may want to explore a false path if, first, you have lots
of problem-solving resources to spare at the moment; second, if the initial
cause for failure is violated by the new problem (in other words, the
reason why that particular attempt at solving the problem failed before
does not hold up under the new problem - the new problem differs enough
from it in that respect); and third, if there is some reason to believe
that that path ought to be the preferred one if it can be taken at the
present time. So you do not block yourself from other qualitatively
different kinds of problem solving processes.

IV.   KNOWLEDGE ACQUISITION

Now let me try to merge problem solving with knowledge acquisition. One
kind of knowledge acquisition that you can do in this analogical reasoning
is just simply to remember all of your experiences. This serves as a form
of knowledge acquisition because now you have a larger basis from which you
can solve new problems. As you accumulate a corpus of experience, a new
problem that comes in is more likely to match one of the previous ones - or
match one more closely, therefore you might solve it more effectively.

Let us worry about a slightly more interesting kind of learning. In
learning from examples, one has a set of positive examples of a concept,
and a set of negative examples. The objective is to find a general concept
description that includes all the positive examples, but none of the negative ones. For instance, if a learning system is shown - and provided with
full information of - a cat and a dog as positive examples, and a barn and
a tree as negative ones, any of a large number of concepts could be
inferred. But, if a horse is included as a positive instance, and a lion
and a sparrow as negative ones, the concept of "domestic animal" begins to
emerge. The more constraining the negative exemplars the better. The
concept is considered acquired if from it you can rederive all the positive
examples but you cannot rederive any of the negative examples. Also, the
concept should be more general than the disjunction of all the positive
examples.

Now, what is the point of learning from examples? Suppose that we had a
problem that we had solved before, and by the analogical process, we
derived a set of other solutions, as in Figure 10. The solved problems
serve as a cluster of positive examples from which you can try to form a
generalized plan. That generalization is a plan for solving problems of
the class which are related by having a common analogical ancestor. These
solutions are similar to each other because they were derived from a common
ancestor, and in fact, if you remember the reasons why you made the
specific transformations, you can come up with contingencies expressed
directly in your general plan. If not, you come only with the smallest
common denominator, in other words, all those steps which held constant
across all solutions. Similarly, if you had any false analogies, you can
use them as negative exemplars to cut down the search, to narrow down the
generalization. For example, suppose that I determined by analogy how to
travel from Boston to Rutgers University and from there to Maryland, and

then also by analogy how to get from there to Pittsburgh; everything is working out just fine. And now I have a new problem: how do I get from there to San Francisco? The distance is bigger, but let me try the same driving plan, and it works, I can get there by car also. Now I need to get from Boston to London; well, that is about the same distance as San Francisco, let me again try the same plan; and I get as far as Cape Cod. Then I realize that something has gone wrong, and now I understand there is a crucial difference between that problem and the previous ones. They are not in the same connected land mass, and that is important. That is a feature that I did not know was important before, but I can put it in now to bound my generalization only to those situations for which that feature is satisfied. So the near-miss negative exemplars, just like the constraining negative exemplars in declarative concept descriptors, work for learning generalized plans.

V.   DESIGNING NEW AI LANGUAGES

What are the desired criteria that an AI language should have in order to be able to do the kind of transformations we have discussed and to be able to represent the kinds of knowledge that we found necessary? I believe in the maximum-inclusion property - when in doubt, put it in. If the language can support more than one kind of reasoning, if it can support a representation that can be interpreted directly or can be reasoned from, if it can serve as a declarative knowledge source, then all the better. A language is a tool kit, and the more extensible and versatile the better. The notion of trying to have the smallest possible language that can do X is an absurd one from the point of view of trying to develop effective AI systems. Mathematical axiomatizations are not analogous to language design criteria; if that were so, we would find ourselves programming Turing machines. The language has to allow for problem space transformation. The closer the language reflects the problem solving process, the better. If you want to be able to solve a problem by analogy, you need to extend your language to be able to support that. That means that solutions must be reasoned from as well as executed. Introspection into processes is an example of a capability the language should support. For example: "Why did I do what I just did, or what I did the last time?" "Is it justified under the new situation?" You have to be able to represent the reason for why you did what you did explicitly, in order to see whether or not that reasoning is justified under a different set of circumstances. You also have to be able to switch problem spaces, as in the example that I used to show that a constraint space occasionally can prove more effective than the object space for solving problems; that ought to be supported in the language. To reiterate: One wants to allow for introspection, interpretation, and reflection, all of them using the same knowledge acquisition, knowledge integration, and reasoning techniques.

Two languages that are currently being developed (actually, these two languages have different philosophies, but they happen to meet some of the same criteria, and that makes them kind of interesting) are SRL+ and SOAR. SRL+ is a frame-based language that has as its general operations the ability to evaluate a piece of the knowledge base, to interpret it as though it were a procedure, and to pass information in an object-oriented technique. This flexibility makes it extremely effective for rapid prototyping of problem solvers and expert systems.

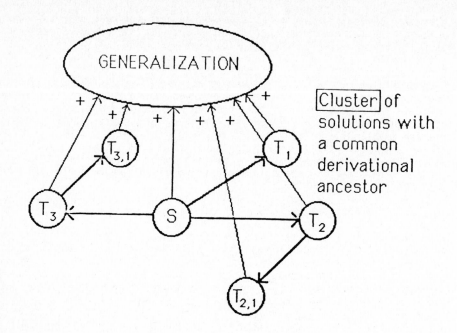

$$\delta(T_i, T_j) \ll \delta(T_i, T_k)$$
$$\forall T_i, T_j T_k \text{ s.t. } T_i \in \mathbf{C} \longleftrightarrow T_j \in \mathbf{C} \ \& $$
$$T_i \in \mathbf{C} \longleftrightarrow T_k \notin \mathbf{C}$$

⊖ Members of a cluster = <u>+ Instances</u>

Members of other clusters = <u>− Instances</u>

(or failed analogies serve as − Instances)

to an <u>Induction Engine</u>

FIGURE 10
ANALOGY PROVIDES DATA FOR INDUCTION

Perhaps a more radical departure from standard languages is SOAR, being developed by Laird, Rosenbloom and Newell, jointly between Stanford and Carnegie-Mellon and Berkeley. SOAR encodes directly the problem-solving state space into the language. It is meant to be a unified problem solving language. Its units are problem states and operators and evaluation functions, and its general operations are things like selecting a problem, selecting operators or applying them or generating new operators or new states or new spaces. For example, to generate a transformation from one space to another, it generates that new space and then applies that transformation to each of the objects in that space. The selection procedure applies to operators, to goals, and to spaces, so it has to select which space this problem will be most directly solvable in. Initial results have proven quite enlightening.

The design of the language ought to be motivated by the kinds of problems you wish to solve in it, and getting a deep understanding of the problem-solving structure itself is crucial in developing a good design of a language, rather than trying to do it the other way around.

Let me close off with a very short story I wanted to mention. Once upon a time, like all stories begin, there was a master, a physics professor, and his very young prodigy. He was only 12, and studying college physics. The professor wanted to show him off to some of his colleagues, saying, "Look at this young kid in the class, he is as smart or even smarter than all of the older students." One of his colleagues said, "Oh yeah, well can I suggest some problems for him to do?" And the physics professor said, "Sure," whereupon the colleague started off with an easy problem. He said, "Here I have a barometer. It is a mercury column barometer and it measures the pressure by telling you how high the column of mercury goes on this metric scale over here. See that tall building over there? Here, use this mercury barometer and tell me how tall that building is." And the student said, "No problem." He ran up to the building, into the elevator, climbed on the roof of the building, dropped the barometer, looked at his watch - one, two, three, four, smash! "D equals minus 1/2 GT squared," and he got the answer. The physics professor was a little bit appalled by this in front of all his colleagues, and when the student came back, he said, "No, no, no, no! We will give you another chance. That is not the way you use a barometer. Here, here is a second barometer. Go use it, but this time, you are not allowed to break it; you are not allowed to drop it. Do you think you know what you are going to do?" The prodigy responds "Sure!", runs off, knocks on the manager's door, and says, "Here, I will trade this nice new valuable mercury barometer if you will only tell me...." "No, no, no!" The physics professor stepped in before the prodigy gave away the barometer; that is the last one they had, after all. "No, Junior, let us try it again, and this time you have to have the barometer with you and intact, after you have finished." The student responds "Sure!" He picks up a ball of twine, measures the ball of twine, and the professor guesses what he is going to do before the student gets to the top of the building and starts to let down the barometer as a weight at the bottom of the twine, and he says, "No, no, no, use the barometer by itself, do not add anything else, and it has to remain intact." And the kid again says, "Sure!" The sun was out; there was a shadow of the building in a nice diagonal; and he aligned the barometer, which was nicely labeled with metric units to measure the column of mercury, with the shadow. By triangulation he figured out the height of the building. At this time the professor's colleague was totally amazed, and said, "This kid is a genius, he solved it in four different ways that we never would have thought of!" And the professor says, "Oh, really?" trying to recover and save some face.

Well, the lesson from this story is that it describes the state of many of the AI problem-solving systems. It is not the fact that they cannot solve a problem, cannot be creative. If we define "creative" as doing unexpected things, things that even the programmers do not expect them to do, these systems are incredibly creative. Examples abound in the literature. The real problem seems to be that most AI systems lack common-sense, an ability that people have, and lack the notion that there exist certain good ways of solving problems. Usually humans diverge from proven solution strategies only with good cause.

The Role of Language in Problem Solving I
R. Jernigan, B.W. Hamill, and D.M. Weintraub (Editors)
 Elsevier Science Publishers B.V. (North-Holland), 1985

PROBLEM SOLVING AND THE EVOLUTION OF PROGRAMMING LANGUAGES

J. C. Boudreaux

Center for Manufacturing Engineering
National Bureau of Standards
Room A-351 Metrology
Washington, D. C. 20234

Backus has observed that von Neumann programming
languages are fat and weak. Though there are current
efforts to provide alternate models of computation, an
examination of the genealogy of programming languages
suggests that it is unlikely that the issues now facing
programming language designers will be resolved by the
simple expedient of replacing one model with another.
What such an examination does suggest is that each
succeeding generation transfers new and more difficult
cognitive functions from the programmer to the computer.
If this is correct, then we can predict that the next
generation will come about not by some revolutionary
advance in computer technology, but by the successful
automation of higher-order cognitive functions which now
require human attention. One ideal solution would be a
cluster of programming languages that are expressive
enough to reflect as nearly as possible the user's own
cognitive framework, i.e., the structured world of
abstract objects which define the user's application
domain, together with the set of transformation rules on
that domain which permit the user to create and/or modify
those objects. In this paper, I will examine existing
programming languages and then show how ideal programming
languages could be realized in practice.

1. INTRODUCTION

It is generally believed that the human species is distinguishable from all
others by its marked capacity to solve symbolic problems. In this sense, a
problem is any one of an indefinitely large class of mental irritants which
are bothersome enough, though possibly just barely so, to trigger a
"solve!" response on the part of otherwise contented humans. Of course
humans are at times confronted by situations that are bothersome enough to
be genuinely problematic and that even have a sharp problematic formulation, but that are not and cannot be accepted as authentic because they do
not admit of practical solution. One is reminded of such metaphysical
questions as "why is there something rather than nothing?" The fact that
this question has no solution, or equivalently that it has very many
different solutions, in no way diminishes its intellectual power or its
seductive emotional appeal.

What we are primarily concerned with in this paper is a very tightly bound
class of problems, namely, those that admit of computational solution, cf.

Amarel [2] and also McCarthy [13]. I conjecture that every member of this
class of problems has the same underlying morphology, though of course I
know of no way to confirm this conjecture. First, the actual situation
which gives rise to the problem is a configuration of objects, and the
totality of all possible configurations may be taken to be the problem
domain. Second, every problem is based on a set of transformations which
allow us to move from one configuration to another by creating, deleting,
or modifying one or more objects in the initial configuration.

In general, this analysis of the essential features of problems is guided
by what may be called a linguistic orientation. That is, there should be
precise combinatorial rules for distinguishing well-formed (admissible)
from ill-formed (inadmissible) configurations of objects. This is intended
to suggest a connection between objects and such lexical items as words,
which in turn implies a connection between configurations and sentences.
This analogy suggests that objects may themselves be decomposable into
finer constituent parts, just as words are decomposable into phonemes.
Though this raises the possibility of a combinatorial explosion, in particular cases it is usually assumed that there is a finite bound on the depth
to which this decomposition need proceed. Of course, one may choose to
speculate about the existence of absolutely indecomposable simples, i.e.,
logical atoms, but this notion poses issues that are far beyond the
intended scope of this paper.

These restrictions rule out a broad class of ill-posed, riddle-like
problems which presuppose that the problem solver has mastered the art of
uncovering the correct plaintext that lies below an ornate, highly metaphorical surface text. For example, it is difficult to imagine how one
could construct an abstract problem class with respect to such problems as
that posed to the Greek hero Oedipus by the Theban sphinx: "What walks on
four legs at dawn, two legs at noon, and three legs at dusk?" Oedipus's
solution, i.e., humankind, reveals an intelligence that is uncommonly
skillful in decoding complex, even nested metaphors: the third "leg"
supporting locomotion during the "dusk" phase of human life is a staff,
which is not an organic part of the human body.

But merely to characterize the class of computational problems negatively
is not enough. The main topic of this paper is the interdependence between
problem solving and the evolution of programming languages. Since representation is clearly the crucial notion, I would like to present a very
obvious and very clear worked example of the way in which a programming
language can support the efforts of the problem solver.

Many successful programming languages have been developed, but none has had
the enormous success or longevity of FORTRAN. Though I will be discussing
more modern languages in the next section, I think that even now the internal coherence and consistency of the FORTRAN world makes this language a
very good paradigm of the complicated connections between a programming
language and human problem solving. The stunning success of FORTRAN may be
attributed to the simple underlying model of computation -- call it the
FORTRAN machine -- which specifies how FORTRAN programs are to be executed.

The transparency of the FORTRAN machine to the scientific programmer can
best be illustrated by considering the solution of a simple problem: given
an arbitrary univariate polynomial of the form

$$A(X) = A_n X^n + \ldots + A_1 X + A_0$$

and an arbitrary real number Y, then evaluate A(X) at Y.

According to the accepted wisdom, the craft of computer programming is divisible into two specialties: selecting appropriate data structures for information-bearing objects in the problem domain, and designing algorithms that manipulate and transform those data structures to provide a state in which the answer is contained. The most direct way to illustrate this folk wisdom is to trace the particular steps which lead to the solution of this problem. First, integral values, including the polynomial's exponents and the indices of its coefficients, should be represented by the FORTRAN data type INTEGER; and all real values, including the polynomial's coefficient, the class of values that may be substituted for its variable, and hence the value returned by the evaluation transform, should be represented by the FORTRAN data type REAL. Second, the univariate polynomial A(X) itself may be represented by a unidimensional FORTRAN array, say POLY(0:N), such that for every INTEGER J between the values 0 and N, which designates the polynomial's degree, the value of POLY(J) is the J'th coefficient of A(X).

Having thus selected data structures to represent the objects in the problem domain, the next step is to define an algorithm to represent the evaluation operation. The most obvious algorithm is based on the following iterative schema:

```
        TMP = POLY(0)

        DO 10, J = 1, DEGREE
        TMP = TMP + POLY(J) * ARG ** J
   10   CONTINUE
```

Though this algorithm is correct, it is not computationally efficient, which means that it cannot be accepted as a fully satisfactory solution to the original problem. In particular, since the FORTRAN expression "ARG ** J" is usually interpreted by the loop

```
        TMPB  =  1.0

        DO 20, K = 1,J
        TMPB  =  TMPB * ARG
   20   CONTINUE
```

the number of covert multiplications increases very rapidly with increases in the degree of the polynomial. A much more efficient algorithm for evaluation may be based on Horner's Rule, which slightly modifies the underlying representation of the polynomial:

$$A(X) = (\ldots(A_n X + A_{n-1}) X + \ldots)X + A_0$$

Since the number of multiplications required by the revised representation is (provably) optimal, the following algorithm can be accepted as a highly satisfying solution to the polynomial evaluation problem:

```
        REAL FUNCTION HORNER(POLY, DEGREE, ARG)
```

```
        INTEGER DEGREE, J
        REAL POLY(0:DEGREE), ARG, TMP

        TMP = POLY(DEGREE)

        DO 10, J = DEGREE - 1, 0, -1
        TMP = TMP * ARG + POLY(J)
   10   CONTINUE

        HORNER = TMP
        END
```

Notice that the change has no effect on the manner in which objects are represented. In a fuller treatment of this subject, we would also have to consider this issue as well. For example, it is obvious that the representation of the polynomial presupposes that it is dense, that is, that almost all of its coefficients are non-zero. Given this presupposition, the representation is good enough, but not if the polynomial is sparse, cf. [1] for further discussion.

In this paper I shall be discussing as accurately as I can some of the more pressing issues in the design of programming languages. Though discussions of this sort are fraught with the dangers of subjective judgments, I will try to avoid these dangers by presenting my remarks within the context of specific languages. For reasons that will be made clear as we proceed, I have chosen to concentrate on the programming language LISP. To insure that this paper is more or less self-contained, I will give the reader as much information about this language as will be needed to follow the discussion in the final section of this paper. This increases the length of the paper, but I believe that this negative side effect is more than offset by the increase in the clarity and rigor of the exposition.

Before resuming the thread of this exposition, let's first entertain an evolutionary hypothesis concerning the origin of programming languages.

2. THE EVOLUTION OF PROGRAMMING LANGUAGES

Strictly speaking, programming languages are human artifacts and thus may be said to evolve only in an analogical or metaphoric sense. A good case point is the development of FORTRAN. In the early 1950's computer programming was an activity that was carried out in the native language of the programmer's machine or possibly in an assembly language which allowed symbolic reference to machine operations and storage locations. Since this programming style requires the programmer to keep track of an enormous amount of unrewarding clerical detail, the need for a less labor intensive method for programming computers was obvious to many reflective observers. In retrospect, it is easy to see that the solution to the problem is to allow programmers to write out polynomial expressions on the right-hand side of assignment statements, and then to define a small cluster of rudimentary features for formatted input/output and program control. Taken together, these "easy" solutions constitute the main technical contributions of FORTRAN, and in my opinion, there is no better evidence of the brilliance of the work done by John Backus and his colleagues at IBM than the apparent inevitability - even banality - of their FORTRAN design.

There are many lessons to be learned from the FORTRAN project, not the
least of which is that it is very difficult to predict future advances in
programming language design by merely reviewing the existing state of
affairs. If we consider programming language design as a kind of problem
solving, then we are immediately struck by its lack of similarity to most
of the problems mentioned above. The problem posed by language design is
not how best to reach some pre-determined goals but what goals are reasonable and prudent to adopt in the existing state of affairs. Judgments of
this sort require language designers to be acutely conscious of the expectations and beliefs of the scientific community for whom the project is
undertaken. The fact that algorithms are written in a language-like
formalism which is the common intellectual property of the entire community
makes it very likely that only those designs which are seen as locally
smooth continuations of existing practice stand much of a chance of being
finally accepted.

One straightforward view of the evolutionary development of programming
languages has been proposed recently by Alan Kay in [11]. Though there are
a few quibbles that can be made about the adequacy of his divisions, it can
serve as a useful guide. Specifically, Kay recognizes four distinct genres:

low-level languages (LLLs), e.g.,machine languages, assembly languages;

high-level languages (HLLs), e.g., FORTRAN, ALGOL, and LISP;

very-high-level languages (VHLLs), e.g., SMALLTALK, PROLOG, and LISP;

ultra-high-level languages (UHLLs), e.g., VISICALC, EURISKO, and LISP.

Notice that LISP appears at every level above LLLs, a fact that Kay
explains in the following way: "The language LISP has changed repeatedly,
each time becoming a new genre."

From a theoretical point of view, this discussion of the genealogy of programming languages is at best a harmless diversion which must not obscure
the fact that any and all differences from one generation to the next are
entirely superficial, stylistic variations which may for a time disguise -
but which can never really hide - the fundamental logical equivalence of
all programming languages.

There is a way to interpret the claim that all programming languages are
identical, which makes the observation trivially true: every programming
language - even the most primitive - is expressive enough to provide us
with both the data structures and algorithms needed to construct an
internal model of the Universal Turing machine.

When I claim that programming languages are structurally identical, what I
mean is that until quite recently, all practical programming languages were
based upon a model of computation that was proposed in the 1940's by John
von Neumann. Though not the first model of computation - the Turing
machine preceded it by almost a decade - it was the first model which
provided secure supports for both the commercial development of computer
equipment and also the design of programming languages. This view of the
structural identity of programming languages was first explicitly formulated in the Turing Award lecture of John Backus (cf. [5]):

In its simplest form a von Neumann computer has three parts: a central
processing unit (or CPU), a store, and a connecting tube that can
transmit a single word between the CPU and the store (and send an
address to the store). I propose to call this tube the von Neumann
bottleneck. The task of a program is to change the contents of store
in some major way; when one considers that this task must be accomplished entirely by pumping single words back and forth through the von
Neumann bottleneck, the reason for its name becomes clear.

Conventional programming languages are basically high level, complex
versions of the von Neumann computer. ... Von Neumann programming
languages use variables to imitate the computer's storage cells;
control statements elaborate its jump and test instructions; and
assignment statements imitate its fetching, storing, and arithmetic.
The assignment statement is the von Neumann bottleneck of programming
languages...

Given this store-centered model of computing and the evident need to
maintain very strict control over the assignment process, von Neumann
languages tend to have rigid syntactic exoskeletons. This rigidity is
noticeable in FORTRAN, and overwhelmingly obvious in more "modern"
languages, such as the programming language Ada which was developed in the
mid-1970's by the U.S. Department of Defense, cf. [4] for details.

The von Neumann architecture and its associated model of program structure
completely dominate the field of general purpose computer organization and
programming language design. Though this dominance is not likely to be
challenged in the near term, recent discussions have stressed two significant weaknesses: first, its lack of useful mathematical properties which
severely complicates the task of program verification, and second, its
inability to represent and utilize parallelism. These weaknesses have
caused many researchers to question the continuing viability of the von
Neumann architecture for present day computing. Current efforts to provide
a more adequate conceptual foundation for computing and programming
language design have led to the development of two alternate paradigms: the
dataflow machine and the reduction machine.

The dataflow machine may be imagined to be a directed graph whose instructions, or "nodes", are interconnected by paths. Each instruction has a
definite number of both input and output paths over which data values, also
called data tokens, may be passed. Unlike the rigid "one-at-a-time" flow
of control of the von Neumann architecture, dataflow machines are expressly
designed for parallel execution: an instruction is executed as soon as a
data token has arrived on each of its input paths. After an instruction is
executed, the resulting value is placed on each of its output paths, thereby making the data token available to other instructions.

The reduction machine represents an even more radical departure from the
prevailing model of computation. Though they differ in important respects,
both von Neumann and dataflow programs are built from fixed-sized instructions which contain operators and operands, and higher level programs are
built by threading these instructions together according to explicitly
defined combinatorial rules. In contrast, reduction programs are built
from nested expressions, and the ultimate goal of computation is to successively simplify an expression, i.e., to "reduce" the degree of nesting, by
replacing compound expressions with their values. That is, a reduction

program is equivalent to its resulting value in the same sense that 3 + 3 is equivalent to 6. The nearest analogue to an instruction in reduction is a function application of the form <function><argument>, which returns its value in place. Such an instruction is executed whenever its value is required by an invoking function.

Though these rival models of computation suggest interesting possibilities, it would be a mistake to suppose that the problems now confronting language designers could be solved by the simple expedient of replacing the von Neumann machine.

A thoughtful examination of the genealogy of programming languages reveals some hypotheses about the evolutionary pattern of programming language development. Though several patterns emerge, one is quite striking: each succeeding generation transfers new and more difficult cognitive skills from the programmer to the computer. That is, as one moves from from one generation to the next, tasks which once required the intellectual atten- tion of humans are reduced to finitary algorithms that can be routinely, and safely, assigned to computers. If this hypothetical pattern has any substantive validity, then the next step will not come about by any revolu- tionary advance in the design of HLLs, but by the successful automation of some of the comparatively high-order cognitive functions which now demand the attention of humans.

Since the user is assumed to be technically expert in his field, the purpose of automation is to provide him with a helpful working environment - that is, an environment which would tend to amplify and extend his field- related capacities. Though there are many possible paradigms that could be invoked here, one very familiar and commonplace mechanism for achieving this end is to surround the user with a computer-based system that he could come to appreciate as a reliable, if not overly intelligent, assistant who could carry out a number of elementary operations without any prompting and who could be trained to undertake more complicated operations if the user were willing to specify step-by-step descriptions of what needed to be done. Since we are imagining future possibilities and not presently exist- ing systems, we might also add that the notational system should be similar in form and content to that which would be used to explain the same opera- tion to a human assistant.

The last observation imposes strict constraints on the design of ideal programming languages. But to avoid confusion, it should be emphasized that it does not imply that an ideal language should be a variant of the user's native language. Though the attractiveness of the natural language paradigm is not lost on me, I am persuaded that it tends to concentrate attention on the wrong set of issues. There is really very little to be gained by building machines that cause users to suffer the delusion that they are addressing members of their own species. Rather we should concen- trate our efforts on the design of programming languages that allow the user to exercise a high degree of predictive control over the machine's behavior. But this is a very difficult task, that demands very careful analysis of subtle, field-dependent variations in cognitive style.

3. IDEAL PROGRAMMING LANGUAGES

If programming languages really did evolve, then there would be no surer or more rational path toward future ideals than simply to wait and observe what happens. But as the earlier comments on FORTRAN suggested, programming languages are really human artifacts and as such need to be first conceived vaguely and in an obscure way in the designer's vision of technical needs that are not being satisfactorily met by existing programming languages. In most cases, the proposed design is really an adaptation of an existing programming language developed by adding new features and then smoothing and rationalizing the results. For example, Ada may be viewed as a development of Pascal, extended to include features to support multitasking and programming in the large, and Pascal itself is a development of Algol. The commonly used expression "Algol-like" is a testament to the dependence of a family of programming languages upon a common ancestor. Though there are no immediately apparent signs that this robust family is near extinction, I believe that the technical problems facing language designers will require the adoption of a new ancestral paradigm and I suggest that the paradigm of choice is none other than LISP!

As Kay has already observed, LISP began its career at the beginning of the age of FORTRAN and has been transforming itself into new genres ever since. In fact, what recommends LISP to our attention in this context is precisely this inherent capacity for evolutionary development. Since I do not assume that LISP is part of the common knowledge of scientific programmers, let me begin with a concise introduction. For those readers who are already acquainted with some variant of LISP, I should remark that the specific dialect to be introduced here is FRANZLISP, as described in [7].

The single common fact which runs through almost all variants of LISP is that it is an interactive language which presents the user with a comparatively simple interface. Thus, from the user's perspective, the LISP system consists of an interpreter which signals its availability by printing a prompt symbol. When the user responds by keying in an expression, the interpreter immediately returns the value of that expression on the very next line. The following example is the transcript of two very simple exchanges between a user and the LISP system:

        -> 12
        12

        -> 23.4567
        23.4567

In this case, the user has keyed in an integer value, which in LISP jargon is called a **fixnum**, and then a floating point number, or **flonum**. As important as the quick response is, there is another more important point, namely, the interpreter is extraordinarily simple. In fact, MacLennan presents a version of a simple statically scoped interpreter written in LISP that takes all of 25 lines! See MacLennan [12], Chapter 11 for a very carefully crafted explanation. The simplicity of the interpreter merely reflects the underlying simplicity and regularity of the language itself.

LISP recognizes only two kinds of objects: atoms and lists. Atoms include such scalar values as integers (**fixnums**), floating point reals (**flonums**), symbols, and strings. A list is an object that may always be resolved into a head component which may either be an atom or a list, and a tail component which must be a list. The accepted notation for lists is to enclose their components within mated parentheses:

                    (this list)

In this example, the head is the symbolic atom **this,** and the tail is the list **(list).** The list which has no components is called the null list and may be represented by the empty-nest expression "**()**" or by the constant symbol "**nil**". The null list is a useful artifact, especially in that it permits us to disambiguate such expressions as **(list),** which can only be interpreted as the list whose head is the symbol **list** and whose tail is the null list.

The only other class of entities recognized by LISP are functions. In order to signal the LISP interpreter that a particular function is to be applied to a (possibly empty) sequence of arguments, the programmer simply presents the interpreter with a list object whose head is the symbolic name of the function and whose tail is the list of the expressions to be passed to the function as arguments. Though there are important exceptions, LISP functions usually have a fixed number of arguments which is identical to the number of components, or length, of the argument list.

LISP programmers work in programming environments that are already supplied with pre-defined LISP functions. In fact, one LISP dialect will differ from another in both the variety and complexity of the pre-defined functions that it makes available. Given the enormous diversity of LISP dialects, it is quite impossible to present even a representative inventory. But to make sure that the reader does indeed understand the points to be discussed and also the examples to be given, let me provide a brief sketch of a few very basic functions.

Two functions, **car** and **cdr**, are used to select the components of any list; in particular, **car** selects the head component and **cdr** selects the tail:

>   -> (car '(this list))
>   this
>
>   -> (cdr '(this list))
>   (list)

The single quote is an important LISP function which is used to signal the LISP interpreter that the LISP object following the occurrence of the single quote is not to be evaluated but treated as a literal. Had the single quote been omitted, the LISP interpreter would have processed the first example as if the programmer intended to apply car to the value obtained by applying a function named **this** to the value associated with the argument **list,** which in this case would have been unintelligible. But in other cases, this is precisely what we have in mind:

>   -> (car (cdr '(this list)))
>   list

Given the ubiquity of these functions, the usual practice of LISP systems is to abbreviate nested **car**'s and **cdr**'s by allowing any finite sequence of a's and d's in the context "c...r". Thus, the example just given could have been written:

>   -> (cadr '(this list))
>   list

There are several important contexts in which the single quote function is used covertly, for example, the LISP function **setq**, which abbreviates "set quote", is used to assign LISP values to symbols:

>   -> (setq my_ex '(this list))
>   (this list)

In general, the effect of the evaluation of this function is the association of the value of the second argument with the symbol in the initial
argument place. This symbol may then be used as a LISP variable. The fact
that **my_ex** has a value is recorded in the otherwise hidden symbol table,
and from this point on, if the interpreter is presented with this name, it
will respond with the assigned value:

>   -> my_ex
>   (this list)

There is one primitive constructor function for lists, namely, **cons**, which
when applied to two arguments, returns a list whose head is equal to the
first argument and whose tail is equal to the second:

>   -> (cons 'this '(list))
>   (this list)

The relationship between the constructor function and the selector
functions is nicely illustrated by the following interactive exchange:

>   -> (equal (cons (car my_ex) (cdr my_ex)) my ex)
>   t

where the interpreter's response **t** represents the Boolean value true as
opposed to **nil** which represents the Boolean value false. The truth of this
test tells us that the effect of **cons**-ing the **car** and the **cdr** of any non-
null list is equal to the list itself.

Symbols may also be used to represent entities with properties where the
property name is itself a symbol and the value of the property is any LISP
expression. Thus to express the fact that Abraham is the father of Isaac,
we would write

>   -> (putprop 'isaac 'abraham 'father)
>   abraham

To retrieve the value of a property, we would write

>   -> (get 'isaac 'father)
>   abraham

To this point, every function has had a fixed adicy. The next function,
cond, which expresses the primitive conditional test method in LISP, may
have any finite number of arguments. Each argument is a list, called a
conditional clause. The interpreter processes each clause in order until
one is discovered whose **car** is non-nil. Then the **cdr** of the successful
clause is interpreted, that is, all of the components of the **cdr** of this
clause are processed and the value returned by the **cond** function is the
final result obtained.

```
-> (cond
    ((equal (car my_ex) 'this) 'that)
    (t 'them))
that
```

Since the LISP interpreter simply ignores white space, i.e., carriage returns, line feeds, blanks, and so on, in response to the prompt, it is possible to organize LISP code in a visually satisfying way. For example, it improves the legibility of the above example to have each of the conditional clauses written on a separate line. In this case, the value returned by **cond** is a consequence of the fact that the car of the first clause has the Boolean value **t**. Since this value is always non-nil, the second conditional clause is really playing the role of an otherwise clause, at least in the sense that whatever results are specified by the **cdr** of this clause will be returned only if no preceding clause has been successful.

All LISP systems share the characteristic that programmers are encouraged to construct new application-specific functions of their own devising. Once written, these functions have the same status as the functions supplied by the LISP system. To illustrate this important feature, let's use the example of the square function:

```
-> (def square (lambda (X) (times X X)))
square
```

The response square tells us that the function is subsequently available as a LISP function, as the following exchanges clearly indicate:

```
-> (square 4)
16

-> (square 2.36)
5.5696
```

Note that **def** is just another LISP function, differing from others that have already been mentioned, primarily with respect to its side effects on the LISP environment. Like **setq**, this function causes a value to be associated with a symbolic name, in this case the symbolic name square. It associates this symbol with the otherwise anonymous function defined by the lambda abstraction formula

```
(lambda (X) (times X X))
```

The variable **X** in this definition is said to be lambda bound, which means that the actual value is dependent upon the argument being passed during the activation of the function. In effect, the lambda expression creates a nested environment in which all lambda bound variables are set to the argument values, then the inner expression is evaluated in the usual way until some resulting value emerges. Once this value has been obtained, the nested environment is deleted and the resulting value is returned in place.

The purpose of this brief tour of LISP has been to acquaint the reader with enough of the folklore to be able to form a reasonable opinion of the claim that future programming languages should harken to LISP as their totemic ancestor. What now needs to be done is to concentrate attention on those

features that are most likely to be the sources of fresh inspiration and
creative growth. Though there are features of LISP that the novice finds
strongly unacceptable, like the mountains of parentheses, I think that such
matters are mere surface-level irregularities that have little bearing on
the technical issues under discussion. In the next subsections, I will
discuss those features of LISP which in my opinion are forward looking and
which do merit more careful consideration. As each point is discussed,
keep in mind that the ultimate ideal is to identify some of the important
linguistic features of a family of programming languages which will be
expressively rich enough to permit technically and scientifically trained
users to articulate domain-specific knowledge and computational skills.

3.1 Universal Typefree Variables

In its starkest and least restricted form, the project of this paper is to
try to imagine a family of programming languages that is expressive enough
to allow scientific communities to state that portion of their shared
knowledge which can be represented by algorithms. Thus conceived, this
project bears a close resemblance to the much older project of defining an
authentic foundational system for pure mathematics. Though this analogy
has been noted by others, especially Backus [5] and [6], I think that it is
important enough to be pushed to its furthest limit.

For obvious reasons, Algol-like languages tend to accept strong typing as
an intrinsically useful discipline. In this respect, they are analogous to
the foundational system first proposed by the British logicians Russell and
Whitehead in their *Principia Mathematica*. This foundational system is
based on a very strictly enforced theory of types in which every significant component of a well-formed expression can belong to one and only one
type. All such systems suffer from a common defect: an unacceptably high
level of redundancy. In the simplest version of the theory, the Principia
types are identified with the members of the following hierarchy: individuals, sets of individuals, sets of sets of individuals, and so on. Since
every well-formed expression has one and only one type, it is obviously
impossible to define a single union operation. To stay within the strict
confines of this doctrine, one would have to define a new union operator
for each member of the hierarchy! At this point, I think that the connections with strongly typed programming languages are beginning to emerge
with particular clarity.

The introduction to LISP given above shows that LISP accepts a radically
different paradigm in this respect, in fact, one which is more closely
analogous to the doctrine of set theory. As in the theory of sets, and
also the calculi of lambda conversion, every variable is understood to
range without restriction over the entire domain of existing entities. To
illustrate the computational effects that can be achieved with universal
typefree variables, I will now define a series of LISP functions that
implement a stack processor. In developing this example, I have borrowed
freely from the axiomatic analysis given in Horowitz and Sahni [10]. The
first group of functions allows stacks to be created and deleted, and also
provides the customary stack-empty test:

```
(def create-stack
     (lambda ()
          (cond
               ((get 'stack$ 'guard)
```

```
                    (print "ERROR: stack reserved"))
            (t (putprop 'stack$ nil 'value)
               (putprop 'stack$ t 'guard)))))

(def delete-stack
     (lambda ()
          (cond

               ((null (get 'stack$ 'guard))
                    (print "ERROR: DELETE unCREATEd stack"))
               (t (putprop 'stack$ nil 'guard)
                  (putprop 'stack$ nil 'value)
                  t))))

(def isempty?
     (lambda ()
          (cond
               ((get 'stack$ 'guard)
                    (null (get 'stack$ 'value)))
               (t (print "ERROR: query to unCREATEd stack")))))
```

Notice that creating a stack is defined as the association of a property
list with the oddly spelled symbolic atom **stack$**. The property list has
two indicators: value and guard. The first indicator allows one to get
the current value of the stack. The second is a Boolean-valued indicator
that will provide both a primitive form of protection to the stack
processor and a certain discipline for its proper use. Thus, once **create-
stack** has been successfully invoked, the resulting value attributed to
guard prevents any successful re-invocation of this function until the
stack is released by invoking **delete-stack**. This protection is important
because of the dramatic effect that **create-stack** has on the current **value**
of the stack, in particular, it sets it to nil, thereby erasing whatever
information had been available beforehand.

The second group of functions defines the customary operations for modifying the current value of the stack:

```
(def push
     (lambda (item)
          (cond
               ((get 'stack$ 'guard)
                    (putprop 'stack$
                             (cons item (get 'stack$ 'value))
                             'value)
                    t)
               (t (print "ERROR: PUSHed unCREATEd stack")))))
(def pop
     (lambda ()
          (cond
               ((get 'stack$ 'guard)
                (cond
                 ((isempty?)
                     (print "ERROR: POPed empty stack"))
                 (t (setq tmp (get 'stack$ 'value))
                    (putprop 'stack$ (cdr tmp) 'value)
                    (car tmp))))
               (t (print "ERROR: POPed unCREATEd stack")))))
```

The following exchange illustrates the use of the stack processor once the file containing the preceding functions has been successfully loaded into the LISP interpreter. The function **plist** will also be used to show the actual contents of the property list of **stack$**.

        -> (create-stack)
        t

        -> (push 2)
        t

        -> (push 'two)
        t

        -> (push '(this list))
        t

        -> (plist 'stack$)
        (guard t value ((this list) two 2))

        -> (pop)
        (this list)

        -> (plist 'stack$)
        (guard t value (two 2))

        -> (delete-stack)
        t

        -> (plist 'stack$)
        (guard nil value nil)

The fundamental lesson here is that LISP variables are strongly type-free and the user does have the capacity to support the abstraction directly. Any LISP object can be pushed on the stack, and subsequently popped. Thus, the idea that one needs to differentiate integer stacks, floating point stacks, and so on, and in particular the idea that one should not push compound objects on stacks, is explicitly revealed as an artifact of the method of representing stacks in strongly typed languages by over-sized arrays.

3.2 Types as Predicates

Of course, the fact that LISP is not strongly typed does not mean that there is no way in LISP to enforce the underlying discipline that is clearly useful in many practical contexts. In LISP, as in set theory, one considers types to be predicates, that is, suitably defined Boolean-valued functions which return the value **t** if the argument is a bonafide member of the type, and **nil** otherwise. As a matter of explicit contrast, consider the following Ada expression which may be used as the type definition of natural numbers:

        type natural is new integer range 0 .. integer'last;

The intent of this definition is clear enough: the name "natural" is to be associated with any integer value between 0 and the implementation-dependent value integer'last. The same intent is captured by the following LISP function:

```
(def natural (lambda (intx)
        (and (fixp intx) (not (lessp intx 0)))))
```

The monadic function fixp returns **t** if its argument is a **fixnum** or an infinite precision integer, called a **bignum**. It is very difficult for me to understand why anyone would find the Ada version any more intuitively appealing or any easier to explain than the corresponding definition in LISP.

Moreover, there are advantages to be had by accepting the types-as-predicates approach which are very difficult or impossible to simulate in the field of strongly typed languages, for example:

```
(def even (lambda (intx)
        (and (natural intx) (equal (mod intx 2) 0))))

(def odd (lambda (intx)
        (and (natural intx) (equal (mod intx 2) 1))))
```

Though not particularly interesting in their own right, these definitions do show that type names can be associated with infinitely large collections of LISP objects. To really appreciate the power that this places in the hands of the programmer, the reader should try to define a more complicated type, like the type of all lists whose atomic constituents are either **fixnums** or **bignums!**

Finally, this approach to the discipline of types, if accepted, has interesting effects on the manner in which programs are built. One way to do type-checking is to surround the body of the LISP definition within a cond clause whose initial Boolean expression tests all relevant argument values:

```
(def TYPE-GUARDED (lambda ( arguments ...)
        (cond ((TYPE-CHECKing arguments ...)

                BODY OF TYPE-GUARDED )

(t (print "TYPE ERROR:")
        (list arguments ...)))))
```

This program scheme permits the execution of the body if the arguments survive the type checking routine, otherwise an error message is printed and a list of the original arguments is returned unchanged. Let's consider a specific example:

```
-> (def smoo (lambda (a b)
        (cond ((and (natural a) (natural b))
                (plus (square a) (square b)))
                (t (print "TYPE ERROR:")
                        (list a b)))))
smoo
```

```
-> (smoo 2 3)
13

-> (smoo 2.0 3.0)
"TYPE ERROR:"(2.0 3.0)
```

Should the type checking test fail, the value returned by smoo is only the list of arguments and not the error message.

This is not an especially elegant solution, but it does illustrate one immediately useful technique.

3.3 Higher Order Functionals

A large part of the power of LISP is its capacity to allow the creation of new kinds of functions that can take functions as arguments. It is this capacity more than any other which explains the fecundity of LISP and its chameleon-like capacity to adapt to new computing environments. Two examples, borrowed from Backus's FP system, will show very clearly the essential properties of this method:

```
(def apply-to-all (lambda (foo listx)
     (cond ((null listx) nil)
           (t (cons (foo (car listx))
                    (apply-to-all foo (cdr listx)))))))
```

The operation of this function is shown by the following example, which uses the LISP function square defined above:

```
-> (apply-to-all 'square '(1 2 3 4 5))
(1 4 9 16 25)
```

In order for this functional to work, the function lambda bound to **foo** must be a monadic function. The corresponding functional in FranzLISP, called **mapcar,** may be applied to a function of any adicy, but in that case the number of lists following the function must be equal to the adicy of the function, and all of the lists must be the same length.

The second functional is closely analogous to the reduce operation in the programming language APL. It inserts a dyadic function between the components of a list and returns a single final value:

```
(def insert (lambda (foo listx)
     (cond ((greaterp (length listx) 1)
            (foo (car listx) (insert foo (cdr listx))))
           ((equal (length listx) 1) (car listx))
           (t (print "ERROR: incorrect argument")))))
```

The following examples illustrate the utility of this functional:

```
-> (insert 'plus '(1 2 3 4))
10

-> (insert 'times '(1 2 3 4))
24
```

Though one can easily imagine interesting uses for both of these functionals, in neither case would one be disposed to say that any specially new linguistic potentialities had been discovered. However, the next functional, also borrowed from Backus, does suggest some very exciting possibilities:

```
(def construct (lambda (lisfoo lisx)
     (cond ((null lisfoo) nil)
           (t (cons (apply (car lisfoo) lisx)
                    (construct (cdr lisfoo) lisx))))))
```

Before discussing what I believe is exciting about this functional, let's look at an exchange with the LISP system into which this definition has been loaded:

```
-> (def incr (lambda (x) (plus x 1)))
incr

-> (def dcr (lambda (x) (plus x -1)))
dcr

-> (construct '(incr dcr) '(10))
(11 9)

-> (construct '(car cdr cddr cdddr) '((1 2 3 4 5)))
(1 (2 3 4 5) (3 4 5) (4 5))

-> (construct '(times plus diff) '(7 8))
(56 15 -1)
```

The initial argument of this functional is a list whose components are either symbols with function bindings or anonymous functions specified by lambda abstractions:

```
-> (construct '(incr (lambda (x) (plus x 2))) '(5))
(6 7)

-> (apply-to-all
      '(lambda (x) (construct '(plus times) x))
      '((1 2) (3 4) (5 6))))
((3 2) (7 12) (11 30))
```

At the beginning of the discussion of this functional, I suggested that it raised the possibility of a new family of programming languages that would mark a radical departure from existing traditions. What I have in mind is a family of languages which is not based on the program-as-text paradigm. This paradigm takes as a fundamental given that all bonafide programs are linearly ordered sequences of words and that the craft of programming language design is to define a suitable context-free grammar for accepting all well-formed sequences and rejecting those that are ill-formed. I suppose that the use of the term "language" has predisposed us to accept this paradigm uncritically and without reservation. However, at this juncture in the development of our science, it seems that the textfile paradigm is retarding our progress.

What the construct functional suggests is that there are very many abstract structures besides linear orderings that can be used to structure algorithms. In my opinion, this is also the primary message of the non-von Neumann machines mentioned in the second section. I admit that this suggestion does go beyond our present evidence. After all, what construct illustrates is that we can imagine a way to put function-valued vectors to work. It does not help us to take the next step, which is to discover the computational utility of more complicated function-valued structures. For example, it is now unclear how we should best imagine the use of function-valued lattices. Though a great deal of work needs to be done, I find it to be a very pleasing and not fully expected result that an inquiry that began as an effort to apply foundational approaches to the design of programming languages should ultimately discover the deep significance of diagrams and other primarily visual icons. But a more thorough development of this theme must await another occasion.

4. CONCLUSIONS

Perhaps I should say that I do not believe that LISP as we now have it is the ideal that we have been seeking. My attitude toward LISP can be aptly expressed by a paraphrase of a remark of the Austrian philosopher Ludwig Wittgenstein: LISP is a ladder and once we've climbed it, we should throw it away! The preceding section has shown us some of the things that LISP has to offer. In future investigations, I will be considering what new possibilities can be brought into focus.

I should emphasize that the origin of this paper is deeply fixed in the investigations that I am now pursuing at the National Bureau of Standards. At this time, most of my efforts are being spent on the design of a programming language system to drive fully automated manufacturing workstations. As can be easily imagined, this is a world in which the success or failure of a computer program is not and cannot be measured by the way in which it affects the contents of store. In fact, the only useful criteria of success are those that may be defined in terms of the actual effect of the program on a certain well-defined portion of the material universe. In essence, my project is to choreograph the worksta-tion so that the final block of shaped metal satisfies all of the product specifications. Until I began to consider the consequences of a change of paradigm, my colleagues and I seemed to be lurching from one conceptual muddle to another. Though we are not yet safely home, several recent successes have encouraged us to believe that we have at last found the beginnings of a homeward-bound trail.

BIBLIOGRAPHY
1. Aho, A.V., Hopcroft, J.E., and Ullman, J.D. The Design and Analysis of
    Computer Algorithms. Addison-Wesley; 1974.

2. Amarel, S. "Representations and Modeling in Problems of Program
    Formation," in B. Metlzer and D. Michie, Machine Intelligence 6,
    Edinburgh University Press, 1971; 411-466.

3. ANSI X3.9-1978, "American National Standard Programming Language
    FORTRAN," (FORTRAN 77) American National Standards Institute Inc.;
    1978.

4. ANSI/MIL-STD 1815A, "American National Standard Reference Manual for the Ada Programming Language," American National Standards Institute Inc.; 1983.

5. Backus, J. "Can Programming be Liberated from the von Neumann Style? A functional style and its algebra of programs," Communications of the ACM, vol 21 (1978), 613-641.

6. Backus, J. "Is Computer Science Based on the Wrong Fundamental Concept of 'Program'? An Extended Concept," in J.W.de Bakker and J.C. van Vliet (eds.), Algorithmic Languages, North Holland, New York, 1981; 133-156.

7. Foderaro, J.K. and Sklower, K.L. "The FRANZLISP Manual," University of California, Berkeley; 1982.

8. Harland, D.M. "User-Defined Types in a Polymorphic Language," The Computer Journal, vol 27 (1984), 47-56.

9. Harland, D.M. Polymorphic Programming Languages: Design and Implementation, John Wiley, 1984.

10. Horowitz, E. and Sahni,S., Fundamentals of Data Structures, Computer Science Press, Inc.; 1976.

11. Kay, A. "Computer Software," Scientific American, vol. 251 (1984), 53-59.

12. MacLennan, B.J. Principles of Programming Languages: Design, Evaluation, and Implementation, Holt, Rinehart and Winston; 1983.

13. McCarthy, J. "Circumscription – A Form of Non-Montonic Reasoning," Artificial Intelligence, vol 13 (1980), 27-39.

14. Pearl, J. Heuristics: Intelligent Search Strategies for Computer Problem Solving. Addison-Wesley, 1984.

15. Rowe, L.A. "Programming Language Issues for the 1980's," SIGPLAN Notices, vol 19 (1984), 51-61.

16. Treleaven, P.C., Brownbridge, D.R., and Hopkins, R.P. "Data-Driven and Demand-Driven Computer Architecture," ACM Computing Surveys, vol 14 (1982), 93-143.

QUESTION AND ANSWER PERIOD

CARBONELL
I had two comments, but I believe he is going to make one of them so I will make the other one. The other one is that with respect to natural language and transparency, one reason why one may want to have natural language in large scale systems, such as expert systems, is for them to be able to explain their behavior, the reasons why they have taken certain actions, to a human user who can then either override that solution, change a description of a problem, send it back to the builder to say it is ignoring this

important piece of knowledge, or accept the diagnosis or the answer or
whatever the case might be. It is a means of making it more transparent by
changing the internal part, although I do agree that imputing additional
intelligence is indeed a danger, just like it is a danger to impute
additional intelligence to hand-held calculators because of course they do
a higher level operation - mathematics - which is more complicated than
language (at least in common folk belief).

BOUDREAUX
I have two points. First, the point that your are making is absolutely
correct, but I distinguish between the language in which I design the
algorithms, the abstract data structures that I'm providing, and the
environmental component of the system which would supply information to the
user in terms of help files, and I think that with respect to those I would
strongly encourage the use of a language that is other than first order
logic. The second observation that I have on your comment is that I deny
that mathematics is more complicated than English. For native speakers of
English the analogy would be to attempt to read medieval English text. If
you do that you will see just how complex something which is called natural
language really, really is.

CARBONELL
I agree absolutely with that. I was just referring to the common folk
belief that mathematics is more difficult....

BARSTOW
I'm curious about why you speak in terms of a single language, rather than
several languages, when you spoke about the great diversity of your target
users. It sounds like you will require technical expertise not only in
their area of expertise but also in the area of computation?

BOUDREAUX
Good. Yes. The single language is the language that I am choosing, in
fact, to implement the system in, that language which that underlies the
work which I'm doing. The language that I have in fact chosen in the
design of this language system for work stations is Franz LISP. The truth
is out. Now, the second point that you made is, to me, the language that
one uses to describe the interface, or the family of languages, I think,
that one uses to describe an interface that a user would have to an underlying system. And I claim that one of the weakest things about our present
theoretical understanding of these issues today is in how to elicit from a
user community that is otherwise naive what kind of language structure they
would find most useful. It is a practical, actual, real issue that I need
help with now, and unhappily I don't know where to turn for it. Except to
you sir, perhaps.

BARSTOW
Let me give you an example that will make it sound even worse. We designed
an equational description language which would allow users to write equations that we thought would be very natural. Unfortunately, some of the
people who used it remembered that the equal sign was also an assignment
statement in FORTRAN, so rather then thinking about pure equations they
started to use them as if they were assignment statements.

BOUDREAUX
Beautiful. That is a great example.

SHNEIDERMAN
I do want to speak in support of your concerns about natural language and
that precise, concise artificial languages often help sharpen thinking - a
view that has been clear to me. Part of the problem with natural language
also is that it makes it difficult to plan ahead a sequence of actions -
you are waiting for the clarification dialog to ensue. That is a further
issue. But I think the underlying principle is this distinction that I
make between syntactic and semantic domains. Natural language only
relieves the user of the need to learn some specific syntactic details.
But the hard part of the use of computers all around is understanding the
semantic domain, the possibilities that are available on the machine, and
in fact as you suggest, I believe natural language only confuses that
issue, whereas a much more sharp language which may tap a user's knowledge
of whatever - set theory, Boolean algebra, predicate calculus - may in fact
be the more productive one in the long run, so I want to give you that
encouragement.

BOUDREAUX
In fact let me give you a concrete case. In the early turn of the century
- not very often computer scientists refer to that - at the turn of the
century a great, enormous amount of effort was being spent by people who
were involved in logic - I'm thinking now of Russell and Whitehead and
later Carnap, and so forth - attempting to give formal explications for the
notions that we in our natural language use. So it was an attempt to use
the mechanisms available in formal logic as it was then understood to make
clear what we say when we say what we ordinarily say. The result was,
after four decades of extremely diligent work by some the ablest prac-
tioners in the field, remarkably slim. It turns out that the intrinsic
complexity of natural language far exceeds our capacity, I think, our
capacity to express it in a reasonable formalism.

SHNEIDERMAN
I think the challenge is, as you have suggested with some of your illustra-
tions, how do we get rid of some of the clutter of syntactic detail and
focus on a notation which expresses the semantics without the additional
clutter. I might note that your FORTRAN example did still have one
syntactic error that I caught.

BOUDREAUX
Did it?

SHNEIDERMAN
I believe so. Which was DO 10 comma I.

BOUDREAUX
Say again?

SHNEIDERMAN
DO 10 comma I.

BOUDREAUX
That is fine. No. On FORTRAN 77.

SHNEIDERMAN
Is that right? OK.

BOUDREAUX
Get with it, man!

SAMMET
Two points. First of all, if you don't understand why ADA is better than
FORTRAN, that is probably because you don't appreciate the need for
languages which are capable of dealing with problems that have millions of
lines of code. FORTRAN is not capable and to the extent that FORTRAN has
improved, it is no longer the original FORTRAN. That is one point. Second
point - with regard to natural language, I take issue with my good friend
Ben Shneiderman. We have argued this point many times. I quite agree that
the semantics is the hard part, but I suggest that one of the difficulties
that users actually have in communicating with the computer is that they do
have to worry about semicolons, whether they be cancerous or otherwise.
And if they can be spared the problem of learning the syntax, and can in
fact communicate with the computer the same way they communicate with each
other, then I suggest they would have more time to concentrate on the
semantics.

BOUDREAUX
Well now, you give me a smorgasbord. Number one, I pointed out that the
notion of a package is one of what I judge to be ADA's signal contributions.

SAMMET
Very generous of you.

BOUDREAUX
Well thank you.

LYON
Since we've talked about programming environments, I just wanted bring that
up, because I think there is something that we can get out of it more than
perhaps what has been mentioned. On one hand we might imagine a puffy
cloud of abstract functions which we agree is what we often want, the
semantic invariants. On the other hand, we're often stuck with a bunch of
circumstances that are going to give us a lot of artifacts - hardware, how
a given group of users wanted to use these language things. Caplan, in a
book on design, mentioned that the designer is someone who has to really
often reconcile the functions that are to be accomplished - namely, if he
has a chair, it is supposed to hold people - with the circumstances of its
use - namely, given anatomical shapes are going sit on this chair - and
that has been a problem in languages because they have been nailed often to
concrete syntax. But a lot of the languages pander more to character set
and deterministic parsing than they should. And we can get off the hook if
the programming environment can help with that, so that you might imagine a
template-driven system where, for given different sets of circumstances
with the same abstract functions, there may be different concrete
languages, the objective realizations. And you would only have to do
recognition, you wouldn't have to memorize all these different concrete
realizations, because when the environment wanted your opinion it would
beat it out of you.

BOUDREAUX
Good point.

BARSTOW
Let me add to that. I think the programming environments, in fact, are the way around the problem of natural language, because what they do is to let you avoid all the syntactic problems, and they can even structure the dialogue and all of that.

BOUDREAUX
I agree.

PIETRZYKOWSKI
That was unfortunate because that was what I wanted to say, but I have something else to say as well. But it is somewhat on a little more serious note. It is indeed important to realize that these days are not the days of the punch cards. That people don't scribble on paper and worry about the thing. People do use the terminals and if they have a decent programming environment, the whole big hoo-ha about syntax is all gone, except in such a monstrosity as ADA, of course. But there are not so many friends of this. Now the other aspect which I would like to raise, we silently assume, and I hope I will have an opportunity to convince you that it is otherwise, that language is a textual phenomenon. That it is a linear chain of symbols which describe.

BARSTOW
My point was that that does not necessarily hold.

PIETRZYKOWSKI
OK. What I wanted to say was that in terms of language, we definitely - particularly in the era where graphics is nothing particularly exotic in computers - we have to consider pictorial contructs on the same level as textual linear contructs, and the interesting point is that that seems to have a lot of bearing on a variety of issues which are otherwise invisible. The two-dimensional representation completely can change the way of approaching problems, completely can change. We should remember that anytime we want to explain something to you, we start scribbling and making little pictures. Nobody writes equations to communicate with each other and nobody can talk incessantly - oh yes some people do, but they don't understand each other.

BOUDREAUX
If I might, just one big comment there. It seems that the problem I have with graphic demonstrations of that form is that the ambiguity level of them is, I think, rather higher than for linear constructs.

LYON
Yes, but in fact with a graphical programming language you may say in fact you can hardly parse it, but that misses the point. That is an artifact; you may not have to parse it if you built it with an environment that was template driven.

BOUDREAUX
Oh, I see.

LYON
That is what I meant by the circumstances. The circumstances are that
maybe we can give a good punt on parsing technology and do something else.

BOUDREAUX
Good, that is it. Thank you very much.

The Role of Language in Problem Solving I
R. Jernigan, B.W. Hamill, and D.M. Weintraub (Editors)
© Elsevier Science Publishers B.V. (North-Holland), 1985

AN ARGUMENT FOR NON-PROCEDURAL LANGUAGES[*]

J. Baron[1], B. Szymanski[2], E. Lock[2] and N. Prywes[2]

University of Pennsylvania
School of Engineering and Applied Science
Philadelphia, Pennsylvania 19104-3897

Programming with a nonprocedural language requires
focusing on analysis of the problem rather than on a
solution in the form of a sequential computer program.
The claim is made that the nonprocedural language fits
better into the natural process of problem solving.
The user of a nonprocedural language specifies a set of
constraints to be satisfied which are derived from the
problem, rather than a sequence of steps to be taken.
On the other hand nonprocedural languages do not permit
the user to make several assignments to one variable,
since the order of these assignments could not be
specified. We argue here that nonprocedural languages
are better suited than procedural languages for most
applications. The discussion is in the context of the
MODEL system, which uses a nonprocedural language and
also provides in depth logical checking and rapid
prototyping. We present data in which subjects learn
very quickly how to deal with the most serious difficulty in MODEL, the need to represent iterations as
vectors. We also describe the development of a large
accounting system by a single programmer using MODEL
which demonstrates that far less code is required for
the MODEL specification than for an equivalent
procedural program, debugging is much faster, and less
time is required than would be expected when using a
procedural language.

I.   INTRODUCTION

Procedural programming languages are those where the order of assignments
of variables is specified by the program. As the computation progresses it
is possible to make several assignments for the same variable, each at a
different time. Nonprocedural languages, in their pure form, specify
constraints that are to be satisfied, usually in the form of equations.
Variables' values cannot be reassigned, for the user does not specify the
order in which the constraints are satisfied. The MODEL nonprocedural

*   Support received from the Office of Naval Research under Contract No.
        N00014-83-K-0560.
1   Department of Psychology
2   Department of Computer and Information Science

language, [Prywes and Pnueli, 1983; Schwartz, 1983, Shi 1984], works by
having the system write computer programs that satisfy the constraints
given in a specification. Thus, it relieves the programmer of the task of
sequencing the commands provided to the computer. As related further, the
MODEL system also interacts with the user in performing in depth checking
to identify missing or inconsistent parts of the specification and provides
a rapid prototyping capability.

There is some tension between procedural and nonprocedural approaches,
especially as "computer literacy" becomes more widespread. Practically all
languages that students now learn are procedural, yet the mathematics that
they learn has strong elements of nonprocedurality. For example, in
algebra, students learn to write down first the constraints of the problem,
in terms of devices such as simultaneous equations to be solved. Students
then use procedures for solving these problems, but the initial representation of the problem is in terms of the constraints, not the procedures.
The same could be said for many applications of mathematics. When students
of these disciplines turn to computers to help them, it would seem most
natural for them to communicate with the computer in terms of the same type
of problem representation.

In this paper, we make a case for the superiority of nonprocedural languages. We cannot make an absolute case. It is, in general, impossible to
demonstrate the absolute superiority of one technology over another in all
conceivable cases. For example, high level compiler languages seem superior
to the assembly languages in most situations, but there are situations in
which lower level assembly languages are still the preferred means of
programming. What we can do is argue that nonprocedural languages are more
useful for the vast bulk of current applications of computers, and for many
applications still in the future.

Section 2 reviews and rebuts some of the previously published arguments in
favor of procedural vs. nonprocedural approaches. However, the main
support for our position favoring the nonprocedural approach comes from the
two experiments. In particular, the results of these experiments allow us
to discuss and evaluate the criteria for superiority of one approach over
the other. These criteria consist of the time required to learn a language,
to program and debug a computation in that language, and to modify a
program in software maintenance.

The first experiment in Section 3, examines programming of iterative computations, which embodies the most fundamental difference between the two
types of languages. It is related to how quickly users learn a programming
language, understand a problem in terms of the language, propose a solution
to it and how much they are prone to make errors. To deal with these
concepts, it has been necessary to experiment with very small problems. As
will be reported, the results of the first experiment favors the nonprocedural approach slightly. The second experiment takes the point of view
that small problems are simple and conveniently within the span of human
cognition. Therefore, the effectiveness of reaching a solution depends
only in a minor way on the type of language used for representing it
[Sheil, 1981]. The second experiment is described in Section 4, including
a brief review of the MODEL system. This experiment is much larger in the
magnitude of the problems programmed. It consists of using MODEL in
development of an accounting system of seven relatively long programs and
nine complex data files. Because of the size of the problem, we were able

to use only one subject. However the experiment allows detailed analysis of the results, showing times attributed to the various programming phases with emphasis on count of various error types and times required to make respective corrections. The advantages of the nonprocedural approach become very significant in terms of length of respective programs, and time required for the overall development.

2. DISCUSSION OF SOME ADVANTAGES AND DISADVANTAGES OF PROCEDURAL AND NONPROCEDURAL APPROACHES

Nonprocedural problem representation has one great advantage in that the user needs to provide only constraints inherent in the problem while the respective procedures are supplied by the compiler. Any problem can be stated in terms of the constraints to be satisfied. This is because a problem (in the true sense of the term) has a goal to be achieved and the constraints can define the goal [Wertheimer, 1945/1959]. Ordinarily, these constraints are part of a user's representation of a problem, and thus - with the help of the computer - the user does not need to provide the detailed procedure. Occasionally, people may learn a mechanical procedure, such as balancing a checkbook, while forgetting the goal, but this happens only because the procedure is routinized. If computers with nonprocedural languages come to do more of our work, there will be less reason for such routinization of procedural thinking. Such computers will be able to translate nonprocedural specifications into procedures, thus eliminating a great deal of work.

Despite the apparent advantages of nonprocedural languages, it has been claimed that by forcing people to think procedurally, we can actually help them solve problems [Soloway, et al, 1982; Erlich, et al, 1982]. The empirical evidence for this claim is essentially that students who have learned programming, compared to students who have not, are less likely to make mistakes in certain problems involving algebra. For example, when asked to write an equation representing the statement, "There are six times as many students as professors at this university," almost 40% of some groups of subjects wrote the equation: 6S = P. Erlich et al. suggest that the subjects might have been tricked by the ambiguity of the representation, since the incorrect answer just given has a plausible analogy, namely: 3 feet = 1 yard. Programmers were less likely to make this error.

Erlich et al. clearly suggest that the learning of procedural languages was crucial to helping them solve problems. (They go so far as to suggest that "Algebra is not a descriptive language but a procedural one." This claim, of course, flies in the face of the fact that algebra can be, and usually is, used to represent constraints on solutions, independently of whether the user knows in advance what procedures will be used to solve the equations thus formed.) However, most of the specific explanations they give for this effect on students' problem-solving performance do not single out procedural languages, and we might expect training in MODEL alone to have much the same effect as training in PASCAL. In particular, Erlich et al. suggest that the relevant experience involved four factors:

(1) The unambiguous semantics of programming language construction. This of course is characteristic of any workable programming language, procedural or nonprocedural. However, Erlich et al. suggest that the use of the equal sign in procedural languages is inherently unambiguous, in that it can mean only to assign

the value on the right to the variable on the left. This may be
true of the equal sign - putting aside its use in Boolean
expressions - but it is surely not true of procedural languages
in general. Soloway and his colleagues have documented, with
respect to the different loop structures in PASCAL, that the
meanings of each loop structure must be painstakingly memorized
(Soloway, et al, 1981).

(2) Explicitness required by the syntax of programming languages.
This also applies to any programming language.

(3) Viewing an "equation" in a programming language as a dynamic
input/output transformation. This characteristic also does not
single out procedural languages. An equation in a nonprocedural
language represents a transformation too, although static, not
dynamic. The fact whether it is dynamic or static seems to be
of little importance. There is no evidence that the transformation characteristic by itself is responsible for the effect
found.

(4) The practice of debugging programs. Although nonprocedural
languages may involve much less debugging, for there is less to
debug, learning to use them would still provide ample opportunity to learn debugging. In general, practice at programming in
any type of language might make people more cautious about how
to represent a problem solution.

(5) The practice of decomposing a problem into explicit steps.
Here, Erlich et al. give the example of a student who wrote the
equations $X=C/4$ and $S=5*X$ before writing his final equation,
$S=5*C/4$. Note that these equations could be interpreted procedurally or nonprocedurally, and that nothing about procedurality
encourages a student to write them this way rather than
directly. If students have learned to decompose a problem, it
was because it helped, not because it was encouraged by the
language itself.

3.   EXPERIMENT 1: THE COST OF NONPROCEDURALITY

The first experiment we report was suggested by the invention of the MODEL
programming language [Prywes and Pnueli, 1983]. This language is entirely
nonprocedural. Its variables cannot be given more than one assignment in a
specification, lest the specification be ambiguous. The MODEL compiler
(actually an automatic program-writing program) chooses an optimal order of
operations, independently of the order in which assignments are made by the
user. For many cases, this is not a source of difficulty for the user.
However, there is one kind of programming problem for which such semantics
may cause difficulty. This is the kind of problem that involves iteration,
the kind that is easily solved in procedural languages with the use of
loops in which one variable (or more) is reassigned with each iteration of
the loop. For example, when we find the sum of an array of numbers, we
usually do this by the use of some variable, say, SUM. It is repeatedly
set equal to its last value plus the next number in the array. When the
last number in the array is reached, the value of SUM is the sum of the
whole array. Another typical example is the estimation of a value by
iteration or successive correction. This case occurs when it is hard to

calculate a value analytically, but relatively easy to correct an estimated value so as to make it a better estimate. Of interest here is that the number of iterations usually cannot be specified in advance.

The first case, in which the number of iterations is known, is handled easily in MODEL by the use of an array in place of a scalar variable (such as SUM). The successive values in the array correspond to the successive values of the scalar variable in the iteration. More generally, any N dimensional matrix is replaced with an N+1 dimensional matrix. The added dimension of the matrix (in the nonprocedural specification) corresponds to successive values of the rest of the matrix (in the procedural program).

For example, to find the sum of an array of number A[1..100], we would say,

IF I=1 THEN SUM[I] = A[I] ELSE SUM[I] = SUM[I-1] + A[I]

SUM[100] is the value we are seeking. (PASCAL-like syntax is used here, rather than MODEL syntax because the subjects in this experiment all knew PASCAL. Also note that this example, and all others to follow, omit declaration required in PASCAL and MODEL, and input-output statements which are required only in PASCAL). The condition in the IF statement serves the function of initializing SUM[.], which must be done somehow - not necessarily the way it is done here - in any language.

In cases in which the number of (procedural) iterations is unknown, the size of the corresponding (nonprocedural) array cannot be specified in advance. In MODEL, this is handled by defining an auxiliary array using the tag END., which has the value FALSE('0'B) for every value but its last. The specification may state that this value is TRUE ('1'B) if some condition is met. The lowest possible position in the END. array for which the condition is met will be assigned a value of TRUE, and this will be the last position in both the END. array and its corresponding actual array. For example, suppose we want to solve the equation X=exp(-X), that is, X equals e to the minus X power, by iteration. Procedurally, we would first set X equal to 1 and then compute successive new values by setting X equal to exp(-X), until the successive values differ by some small amount, say, .001. Nonprocedurally, we would say,

IF I=1 THEN X[I] = 1 ELSE X[I] = EXP(-X[I-1]).

However, our specification must include an additional statement:

END.X[I] = ABS(X[I]-X[I-1])<.001

The value we seek is now the value of X[I] for which END.X[I] is TRUE. There is only one such value, because END.X[I] is not defined for values of I greater than the first such value for which END.X[I] is TRUE. Note, that the very same condition has to be used in procedural languages in a loop head statement with UNTIL construct (or negation of this condition if WHILE is used). Thus the necessity of END specification does not add complexity to the solution.

Most of those who have familiarized themselves with MODEL have found the awkwardness of the approach to otherwise simple iteration problems to be the language's greatest drawback. However, it would seem to be a necessary drawback, for the reasons already discussed. Thus, it is important to find

out how serious a problem it is to convert iterations into arrays. In
fact, this problem seems to be a crucial one for any assessment of the
advantages and disadvantages of nonprocedural languages such as MODEL.

In asking a question like this, we can expect no more than a rough answer.
There is no scale of "seriousness" to which we can appeal. Likewise, the
problem will be more serious for some applications than for others, and
more serious for some users than for others. There is no way of randomly
sampling users and applications. Further, the seriousness of the problem
will depend on how skillful we are in trying to overcome it. Thus, the
best we can do is to examine the effectiveness of a particular kind of
training, for a particular sample of users and problems. In the present
experiment, this is done for what is roughly a worst-case for MODEL. The
problems are those for which iteration is required, and the programmers are
those who are already experienced in using a procedural language, PASCAL.
We ask whether a small amount of training and practice can essentially
eliminate the extra difficulty involved in the use of arrays.

3.1   METHOD

Subjects were 15 undergraduates at the University of Pennsylvania, all of
whom had taken and passed the introductory programming course in PASCAL.
Most were enrolled in an advanced course on database management. (One
additional subject was omitted from the analysis to be reported because,
apparently, he or she did not take the experiment seriously).

Subjects were given a written instruction, followed by six simple problems,
half to be solved in a procedural language, half in a nonprocedural language. Subjects were asked to time themselves on reading the instructions,
and on each problem. They were also asked several questions about their
impressions of the languages. They were paid $5 per hour for all the time
they spent, but in no case more than $10. The problems were done at the
subjects' leisure, and were turned in to the experimenter a couple of days
after being handed out.

The instructions took up 2 1/3 single-spaced pages. There was a brief
introduction to the idea of nonprocedural languages and the procedure of
the experiment. It was pointed out that nonprocedural languages can
simplify the users task for many problems, but there were other problems
that might be awkward because they were usually thought of in terms of
procedures with successive assignments of the same variable. Then, the
main idea of using arrays instead of iteration was introduced. The following statement was put in upper-case letters as the main idea to be learned:
"Whenever a scalar variable takes on different values in the course of a
procedure, replace that variable with an array; if the index of the array
is I, the I'th value in the array corresponds to the I'th successive value
of the variable." Then, subjects were told what syntax they were to use in
solving the problems: they were to use the syntax of PASCAL but without
declarations, input-output statements, or flow-of-control instructions such
as BEGIN, END, DO, or REPEAT. It was pointed out that IF THEN ELSE and
CASE could be used, but they would indicate conditional assignments, not
steps in an order. Assignments were to be made with "=", not ":=". The
use of arrays of unspecified size was explained, and the two examples given
above were presented and explained. In summary, it was suggested that the
trick was to convert a procedure "in time" to an array "in space." It was
pointed out that the user did not have to worry about using up memory,

because the arrays used were only imaginary; specifically, the nonprocedural language compiler would translate a nonprocedural specification into an efficient procedural program.

Subjects were then asked to do the following six problems in order, the odd ones in PASCAL (without declarations and input-output) and the even ones in the corresponding nonprocedural language, or the reverse, depending on whether their birthday fell on an odd or even day of the month:

(1) Given an array A[1..N], find the mean (average) of all the numbers in the array.

(2) Given an array A[1..N], find the mean absolute difference between adjacent numbers.

(3) Given an array A[1..N], find the number of times that a number in the array is equal to the next number in the array.

(4) Given an array A[1..N], find the biggest number in the array.

(5) Solve the equation x=exp(1/x) by iteration, until successive values differ by .001 or less.

(6) Solve the equation x=log(x+5) by iteration, until successive values differ by .001 or less.

Subjects were also asked to comment on what they think the nonprocedural language would be good for, what it would not be good for, and what would make it hard or easy to learn.

3.2   RESULTS AND DISCUSSION

The mean time to read the instructions was 14 minutes. This is probably a reasonable estimate of the time to read a section on the topic included in a manual for MODEL. On the one hand, much of the instructions would be included elsewhere in the manual. On the other, some attention would have to be paid to adding a dimension to arrays and matrices.

The problems were analyzed by pairs: the first pair, the second pair, and the third. Members of a pair were desinged to be closely matched in relevant respects. Thus, it should matter little that some subjects did the first member of each pair in the procedural language while others did the second.

Figure 1 shows the mean reported solution times for the procedural and nonprocedural problem in each pair. (Three subjects did not report times by problem; these means are based on five subjects who did the nonprocedural language first and seven who did the procedural language first.) For the first pair, the nonprocedural language took longer for every subject. This is to be expected, as this language is less familiar to these subjects. All subjects took longer on the nonprocedural language for the second pair as well, although the difference was smaller. However, by the third pair, most subjects took longer with the procedural language: 3 out of five in the group who did the procedural language first (with one subject having equal times on both languages), and five out of seven (with one equal) in the group who did the nonprocedural language first (t(11)=2.53, p>.05 two tailed). Note that this third pair of problems was also the most difficult to solve nonprocedurally in the sense that only this pair required the use of arrays of unspecified size.

Problems	Procedural	Nonprocedural
First pair	5.3 (0)	15.1 (.94)
Second pair	4.7 (0)	9.3 (.72)
Third pair	13.2 (.22)	8.9 (.39)

FIGURE 1
**MEAN TIME IN MINUTES AND (IN PARENTHESES) AVERAGE NUMBER
OF ERRORS PER PROBLEM, BY PAIR OF PROBLEMS**

In sum, for the third pair of problems the difference between procedural and nonprocedural forms had disappeared. The fact that the nonprocedural language was actually easier can probably not be taken seriously, as a few more errors were made (Figure 1) on the nonprocedural language. However, it seems safe to say that the two forms were essentially equal. Thus, a few minutes of instructions plus a small amount of practice - two problems - seems to eliminate the most serious drawback of nonprocedural languages. Of course, it is possible that the difficulty with iteration would reassert itself in some other situations not examined, but there is little reason to think so. (The reader may get some feeling for the surprising ease of converting iteration to arrays by just thinking about how it would be done in his or her favorite problems.) Once we believe that the biggest disadvantage of nonprocedural languages can be overcome, it is more productive to focus on their advantages, which, for some applications, are so obvious as not to require psychological investigation at all.

Figure 1 also show error rates for the different conditions. The procedural language produced fewer errors throughout, although this difference disappeared by the third pair of problems.

Errors were classified according to the following scheme:

Translation - The problem would not work under any sensible interpretation. The subject had wrongly translated the problem as given into a specification, putting syntax aside.

Ambiguous syntax - Any clear syntax was accepted, but often the syntax was ambiguous.

Wrong language - Procedural statements were included in nonprocedural problems or vice versa.

Undefined variable - The most common form of this was actually an out-of-bounds reference to a member of an array, for example, X[I-1] used when I=0.

Extra statement - An unnecessary statement was included, such as
BEGIN or END in a nonprocedural language.

For the procedural language, three of the four errors were translation
errors, and one, in the last problem, involved the use of nonprocedural
statements in the procedural language. In particular, this subject used an
array instead of iterated variable reassignment in problem 6. (He maintained, however, other features of PASCAL, such as ":=," BEGIN, and END.
He had not used these in the nonprocedural language.) For the procedural
language, there was 11 translation errors, 4 ambiguity errors, 4 wrong-
language errors, 8 extra-statement errors, and 7 undefined-variable errors.
(There were also a couple of miscellaneous errors, such as making up
functions that did the job required and failing to initialize.)

All the errors on the last pair of problems were translation errors, (2,
one less that the procedural language) and undefined variable errors (5).
Of course, undefined-variable errors would be caught quickly by such
compilers as MODEL. Still, it would seem that these errors are a natural
consequence of the use of arrays in which each value is a function of the
next lowest value. The user simply forgets that the lowest value in the
array cannot be defined this way. Even remembering to initialize the array
won't solve this problem (and, indeed, failure to initialize occurred only
once). It is also necessary to make sure that the definition of successive
values in terms of lower values excludes the lowest value itself. This
problem might be handled by extra warnings in the manual or by "intelligent" error messages in the compiler, of the sort, "You have defined the
first value of this array as __, and later you define each value, including
the first, as __. I assume you mean this definition to exclude the first."

In sum, this experiment indicates that it is fairly easy to teach
programmers the trick of using arrays instead of iterated reassignments.
One problem that seems to persist is the failure to exclude the first
position in the array from this treatment. It remains to study this issue
in nonprogrammers.

As a result of the present experiment, instruction like those given to our
subject have been incorporated (with improvements) into the MODEL manual
[Schwartz, 1983].

4.   EXPERIMENT 2: SOFTWARE DEVELOPMENT USING NONPROCEDURAL LANGUAGES

4.1   BASIS FOR COMPARISON

The second experiment was undertaken to evaluate the times required for
software development while using the MODEL language. We focused on the
realism and extensive features of the software system that has been
developed in the experiment. The intention was to compare the overall
productivity and error rates in use of MODEL with those when using procedural languages. Ample statistics have been published [Boehm, 1981;
Cougar, 1975; Endes, 1979; Horowitz, 1975; Perlis et al, 1981; Rubey et al,
1975; Schneidewind et al, 1979; Walston et al, 1977; Wolverton, 1974] on
productivity in using high level procedural languages for large software
projects. The number of lines, or statements in a procedural language
generated by the MODEL system serves as a basis to compare the time taken,
errors made, etc., with those reported for similiar length procedural
programs produced manually.

4.2   A BRIEF DESCRIPTION OF THE MODEL SYSTEM

As noted, the MODEL system translates a nonprocedural specification into a
set of procedural programs. The system aids in the composition of a specification by conducting in depth checking of completeness and consistency
and by providing rapid prototyping. The composition of a specification is
an incremental process, at the completion of which the entire produced
software becomes operational. Thus, the traditional multi-phase approach
of software development, typically consisting of requirements, design,
coding and debugging phases, is replaced by a single specification phase.

A large MODEL specification may be divided into module units. Typically
modules corresponds to individual functions or subfunctions. The overall
system consists of two subsystems: a configurator- for translating inter-
module specifications, and a MODEL compiler- for translating single module
specifications. The role of these systems is explained in conjunction with
the procedure for their use below.

The overall procedure in using this methodology is illustrated schematic-
ally in Figure 2. It starts (at the top) with existence of problem. The
human users have to partition the problem into modules. There are two
parallel paths in Figure 2 for module definition and for inter-module
synthesis. They merge at the bottom of the diagram to produce the computation.

The path on the left is followed for each module in a configuration. In
case of system modification, only the specifications of affected modules
need to be added, deleted or changed. This path consists of composing a
specification of a module in the MODEL language, and submitting it to the
MODEL Compiler. The MODEL Compiler constructs a dataflow graph for the
module specification. This graph is used for analysis of consistency and
completenes of definitions, to discover errors, and for optimization of the
generated program. The user must then make corrections to respond to error
and warning messages issued by the MODEL Compiler. Finally, a program is
generated, in our case in PL/1. The program can then be executed as a
process, by itself for testing, and in concurrent operation with other
modules as described in a configuration specification.

The path on the right of Figure 1 is used to integrate modules into a
concurrent computation. A specification of a network of modules and files
is submitted to the Configurator. The Configurator constructs a dataflow
graph of the configuration and analyzes the graph for compatability of the
interconnections and completeness. The user must make appropriate changes
to respond to error or warning messages. The Configurator produces then an
overall customized design to maximize the parallelism in execution of
modules, and generates a set of command language programs for executing the
network of modules in a chosen environment of computers, communications and
operating systems. The Configurator also performs system wide
documentation.

4.3   THE EXPERIMENT

The investigation consisted of progressive development of an accounting
system. The experiment addresses also the question whether it is feasible
and advantageous for the management accountant to express precisely his
processing needs in a nonprocedural language that will be both implemented

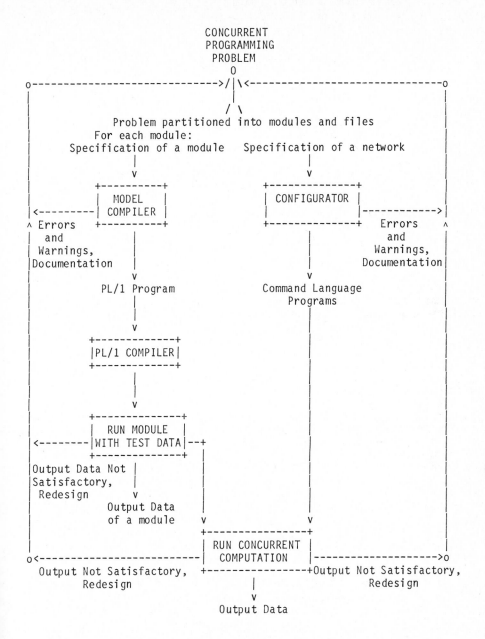

FIGURE 2
SCHEMATIC DIAGRAM OF NONPROCEDURAL PROGRAMMING PROCEDURE

by the underlying language system and understood by the accounting community. The objective is to reduce accountants' dependence on the data processing department and to provide them with full control over the financial processing. This is particularly important when accounting

standards are being changed. To reflect such a situation the experiment
included introducing Financial Accounting Standard 33 (FAS 33) [Ernst &
Whinney, 1979], which requires financial statements with supplementary
information to reflect changing prices. Financial reporting of changing
prices is complex mainly because of measurement problems. Significant
difficulties surround the choice of an index (consumer prices, general
prices, or specific prices) and a price (expert assessments, internal
pricing systems, buying, selling, or yet others) used for calculating the
appropriate adjustments. It has a broad impact so that to incorporate it
into an accounting system in a large corporation would require extensive
funds and time [Martinelli, 1982]. Direct expression of FAS 33 standards
by an accountant can lead to much shorter and easier implementation than
one done by programmers guided by accountants.

Conducting experiments to obtain software development statistics is
expensive and difficult to manage. Therefore, we limited the scope of the
experiment to one developer, who was an MBA student majoring in accounting
with limited programming background. Our study simulated the development
of a typical application and consisted of the following steps:

(1) setting the data programming requirements,
(2) defining files and major program units,
(3) using MODEL to generate the programs and
(4) modifying the MODEL specifications to adopt a new accounting
     standard [Cheng et al, 1983].

The developed accounting system compared favorably with commercially
offered accounting systems and incorporated state-of-the-art features,
including current price valuation of assets.

The development was carried out in three stages, with interruptions of 3-6
months between stages. The three stages were selected to determine vari-
ability in software development times required for (a) initial development,
(b) extension and (c) modification. The system eventually consisted of 7
modules: 2 developed in the first stage for general ledger, 5 implemented
in the second stage for inventory management and 1 of the general ledger
modules modified in the third stage for incorporating valuation of assets
at current prices (as opposed to cost basis used in the first stage).
Development times included the research necessary prior to each stage to
determine the capabilities of the accounting system as competitive with
existing or proposed state-of-the-art accounting systems. It was not
practical to develop the same accounting system using a procedural language
and compare achieved software productivities for two reasons. First, we
assumed that the developer need not have the procedural programming skills.
Second, determining the capabilities of the accounting system has been an
integral part of the software development, and it is unlikely that two
developers, one using a nonprocedural language and the other a procedural
one, would select the same capabilities, necessary to make a comparison
meaningful.

4.4   RESULT AND DISCUSSION

Figures 3 and 4 show overall statistics of program module sizes and
development times.

Stages in Development

	1	2	3	
Steps in programming	General Ledger	Inventory	Valuation at Current Price Modification	Total
Requirements	40 (28)	30 (12)	8 (11)	78 (17)
Programming	23 (16)	49 (20)	13 (17)	85 (19)
Compiler debug.	44 (31)	93 (38)	31 (41)	168 (36)
Run-time debug.	36 (25)	72 (30)	23 (31)	131 (280)
System Total	143 (100)	244 (100)	75 (100)	462 (100)

Units are hours followed by columns (stage) percentage in parentheses

FIGURE 3
TIME BREAKDOWN FOR 3 STAGE DEVELOPMENT OF ACCOUNTING SYSTEM

The first step (requirements) in Figure 3 included the research time for determining state-of-the-art capabilities of accounting systems. In the general ledgers and inventory development stages it involved primarily reviewing commercial systems to select competitive capabilities. In the last stage it involved reviewing Financial Accounting Standards Board [FAS 33, 1979] and other publications on financial reporting and particularly on reporting effects of inflation [Ernst & Whinney, 1979]. The differences in the percentages for this step in the 3 stages are explainable by the different number and sizes of the respective modules. In developing a larger system, a smaller percentage of the time is used in the first step and larger percentages in the debugging. The row for the second and third steps included initial composition of the specification and correction of syntax errors. Time statistics on syntax errors, which are simple and easy to correct, were retained separately from other types of errors. They are not considered as indicative of the software development methodology. Errors discovered by the MODEL compiler (static debugging) were considered separately from errors discovered in dynamic debugging (by running programs with sample data). Stages 1 and 2, i.e initial general ledger development and the extension for inventory management, had very similar relative development times. This is because the extension of inventory management in stage 2 was relatively independent of stage 1. The communications between the general ledger and inventory management was only by sharing the general ledger files. Stage 3 involved a profound modification of the program module which produced the financial statements. It involved addition of input files, changes in reports and a new computational methodology. Figures 3 and 4 show that the MODEL specification is consistently shorter in number of statements than the produced PL/1 program by a ratio of 1 to 7-11 and that .7-1.0 hours of development was required per MODEL statement, i.e. productivity is 7-12 PL/1 statements per hour. Figure 4 shows the statistics for the number of lines for MODEL specifications and respective PL/1 program, but this comparison is somewhat unreliable, as the MODEL specification includes comments while there are no comments in the generated PL/1 program. Static debugging time (excluding syntax) exceeds the dynamic debugging time. Even though there is a considerable improvement over procedural programming, further future improvements are possible through use of additional checking methods in the compiler.

Specification	Stage	MODEL lines	PL/I lines	Ratio (1)	MODEL Stmts.	PL/I Stmts.	Ratio (2)
Financial Stmt	1/3	335	1340	4.00	88	932	10.6
Gen. Ledger	1	280	575	2.05	65	457	7.03
Inv. Order	2	161	396	2.46	33	258	7.82
Inv. Receipt	2	195	792	4.06	45	512	11.38
Inv. Sale	2	264	888	3.36	60	651	10.85
Inv. Status	2	177	698	3.94	41	385	9.39
Inv. On-Order	2	119	426	3.58	31	275	8.87
Total		1531	5115	3.34	363	3470	9.56

```
MODEL Lines Produced per Hour        3.31
PL/I  Lines Produced per Hour       11.07

MODEL Statements Produced per Hour   .786
PL/I  Statement Produced per Hour   7.5
```

Ratio (1) PL/I Lines per MODEL Lines
Ratio (2) PL/I Statements per MODEL Statements

FIGURE 4
PROGRAM MODULES, STAGES DEVELOPED/MODIFIED AND SIZES

The focus in Figure 5 is on the debugging process. Figure 5 shows average times per module to correct each type of error. Also shown are average numbers of errors per MODEL and PL/1 statements (including stage 3). There were no distinct differences between the 3 stages in the error correction times or in the percentages of time expanded for correcting various types of errors. The error count includes changes motivated by a desire to improve the system (e.g. better reporting) and not only errors made inadvertently.

Steps in programming	Error Type	Average Occurrence/ Module	Average Cycles to correct	Average Cycles/ Module	Average Hours/ Cycle	Total Hours/%
Requirements	NA	NA	NA	NA	NA	11/17%
Programming	Syntax	5.5	1.1	6.0	NA	12/19%
Compiler debugging	Ambiguity	3.2	1.4	4.5	1.5	24/36%
	Incompleteness	1.8	2.5	4.5		
	Inconsistency	2.1	3.3	6.9		
Run Time debugging	Test with Data	1.6	2.6	4.2	4.5	19/28%
Total Errors (excl. syntax)		8.7		20.1		
Errors per MODEL stmt		.16		.37		
per PL/1 stmt		.017		.039		

Legend:

Error Type	Description
Syntax	- Within single statement.
Ambiguity	- Use of same name for different purposes
Incompleteness	- Omission of entire statement
Inconsistency	- Found between statements (ambiguity, datatype, dimensionality, range, circular logic)
Test With Data	- Based on examining output reports or data.

FIGURE 5
FREQUENCY OF ERRORS AND CORRECTION TIMES, PER MODULE

Figure 6 shows the extent of the third stage modification by giving the sizes of the effected specification and program before and after the modification as well as the number of statements modified. Figure 6 also shows how the produced program has become more complex through statistics on the numbers of loops and variables used in the produced program. Note the effect of the optimization which changed MODEL array variables into scalar variables in the produced PL/1 program.

The study and statistics reported above provide insight into the development steps and their relative development times using nonprocedural language. The central question of how much more effective are nonprocedural languages over traditional, procedural ones is very difficult to answer. Published rates of statements per hour and errors per statement for using procedural languages can be compared with similar rates for procedural statements generated by the MODEL compiler. On this basis the MODEL methodology productivity reported here is in the ball park of three-fold improvement in statements per hour. However the number of changes made (excluding syntax errors) is only slightly lower than in using procedural methodology.

Complexity Measure	MODEL Before	PL/I Before	MODEL After	PL/I After
Lines of code	198	760	335	1340
Statements	60	533	88	932
Loop Structures	0	21	0	33
# of Variables	50	169	99	262
# of Arrays	46	19	94	41

FIGURE 6
PROGRAM MODIFICATION COMPARISONS

5.   CONCLUSION

If MODEL, or a language like it, were to become widely used, much of what
now passes for "education in computer literacy" would need to change. Much
of this education is in fact education in procedures, in translating one's
knowledge into unambiguous and correct procedures. With a nonprocedural
language such as MODEL, "computer literacy" would have more to do with
mathematical literacy. To take one concrete example, there is a tension in
the teaching of high-school algebra between those who want to treat
functions as procedures and those who want to treat them as ordered pairs.
The former group often argues that procedures, while less mathematically
elegant, are closer to computer programs, hence more useful. Whatever else
may be said for this group, this particular argument could lose its force
if computer languages were themselves to change. On the other hand, if
this group were to win its case, this fact alone would render nonprocedural
languages less useful, because they would not fit so well with the mathematical training of those who would use them.

In the first experiment, we also showed that the one major disadvantage of
MODEL in having to represent iteration by an array is easily overcome by
very short training. MODEL could be used as a general-purpose mathematical
language, for the same sorts of problems for which PASCAL, PL/1, and APL
are now used. For certain application, MODEL would have major advantages,
increasing software productivity and enabling larger groups of users to
formulate their problems in computer language.

In the second experiment, we showed that a complex problem may be more
easily solved using MODEL than by using a procedural language. The nonprocedural specification of a problem is shorter and free from implementation
details (i.e. input/output operations, flow of control) which require less
computer proficiency from the users. Also checking by the compiler is much
more comprehensive than is the case with compilers of procedural languages.
As a result, the use of MODEL language leads to higher productivity and
shorter development times for medium size compiler software systems.

REFERENCES

1.   Boehm, B.W., "Software Engineering Economics," Prenice-Hull, 1981.

2.   Cheng, T.T., Lock, E.D. and Prywes, N.S., "Use of Program Generation by Accountants In the Evolution of Accounting Systems: The Case of Financial Reporting of Changing Prices," Proceedings of the Workshop On Reuseabilty In Programming, ITT Programming, Stratford, CT, September 1983, pp. 17-28.

3.   Cougar, J.D., "Evaluation of Business Systems Analysis Techniques," Computing Surveys, (March 1975), pp. 167-236.

4.   Endes, A., "An Analysis of Errors and Their Causes in System Programs," IEEE Transactions on Software Engineering, June 1979, pp. 140-149.

5.   Erlich, K., Soloway, E., and Abbott, V., "Transfer Effects From Programming To Algebra Word Problems," research Report 252, Computer Science Department, Yale University, December 1982.

6.   Ernst & Whinney, "Financial Reporting Developments: Inflation Accounting," December 1979.

7.   Financial Accounting Standards Board, Statement of Financial Accounting Standards No. 33, "Financial Reporting and Changing Prices," (September 1979).

8.   Horowitz, E., ed., "Practical Strategies for Developing Large Software Systems," Addison-Wesley, 1975.

9.   Martinelli, W.P., "Unique Application Needs Force In-House Creation of Financial Programs," Software Focus, (March 1982), pp. 21-22.

10. Perlis, A.J., Sayward, F.G. and Shaw, M., eds., Software Metrics, MIT Press, Cambridge, MA, 1981.

11. Prywes, N.S., and Pneuli, A., "Compilation of Nonprocedural Specifications into Computer Programs," IEEE Transactions on Software Engineering, (May 1983), Vol. SE-9, No. 3, pp. 267-279.

12. Rubey, R.J., Dana, J.A., and Biche, P.W., "Quantitative Aspects of Software Validation," IEEE Transactions on Software Engineering, (June 1975), pp.150-155.

13. Schneidewind, N.F., Hoffman, H.M., "An experiment in Software Error Data Collection and Analysis," IEEE Transactions on Software Engineering, (June 1979), pp. 276-286.

14. Schwartz, S., "The MODEL Concept: Nonprocedural Programming For Nonprogrammers," Moore School Report, University of Pennsylvania, Summer 1983.

15. Sheil, B.A., "The Psychological Study of Programming," Computer Surveys, Vol.13, No.1, March 1981, pp. 101-120.

16.   Shi, Y., "Very High Level Concurrent Programming," Ph.D. Dissertation
       in Computer and Information Science, University of Pennsylvania,
       1984.

17.   Soloway, E., Bonar, J., Woolf, B., Barth, P., Rubin, E. and Erlich,
       K., "Cognition and Programming: Why Your Students Write Those Crazy
       Programs," Computer and Information Science Department, University of
       Massachusetts, Amherst, 1981.

18.   Soloway, E., Lockhead, J. and Clement, J., "Does Computer Programming
       Enhance Problem Solving Ability? some Positive Evidence on Algebra
       Word Problems," Computer and Information Science Department,
       University of Massachusetts, Amherst, 1982.

19.   Walston, C.E. and Felix, C.P., "A Method of Programming Measurement
       and Estimation," IBM Systems Journal, Vol.16, January 1977, pp. 54-
       73.

20.   Wertheimer, M., Productive Thinking, New York: Harper and Row, 1945
       (revised ed. 1959).

21.   Wolverton, R.W., "The Cost of Developing Large Scale Software," IEEE
       Transactions on Computing, (June 1974), pp. 615-636.

QUESTION AND ANSWER PERIOD

JERNIGAN
Do we have any questions?

UNIDENTIFIED MAN
You said you were going to look at the real-time programs?

PRYWES
Yes, we are running and generating real time programs. I just didn't have
a chance to express that, but there are really two systems. There is the
MODEL compiler, that I've talked about, and another one that is called the
Configurator. Configurator is on a higher level, where you say: here are
all the modules and all the files that I have. This is similar to the
notion of data flow. You show how data is communicated, what are the files
that are produced and consumed by the various modules. Then the system
uses concurrent programming techniques and tries to run the programs
concurrently. There is a new system being developed that tells you what is
the delay in any one module, so that you can actually check if time constraints are satisfied. We don't only generate the PL1 that I've shown
you, but we also generate the job control language programs which establish
communication between modules and also ensure proper synchronization.

JERNIGAN
One other question.

UNIDENTIFIED MAN
Would you say more about the interpreter....

PRYWES
There is no interpreter. Highly optimal programs are being generated. We
have spent the majority of the last eight years trying to think how do you
get programs that are highly efficient. By the way, we don't allow the
user to tell us anything about efficiency because we think that our
program, our optimizer, is smarter than most people, considering the amount
of time that they want to put in. We produce highly optimal programs in
terms of the time and in terms of memory that is being expended.

AMAREL
There is also something else. You mentioned also the Configurator, I
think, that produces a model of flow and so on. You mentioned optimizing,
so there is a considerable amount of software that was developed to take as
input those statements in the MODEL language and produce the porgram. Can
you say something more about that system?

PRYWES
The system has 40,000 of statements of PL1. Everything is in PL1. It runs
on the VAX and on IBM under CMS. I don't know if any of you have seen some
of our publications, but we have published it widely. There was a paper,
it must have been in May '83 IEEE <u>Transactions on Software Engineering</u> that
discussed the compilation process and the checking process. We had the
paper in the last ACM annual conference which went a great deal into the
checking.

BOUDREAUX
To what extent do you use your system, your external system to bootstrap
the underlying system, if at all?

PRYWES
We haven't. That just shows how stupid we were. This whole idea that you
can define anything in a nonprocedural language came down to us all so
slowly. So that when we started developing it we were going to write it in
a procedural language. I think it took about two or three years down the
road before we realized that we should have written the system in MODEL.
So this was a process of self education all the way through. The notion
that you can nonprocedurally describe any problem and produce from that a
highly efficient program, came very slowly, considering that we had
procedural bias and background. You could see that we had to go through a
self psychoanalysis to be able to do that.

JERNIGAN
We must move on. Thank you very much.

The Role of Language in Problem Solving I
R. Jernigan, B.W. Hamill, and D.M. Weintraub (Editors)
© Elsevier Science Publishers B.V. (North-Holland), 1985

THE USE OF THE SUBJUNCTIVE IN PROBLEM SOLVING .

Thomas Strothotte[†]

Department of Computer Science
University of Waterloo
Waterloo, Ontario, Canada N2L 3G1

Natural languages have numerous tenses and moods which
are used in personal communication and in conscious
problem solving, but programming languages are generally
restricted to the present tense. In this paper, the
role of the subjunctive in problem solving is studied.
A programming language feature modelling the subjunctive
of natural languages is presented. The feature is
implemented as a new primitive for the programming
language Pascal. It is demonstrated that some of the
awkwardness of algorithmic languages can be removed by
using the subjunctive.

1.   INTRODUCTION

One of the fundamental differences between human problem solving and
problem solving by computer is the way a human uses his imagination. When
confronted with a problem, he is often able to predict the probable outcome
for each of the possible solutions by simply imagining what **would** happen.
For example, suppose a hiker wishes to get as close as safely possible to
the edge of a cliff so as to enjoy the view below. A typical **stepping**
algorithm might be,

```
Would another step be safe?
    if yes,
        take another step toward cliff and
        repeat the stepping process.
    if no, stop here
```

This type of process is very different from a typical computer algorithm:

```
Take another step toward cliff
Did I fall off?
    if yes, back up one step
    if no, repeat the stepping process.
```

---

† Present Address: INRIA, Rocquencourt BP 105, F-78153 Le Chesnay
                    Cedex, France

Computers cannot predict what would happen, they must do and examine the consequences. As a result, processing often overshoots the desired target.

This idea of overshooting the desired target appears repeatedly in algorithm design. An example is the deletion of an item from a singly-linked list in Pascal. The usual algorithm makes use of a pointer to chain down the list until the unwanted item is found. By this time, however, the previous item, whose link field must be updated, is no longer accessible. In an effort to find the unwanted item, processing has chained one step too far. In this case maintaining a lagging pointer fixes the problem, but in other circumstances it is more difficult to back up the unwanted extra step.

A human uses his imagination to avoid having to overshoot the target. The tool we have in our language to support the process is the **subjunctive** [5]. Instead of taking that next step to the cliff, we ask, "what **would** happen?"

This paper reports on a new language feature that models the way a human uses the subjunctive in natural languages. First the subjunctive feature, as designed for Pascal, is introduced and a compiler for the language extensions is presented. Next, examples are given that show that some of the awkwardness of algorithms expressed in an algorithmic language can be removed 3ith the help of the subjunctive.

2.   LANGUAGE PRIMITIVE

Algorithmic programming languages such as Algol [8] or Pascal [6] deal exclusively in the present tense. Other tenses as they are known in natural languages are not used. Control structures such as the if-statement base their decisions only on the current values of variables. In order to model the handling of uncertainty of the future as it is handled in human problem solving, a subjunctive primitive has been disigned for algorithmic languages. Pascal was chosen because of its simplicity and general avail-ability. A subjunctive is introduced into Pascal by allowing expressions to be qualified by the **wouldbe** keyword:

        <subjunctive expression> ::= **wouldbe** (<expr>)

A subjunctive expression is meaningful in situations where it is used with a control structure such as in an **if-then-else**, or a **while**. To illustrate the semantics of the subjunctive conditional in an if-statement, consider a general statement of the form,

        if wouldbe(c) then S else S'

where c itself does not contain a subjunctive. The translation of the construct is illustrated in the flow chart in Figure 1.

First, the state of the program is saved. This can be thought of as storing the values of all variables and data structures. Next, the code S is **tentatively** executed. The term **tentative** execution is used to reflect the uncertainty about the execution. So far, it has not yet been established whether the code should actually be executed or not, but it must first be tried in order to find out.

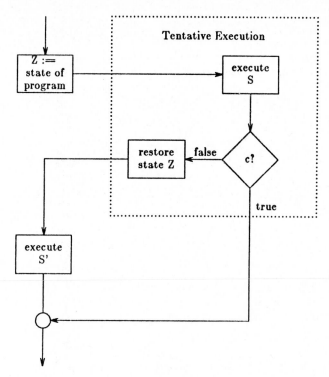

FIGURE 1
TRANSLATION OF IT WOULD BE(c) THEN S ELSE S'

Now the condition c is evaluated. If it is true, it means that the code
should have been executed, as it was, and execution simply continues with
the following statement. If c is false, it indicates that the S code
should never have been executed. The previously saved state of the program
is restored, and the code S' is executed.

The subjunctive feature separates the time of evaluation of an expression
from the time the result affects the flow of control. In figurative terms,
it allows the projected events of **tomorrow** to decide what to do **today**. An
example is the statement,

```
if wouldbe(bet < house_limit) then
    begin
        bet := bet * 2;
        n := n + 1
    end
```

which executes the body of the **then** clause only if at its conclusion the
condition (bet < house_limit) holds true.

The subjuntive can also be used in the

```
while <expr> do <stmt>
```

control structure. Recall that under normal circumstances, <expr> is evaluated first. If it is true, <stmt> is executed, otherwise control passes to the next statement. If <expr> involves a subjunctive, then the body of the loop, <stmt>, is tentatively executed. Then the subjunctive portion of <expr> is evaluated. From here control is handled as in the **if**-statement: if the expression evaluates to true, the side-effects of <stmt> are kept and the process repeated. Otherwise the results of the tentative execution are discarded and control passes to the next statement after the loop.

3.   EXTERNAL COMMUNICATION IN TENTATIVE EXECUTION

The code on which the subjunctive depends must be tentatively executed in order to evaluate the subjunctive. If the subjunctive conditional turns out to be false, the state of the program should be restored to its former state. If the code communicates with the outside world, ie. it countains input or output, this may not be possible. Communication is classified as follows:

- **Immediate actions.** Carry out the I/O, even if the action calls for communication with the outside world. For example, if the code contains output to an interactive user, carry it out irrespective of whether execution is tantative or not. This form of I/O is irreversible. Immediate I/O uses special functions **iread, ireadln, iwrite** and **iwriteln.**

- **Delayed actions.** Delay the action until it is established for certain that the code should really be executed. For example, if a **write** statement is encountered, the output is buffered and only actually generated if the code should be executed. Input is "unread" by retracting the input file pointer, and output is "unwritten" by retracting the output file points. The conventional Pascal I/O functions are treated as being **delayed.**

Section 5 below illustrates uses for the different types of I/O.

4.   IMPLEMENTATION

A compiler to translate a program using the above subjunctive feature into conventional code has been written on a VAX 11/780 running UNIX4.2BSD [7]. The compiler generates code calling Unix's **fork** function to save the state of the program. This function makes an exact duplicate copy of the process which calls it. Whenever a subjunctive clause is encountered, the current process, referred to as the **parent**, creates a duplicate, referred to as the **child** process. The child process executes the body of the then-clause, while the parent simply waits for the result. If the child is successful, that is, the subjunctive conditional turns out to be true, the parent terminates its processing and the child continues with the next statement. If, instead, the subjunctive conditional fails, the child process terminates its processing, and the parent continues on as though nothing had happened.

The translation using **fork** is conceptually simple but can be expensive. The compiler which has been implemented computes the set of possible side-effects of the code which is to be tentatively executed. The values of each of the variables in this set is then copied before entering tentative

execution, and restored if the code should not actually be executed. For
example, consider the following code which uses simple real variables **bet**
and **house_limit**:

```
if wouldbe(bet<house_limit)then
    bet := bet * 2;
```

The set of all possible side-effects of the **then**-clause is {bet}, so only
the value of this variable must be copied. The following translation is
generated by the compiler:

```
temp := bet;
bet := bet * 2;
if not(bet<house_limit)then
    bet := temp;
```

Care must be taken in the analysis so that if the code to be tentatively
executed calls procedures, the values any variables which they change as
side-effects are also copied; the algorithms are further complicated when
procedures are passed as parameters.

The set of possible side-effects of code to be tentatively executed may
include dynamic data structures. In general it cannot be computed
beforehand which parts of those structures may be modified and which ones
not. The approach taken by the compiler to make a copy of dynamic data
just before it is re-assigned a value, and storing the addresses of the
original and the copy in a table; restoring the original values is
accomplished by traversing the table in reverse order. This approach is
similar to an implementation technique used in the Interlisp environment
[9].

5.   EXAMPLES

In this section, three examples will be given illustrating uses for the
subjunctive. The first concerns an end-user undo command in an interactive
user interface. The second example demonstrates removing an element from a
linked list, while the third involves searching a tree.

5.1   USER INTERFACE

The tread in user interfaces is toward more flexibility and control. One
way of making a user-interface more flexible is by allowing an **undo** command
[1]. One of the problems with this is, however, that it can be difficult
to implement. The values of many program variables could potentially be
modified, and these must be reset. Using conventional languages, the
programmer must write code to restore the values of program variables to
their original state. This operation is particularly tedious for dynamic
data structures. In this section, it will be shown how the subjunctive can
be used to program an **undo** in a simple way.

Consider the following problem: a program which prompts a user for
commands, processes the commands and displays the results is required. The
application is an interactive graphics package performing rotations,
projections, etc. on an object which is then displayed on the screen. Once
the object is displayed, the user is asked for a verification. If he types
in 'n', the object is restored as it was before the manipulations.

Otherwise processing continues. Using the subjunctive, the code can be written as follows:

```
if wouldbe(verify ='y') then
    begin
        write('Enter your command');
        read(command);
        process(command);
        display;
        write('Is that ok? Enter y or n');
        read(verify)
    end
```

In essence this code says, "if at the conclusion of the **then** clause the variable **verify** has the value 'y', then execute the code, otherwise just go on as though nothing had happened". The machine begins by asking the user for his command, processing it and showing him the results. The user is then prompted for a verification. If the user types 'y', then the subjunctive conditional is satisfied and execution continues with the next statement. If not, the program **undoes** all side-effects of the then code, so that as far as it is concerned it **never executed the code**, and then continues with the next statement. Placing code analogous to the above within a recursive routine allows undoing any number of commands, one at a time.

The **wouldbe** clause tells the machine to delay the decision as to whether or not to execute the code until the end of the then clause. The machine still has to execute the **then** portion of the clause, but it provides for the possibility of backing up to the state at the beginning of the if, if the subjunctive conditional turns out to be false.

5.2   REMOVING AN ELEMENT FROM A LINKED LIST

Consider the problem of removing an element from a singly-linked list of names. The solution is conceptually simple: chain down the elements of the list until the desired element is found. Then set the link field of the previous element to the link field of the unwanted element. This algorithm works well when executed by hand with pencil and paper.

The problem comes, of course, in obtaining the previous element once the one to be removed has been identified. Once the pointer used to chain down the list points to the one to be removed, the previous one is no longer accessible. This situation is an example of one where an algorithm has run too far and would now like to back up. (Recall the motivating example from the Introduction: It has fallen off the cliff with no way of getting back up). The conventional solution is to maintain a **previous** point while chaining through the list. (Make someone else (a clone) walk toward to cliff ahead of you. When he falls off, you know not to go on). The following code illustrates this typical approach to the problem, identifying the element **prev** lying before the element with the name **findname**:

```
type
    string = array[1..40]of char;
    node = record
            name : string;
            next : ^node
        end;
var
    cur,prev,head : node;
    findname : string;
    . . .
    {assume the list is non-empty and does contain findname}
    {first take care of special case of findname being at start}
    if head^.name = findname then
        head := head ^.next
    else
        begin
            prev := nil;
            cur := head;
            while(cur^.name <>findname)do
                begin
                    prev :cur;
                    cur :cur^.next
                end;
            {now remove unwanted element}
            prev^.next := cur^.next;
        end
. . .
```

The subjunctive used in a loop control allows the program to look ahead at
what **would** happen in the next iteration, if it were to be executed. What
we want in this case is to avoid taking that irreversible step to the
element of the desired name. The following code solves the problem:

```
var
    p,head : node;
    . . .
    {take care of special case of findname being at start}
    if head^.name = findname then
        head := head^.next
    else
        begin
            p := head;
            while wouldbe (p^.name <> findname)do
                p:=p ^.next'
            p^.next := (p^.next)^.next
        end;
    . . .
```

It is interesting to note that the construct as used here performs a **look
ahead**. The next element is still examined to see if it corresponds to the
one with the undesired name, but the currently executing process doesn't
itself take the step until it is sure that it will not overshoot. From the

point of view of problem solving, it is significant that the programmer
must not handle the backing up step. Indeed, although the subjunctive
feature is defined in terms of backtracking, the programmer may conceive of
it in terms of looking ahead.

5.3   TREE SEARCHING

Consider the task of searching for the leaf that corresponds to a solution
to a problem represented as a tree. The desired result is to print out the
labels of all intermediate nodes along the path from the root of the tree
to the solution. A conventional algorithm is as follows:

>   (1)   Search the tree using a recursive function, which is applied
>   to nodes in the tree and returns the value **nil** if the solution
>   does not lie on or below the node, and otherwise returns a
>   list giving the path from the node to the solution.

>   (2)   Traverse the list of nodes obtained from Step (1), printing
>   out the labels of the nodes.

This is a two-step process: first find the path through the tree, then
traverse that path to print out the elements.

A one-pass algorithm can be written using the subjunctive conditional. The
algorithm is again a recursive routine, but it returns only a bit saying
whether or not the node lies on the path to the solution. Using the
delayed output feature discussed previously, the algorithm, in pseudo-code,
is as follows:

```
procedure printout(s : string);
   begin
      {procedure to print out a character string}
      {output uses the "write" and "writeln" functions}
   end

function search (root : tree) : boolean;
var
   s : tree;
   is_on_path : boolean;
begin
   printout(root ^.label);
   if leaf(root)then
      search := true
   else
      search := false
   else
      begin
         is_on_path := false;
         while(s<>nil) and not is_on_path do
            if wouldbe(is_on_path) then
               is_on_path := search(s);
         search := is_on_path
      end
end
```

The code uses the function **write**, which acts as delayed output when in tentative execution. The effect is to produce the output only for those nodes for which the code is actually (not just tentatively) executed. The output corresponds to the nodes lying on the path, as desired.

This technique has wide applicablity as many problems can be expressed as a decision tree. For example, it can be used to perform syntax-directed translation from within a backtracking recognizer. Under conventional circumstances, it would be necessary to first build a parse-tree and then in a second step, traverse the parse-tree to produce the code. The subjunctive eliminates the need for building a parse-tree to produce the code.

6.   RELATED WORK

The subjunctive presented in this paper is a way of expressing **automatic backtracking** algorithms. A number of other languages support this feature. Snobol [3] has backtracking as a special-purpose feature for pattern-matching. Icon [4] has more general backtracking, but is limited to backtracking over one expression. Prolog [2] algorithms are specified in terms of a series of rules forming a database. Problems are solved by searching the database, attempting to satisfy **goals.** Prolog programs have a different flavour from those of algorithmic languages, in that they specify what the solution of the problem is, rather than how to compute the solution: the solution is found by searching the database.

The subjunctive is closely related to Zelkowitz's construct [11] for backtracking in PL/C. His construct allows backtracking to essentially any point in the program. By contrast, the subjunctive imposes structure on the backtracking, forcing the programmer to specify to which points backtracking may go. Further, the programmer using the subjunctive specifies the condition on which backtracking depends at the point to which backtracking will go back to, making it clear to the reader of a program that side-effects the code which is to be executed may be reversed. The subjunctive can be written in terms of Zelkowitz's construct as follows:

```
    Subjunctive:        Zelkowitz's Construct:

                        L: S
                            if not c then
    if wouldbe(c) then          retrace to L and
        S                       begin
    else                            S'
        S'                          goto L2;
                            end
                        L2: . . .
```

7.   CONCLUDING REMARKS

This paper has presented a language construct modeling the subjunctive of natural languages. This construct allows a number of algorithms to be expressed in a manner more like those of everyday situations. In fact, the subjunctive feature can lead to algorithms whose nature differs from algorithms expressed with conventional language constructs. While providing a significant expressive power, the subjunctive also has a realistically efficient implementation.

The results show that the subjunctive is a powerful new tool for algorithmic languages, but there are numerous open questions. One of the underlying assumptions of the subjunctive is the notion of time and sequence. Algorithmic languages are based on the principle that problem solving involves changing and advancing the world until a desired result is attained. This is in sharp contrast to the Prolog view of problem solving, which considers the world to be a static state in which answers are sought. Which of these views most closely matches the human problem solving process? This is an open question inviting experimentation.

An area which is being explored is the use of additional tenses other than the subjunctive. A new language, T-Pascal, has been designed and implemented. The work is outlined in detail in [10]. T-Pascal contains the features described in this paper, as well as a past tense and a future tense. A past tense is used to **rememeber** facts which occurred in the course of the past execution of the program. This feature is useful in replacing the common practice of setting bits to remember facts. The future tense is used to monitor the values of variables over a segment of code. It is hoped that this work shed new light on the process of problem solving by computer.

REFERENCES

1.   Archer, J. The Design and Implementation of a Cooperative Program Development Environment, Ph.D. dissertation, Computer Science Department, Cornell University, Ithaca, N.Y., August, 1981.

2.   Clocksin, W. F. and Mellish, C. S., Programming in Prolog, Springer-Verlag, berlin, 1981.

3.   Griswold, R. E. and Griswold, M. T., A Snobol4 Primer, Prentice-Hall, Englewood Clif, N.J., 1973.

4.   Griswold, R. E. and Griswold, M. T., The Icon Programming Language, Prentice-Hall, Englewood Cliffs, N.J., 1983.

5.   Gschossmann, E., Schaum's Outline of German Grammer, 2nd Edition, Schaum, New York, 1983.

6.   Jensen, K. and Wirth, N., Pascal User Manual and Report, Springer Verlag, Berlin, 1975.

7.   Leffler, S. J., Joy, W. N., and McKusick, M. K., Unix Programmer's Manual, 4.2, Berkeley Software Distribution, Virtual VAX-11 Version, Computer Science Division, Department of Electrical Engineering and Computer Science, University of California, Berkeley, CA 94720, August, 1983.

8.   Naur, P., "Algo160," ACM-SIGPLAN History of Programming Languages Conference, SIGPLAN Notices 13, August, 1978.

9.   Sandewall, E., "Programming in the Interactive Environment," Computing Surveys 10, (1978), 35-71.

10.   Strothotte, T., "Temporal Constucts for an Algorithmic Language,"
       Ph.D. dissertation, Department of Computer Science, McGill
       University, Montreal, Quebec, December, 1984.

11.   Zelkowitz, M. V., "Reverisble Execution," <u>Comm. ACM</u> 16, 9,
       (September 1973), 566.

QUESTION AND ANSWER PERIOD

JERNIGAN
Do we have any questions?

GUIER
Have you explored this completely on recursive functions?

STROTHOTTE
You ask if this has been completely explored on recursive functions, yes it
has.

REMARKS

The Role of Language in Problem Solving I
R. Jernigan, B.W. Hamill, and D.M. Weintraub (Editors)
© Elsevier Science Publishers B.V. (North-Holland), 1985

SPACE AGE DEMANDS FOR POWERFUL COMPUTER LANGUAGES

George C. Weiffenbach

The Johns Hopkins University
Applied Physics Laboratory
Laurel, Maryland 20707

I will present a quite different point of view from the speakers I heard
yesterday. In the first place, I have very carefully abstained from learning how to do any programming or anything of that kind.

The title of my talk, I suspect, is somewhat more grandiose than is appropriate and my real intention is somewhat more modest. In a sense I
represent the user community, and more particularly, I would view myself as
a broker for users, whether they be for scientific applications or whatever
else in the space arena. I'll try to restrict most of my remarks to things
I am more familiar with, activities we are involved in now, and then I will
take a flyer and do a little speculating at the end.

I would like to address four general areas; they look somewhat different,
but there are relationships. The first I will address will be implantable
medical devices, then scientific computing, spacecraft satellite design and
manufacturing, and then last, smart spacecraft.

Under the NASA technology utilization program, we have been involved in a
number of projects that involve devices that are implanted in human beings.
That has been done in conjunction with The Johns Hopkins Medical School,
with Massachusetts General, the Mayo Clinic, and a number of other organizations. Some years ago we started with pacemakers; they essentially had
no smarts, but more recently we have gotten involved in a number of other
kinds of things. As a sampling, there are defibrillators with very primitive intelligence; that is, they can observe the heart rate and note when
it goes into defibrillation, when it is arrested in effect, and then shock
the heart, all internally. Hypertension currently looks like it's going to
be the first closed loop system that we will be able to establish. Closed
loop in the sense that it will make a measurement of a human being's blood
pressure, and in response to that, will inject medicine into the blood
stream. An artificial sphincter, an insulin pump (we actually have insulin
pumps in several dogs in clinical trials with an implantable device that is
not closed loop, but that has a fairly competent computer), and pumps to
inject morphine into the spinal column. The latter may very well be our
first human implant, because it is hard to test those on animals. In all
of these cases, without going into a lot of detail, there are opportunities
for some very smart implantable devices. I don't think it takes too much
imagination to think of the things that you would like to be able to do
that would involve expert systems programmed in Very Large Scale Integrated
circuits.

Now in scientific computing, information processing if you will, we are
involved in three areas of research, one in space physics (magnetohydro-

dynamics or plasma physics), physical oceanography, and solid earth geophysics. These are areas that share some common characteristics. (1) they deal with very complex and heterogeneous systems, (2) there already exist enormous data bases and those data bases are being filled at an accelerating rate (e.g., imaging devices for both physical oceanography and solid earth geophysics put in orbit produce prodigious data rates). Another interesting characteristic is that all three areas are at the point now where the current theory is not adequate, and not adequate in very important ways, because we have already seen a number of phenomena that are predicted only through the introduction of non-linearities.

Furthermore, we have a data glut that you would not believe. As one example, for the imaging radar that was put into orbit on the SEASAT satellite in 1979 only 40% of the data has been processed, no more than 10% has been analyzed, and that satellite lived for only 90 days. What we do not have is an information glut; I think the role of the techniques you are talking about here is obvious.

Now, in spacecraft design and fabrication today we have all the normal activities that you have when you run a business: accounting, management information systems, and the like. It is already quite clear that even in this rather normal kind of enterprise, we come up short in the computing systems that we have to address these issues. But in the satellite business, we have another routine kind of thing, inventory systems, that are not quite normal. Because of the enormous cost of designing, building and launching satellites, and because you really want to get to the most reliable possible system, we keep detailed birth-to-death records on each of the many thousands of components on the satellite. Obviously, record keeping is a serious problem for us. Bob Jernigan and some other people here have put together an inventory system; they have not quite finished putting it altogether - mainly a funding limitation - but it did pull us out of a real choke point. We were so backed up that we were losing schedule, and that is a very costly thing.

Looking at the space hardware design process, we have a shortfall in computer aided design, computer aided manufacturing, and computer aided engineering. In the satellite business, more emphasis has got to be placed on a correct design than in almost any other case. The reason is that there really is no mass production. We cannot rely on a million customers sitting out there someplace to debug the things that we make. There is no feedback through large production. There is no graceful way to see what you have produced, how it functions, and then correct it. Every time you miss the boat there are enormous cost and schedule problems. We are not nearly where we would like to be in space hardware design tools. That is the existing situation; I hope that I am getting across the message that we need the tools that you people can provide us - yesterday.

Looking into the future, I get even more interested in pushing your art. We have already seen the past impact of computers throughout our society. In space technology we are going to see a total turn-around. Satellites have been kind of a "gee-whiz" business; there is a lot of glamour attached to it. You see the enormous rockets go off and it is very impressive. A less spectacular but more impressive future is in store. In ten years I am convinced that we will have the ability to design an information processor, a computer, that can be put into a satellite, that will easily have all the computing power of a CRAY, and will almost certainly be even smarter. One of the drivers that I have noted (and you do not have to be very sharp to see this happening) is the enormous progress that is being made in the hardware side of the computing business. We already have chips that you can buy with half a million components on them, and there does not seem to be any fundamental limitation to increasing the number of components you can put on a silicon chip by a couple of orders of magnitude. The hardware is coming. (It is fascinating that the Japanese have people working on chips of this kind, but with perhaps 40 layers on them.) The hardware is coming. It has got an enormous impetus behind it for all kinds of reasons.

What I have not yet seen is comparable progress in our ability to exploit the hardware, and that clearly means computer architectures and programming. Some very simple arithmetic tells you (verified by a lot of experience) that if you are going to sit down and design a CRAY for a satellite, the number of man-years needed to design the architecture and to turn this into a useful device, is enormous. If you look further downstream as the hardware gets more potent, conventional approaches are simply going to be totally impossible. I do not mean only from the standpoint of the length of time and the number of people; if we do not find more effective ways to carry out this process, you will never get there. I think everybody here must be very much aware of the difficulty of adding and adding people into a programming task of any kind (and designing the architecture of a smart computer has got to be one of the most sophisticated things). Sooner or later you reach a stage where you simply never arrive at the end point. The potential that I can see in future smart, autonomous satellites is mind-boggling, quite literally. I doubt that anybody can really predict the manifold uses that future satellite-borne computers can be put to. But I am convinced that it will, in fact, allow you to do things that we are totally unable to do today in terms of satellite autonomy, reliability, and information processing. I count on this community to provide the means by which we can bring this about.

SESSION:
OVERCOMING LIMITATIONS IMPOSED BY
CURRENT PROGRAMMING LANGUAGES

The Role of Language in Problem Solving 1
R. Jernigan, B.W. Hamill, and D.M. Weintraub (Editors)
© Elsevier Science Publishers B.V. (North-Holland), 1985

XIMM - AN EXPERT SYSTEM FOR IDLE MATERIALS MANAGEMENT:
LOGIC PROGRAMMING FOR CORPORATE STRATEGIES

Robert Jernigan
Decision Resource Systems
5595 Vantage Point Road
Columbia, Maryland 21044

Anand Desai
Digital Equipment Corporation
40 Old Bolton Road
Stow, Massachusetts 01775

The amount of inventory and capital equipment on corporate books that is not being utilized can be measured in
billions of dollars. A major corporate strategy is to
recover this idle investment through redistribution,
upgrading, and dismantling for parts. The experience of
the authors in implementing XIMM for Digital Equipment
Corporation is detailed. This paper focuses on the
problems of integrating multiple data bases, decision
support systems, and management strategies into an expert
system. Particular emphasis is placed on how APLLOG, an
integration of PROLOG and APL, has been used to represent
the rule base of the system.

1.   INTRODUCTION

Given a corporate environment with multiple data bases on computers at
several locations, a problem that requires a synthesis of those data bases,
and a required solution that is based on both an analysis of the data and a
set of rules, it is immediately obvious that a powerful computer language
is required. That language must be capable of the following:

  1.   Expressing the rules that describe how experts wish to
       implement corporate strategy;

  2.   Expressing the rules that describe how objects in the system
       are to be handled;

  3.   Implementing the analytic algorithms used to define the
       decision support system;

  4.   Accessing multiple data bases;

  5.   Adapting to changing requirements.

The language that was chosen to implement this system has APL as a base.
It would be an oversimplification to state that the system is implemented
in APL because of the number of things that have been added. While APL is
a language in itself, it is considered by the authors to be merely the

starting point, i.e., a foundation, for higher level languages and
abilities.

2.   THE PROBLEM

Digital Equipment Corporation is typical of any large organization that
uses or manufactures equipment. Some of the equipment on the books is not
being utilized. The reason are various - overproduction of particular
items, obsolesence, returns, and reorganization are but a few. A corporate
data base has been established to track all idle materials in the organization. Anyone having, or needing, any equipment or material can use the
data base to list the equipment or to inquire about availability.

Equipment may not be reusable in the form in which it exists. If it is,
and if there is a demand, it can be readily picked up by whoever needs
it. What is often the case, the equipment contains parts for which there
is a demand, or conversely, may be used as a part in a larger component
that is in demand.

XIMM is designed to address all three categories.

3.   THE SOLUTION

Three classes of data bases are of interest:

        a.   The demand data bases that forecast equipment requirements
              for both field service and sales;
        b.   The idle materials data base that list all idle materials and
              equipment;
        c.   The Bill of Materials (BOM) data bases that list both the
              materials required to build a component and the materials
              that can be reasonably salvaged from a piece of equipment.

The decision support algorithms are concerned with calculating the optimum
amount of effort that should be expended in regrading, or reconfiguring,
equipment. It is obvious that some parts are not worth the effort it takes
to remove them. Also, it would be pointless to remove parts for which
there is no demand.

The rule base implements the rules that will be used to establish the
priorities and procedures for the distribution , upgrading or dismantling
of the idle equipment. Each component that can be upgraded or dismantled
will have a set of procedures that must be followed, those procedures vary
with the version of the component required or on hand. The rules for
dismantling can also vary with the demand for the parts on the component,
i.e., each dismantling run for a component may not be the same as another
because of varying demand for the parts.

4.   THE LANGUAGE

This section will discuss the language that is being used to implement
XIMM. Known variously as APL/PROLOG, APLLOG, and ALPS (for APL Logic
Programming System), it essentially APL with various extensions added in
the APL language. These extensions have been described elsewhere in detail
[Jernigan 84] and can be summarized as follows:

a.   an interactive editor, which also serves as the primary user
      interface, provides programmatic control of all user
      commands, editing of the rule base, data bases and APL
      functions;
b.   a relational data base system;
c.   a modelling system, APLDOT [Kruba 83];
d.   PROLOG-like Horn clauses; [Clark 82]
e.   full access to the APL system for execution of APL functions.

The intent of these extentions is to provide flexibility in problem definition. This flexibility allows the system implementers to choose from a
variety of representation formats. Another goal is have each part of the
system functional as it is implemented.

The balance of this paper will be devoted to examining some of the
fundamental concepts that have been used in APLLOG and how those concepts
facilitate solutions. When these conceptual extensions are added, they are
added as layers, each with its own structures are execution control functions. Each layer forms part of the foundation upon which a application is
based. (See Figure 1)

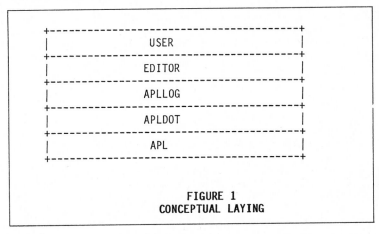

```
+-------------------------------------+
|                USER                 |
+-------------------------------------+
|                EDITOR               |
+-------------------------------------+
|                APLLOG               |
+-------------------------------------+
|                APLDOT               |
+-------------------------------------+
|                 APL                 |
+-------------------------------------+
```

FIGURE 1
CONCEPTUAL LAYING

Each conceptual layer adds conceptual mechanisms for application development. Each layer has full access to the other layers.

4.1   EDITOR

The editor, as the primary interface, has primary control of user input.
Through the editor, data can be added to the data bases, rules can be
written and tried before they are added to the rule base, APL functions can
be written, and any of the existing structures in the system can be
changed. Temporary structures, called SCRIPTS, can be maintained and used
as command sequences, embryos of Horn clauses, or sequences of APL
statements.

4.2   APLLOG

Prolog has been described as a "declarative language whose design includes
a relational data base. Prolog computes relations, just as Lisp computes

functions." [Parsaye 1983] Prolog, as implemented here, is an implementation of a restricted set of predicate calculus. The primary restriction
being that the clauses are in a form known as Horn clauses. Horn clauses
consist of predicates, one of which occurs to the left of a logical implication symbol, "←". This predicate is called the HEAD of the clause. To
the right of the implication symbol are a series of predicates, called
goals. These goals, together with their connectives, the logical symbols
for "and" and "or", are collectively called the TAIL of the clause.
Because of the symbols available in APL, the symbols used are "←" for
implication, "∧" for logical-and and "∧" for logical-or.

Each predicate is the name of a relation or of a clause, the name of a
clause being the predicate symbol used in the HEAD of the clause. A
typical clause that occurs in literature discussing PROLOG is the one
expressing the definition of GRANDFATHER.

$$GRANDFATHER(\Delta A, \Delta B) \leftarrow FATHER(\Delta A, \Delta C) \wedge FATHER(\Delta C, \Delta B)$$

APLLOG uses the "Δ" symbol to distinguish variables from constants. Thus
ΔA, ΔB and ΔC are variables in the above example. The clause can be
interpreted as "A is the grandfather of B if it is true that A is the
father of C and C is the father of B." If in APLLOG, FATHER is defined as
a relation in a data base, the query

GRANDFATHER(ΔA, ΔB)?

represents a relation of all tuples, as established by the FATHER data base
and the GRANDFATHER clause. The evaluation is carried out as follows:

   a.   The rule base and data base is scanned to find a rule or
        relation that matches the name of the predicate in the query,
        in this case "GRANDFATHER".

   b.   The variables in the query are checked to see if they have
        values, in this case, they do not.

   c.   The first goal in the TAIL is tried. Since the variables are
        not assigned before the first goal (FATHER) is tried, the ΔA
        and ΔC variables are assigned all the values in the
        corresponding fields from the FATHER relation.

   d.   The second goal is tried with corresponding assignment of the
        ΔC and ΔB variables. Note that at this point there are two
        ΔC variables, one from the first goal and one from the
        second. There are actually two relations in the TAIL part of
        the clause, both of which are the FATHER relation.

   e.   A relational JOIN operation is performed on the two
        relations in the TAIL. The JOIN uses as its key, variables
        having identical names, i.e. the ΔC variable.

   f.   The resulting relation is passed back to the HEAD and then
        back to the original query.

This explanation oversimplifies the behavior but it does demonstrate the
relational data language capabilities of the Horn clause construction. Of

course, in the XIMM the concern is not with the GRANDFATHER relation but a
similar construct occurs when constructing the definition of the Bill of
Materials (BOM) relation, which is made up of the "options" data set containing costs estimates for dismantling a component and the "parts" data
set containing detailed cost data for the parts belonging to the component.
Using the OPTIONS and PARTS successively as goals in a Horn clause performs
a relational JOIN of those two sets. When the HEAD of the clause is given
a particular OPTION, the resulting relation data set that, when referenced
as a goal in a Horn clause, will be the functional equivalent of a data
base. (Figure 2)

```
OPTIONS                PARTS
+------+------+         +------+------+------+------+
| NAME | COST |         | OPTN | PART | QNTY | COST |
+------+------+         +------+------+------+------+
| PDPX | 1000 |         | PDPX | BULB |   2  |  .10 |
| LAXX |  700 |         | PDPX | DISK |   1  | 4000 |
+------+------+         | LAXX | KEY  |  56  |  .04 |
                        | LAXX | SYO5 |   3  |   99 |
                        +------+------+------+------+


      BOM(ΔOPTION;ΔOPTCOST;ΔPARTS;ΔPTQNTY;ΔPTCOST)←
          OPTIONS(ΔOPTION;ΔOPTCOST)
          ∧ PARTS(ΔOPTION;ΔPARTS;ΔPTQNTY;ΔPTCOST)


   BOM
   +--------+--------+-------+--------+--------+
   | OPTION |OPTCOST | PARTS | PTQNTY | PTCOST |
   +--------+--------+-------+--------+--------+
   | PDPX   |  1000  | BULB  |   2    |  .10   |
   | PDPX   |  1000  | DISK  |   1    | 4000   |
   | LAXX   |   700  | KEY   |  56    |  .04   |
   | LAXX   |   700  | SYO5  |   3    |   99   |
   +--------+--------+-------+--------+--------+
```

FIGURE 2
HORN CLAUSE VARIABLES AS A RELATIONAL DATA BASE

APLLOG finds all data satisfying a goal or query in a single retrieval or
pass through a Horn clause. It is forward searching with no back tracking
required. To accomplish this, the clauses must be specified in disjunctive
normal form. [Robinson 79] In addition to performing a relational JOIN
between the relational sets of each goal in conjunctive subclauses, a set-
union operation is performed on the relational sets resulting from the
conjunctive subclauses. Forward searching algorithms for Prolog have also
been investigated by Popplestone. [Popplestone 79]

4.3   APLDOT

APLDOT is a modelling system that uses relational data sets to store data
and formulas. [Date 75] These data sets may be moderate size data bases,
report definitions, or sets of APL functions that act like data bases.

(One application accessed a large proprietary data base to provide the data
base to the user in a relational format.)

A distinctive feature of APLDOT is the reference function, "Δ", which
provides a non-procedural operational characteristic. With the reference
function, any data element in the system can be accessed as if it were a
named variable. Its function is similar to a SELECT operation in data base
query languages except that it is used in formulas and has the value of the
referenced item. If the referenced item happens to be a formula, the
formula is executed and the result is passed to the calling formula.
Whenever a formula is executed, its value is stored for future references. This prevents needless re-execution of formulas. (See Figure 3)

```
    FORECAST
    +---------------+-----------------------------------+
    | NAME          | DEFINITION                        |
    +---------------+-----------------------------------+
    | CPI           | 4 4 8 7 3.2 9                     |
    | GROWTH(CPI)   | ×\1+(FORECAST Δ'CPI')÷100         |
    | DISC RATE     | FRB Δ'FED DISC RATE'              |
    +---------------+-----------------------------------+

        FORECAST Δ'CPI'
4 4 8 7 3.2 9

    ×\1+(FORECAST Δ'CPI')÷100
1.04 1.0816 1.168128 1.24989696 1.289893663 1.405984092

    FORECAST Δ'GROWTH'
1.04 1.0816 1.168128 1.24989696 1.289893663 1.405984092


                        FIGURE 3
                APLDOT FORMULAS AND DATA
```

This formula would compute a compounded Consumer Price Index. FORECAST is
the name of the data set where CPI is stored. The actual entry for CPI
might be a sequence of numbers or it might be a formula.

All elements in any APLDOT model are available to the user in the APL mode
and in the APLLOG mode.

4.4   APL

By using APL as an implementation language for higher-level languages the
option of dropping back to the functionally of APL is always available.
The most common way of programming in APL is to write user-defined
functions that extend the functionality provided by the system-supplied
primitives. Once defined, these user functions are available at all higher
levels.

One particular advantage of APL is the availability of the parser. A
principle feature of APLDOT is the use of formulas to define concepts. By
relying on the APL syntax for the expression of these formulas there is no

need to write a separate parser for execution of the formulas. (In fact, APLDOT compiles its data sets, which can contain formulas, into APL functions).

The user can also use APL to define pseudo-predicates for APLLOG as APL functions. These APL functions can access the local variable list in Horn clauses or cause side effects such as making terminal queries or printing values of variables. If they modify the local Horn clause variables, care must be taken to maintain the shape restrictions on the variables, i.e., the Horn clause variables always constitute a relation with a one-for-one correspondence between the elements of the variables. An element of any vector for a tuple with corresponding elements of other vectors or rows of matrices. An element of a tuple may be a vector of numbers, which is, itself, a row of a matrix.

Figure 4 shows examples of some of the pseudo-predicates that can be used in Horn clauses.

PEDIT	Formats Horn clauses and data sets and calls the editor.
S	Sorts the local variables of a clause.
$\overline{P}$RINT	Formats and prints data set and clauses.
$\underline{U}$	Uses local clauses to update data sets.
$\underline{P}$	Forces a relation "projection" operation on the relation formed by local variables.
$\underline{R}$	Establishes a range for the local variables, i.e., variables not wanted in the local variable list are discarded.
$\underline{Q}$	Qualifies local variables, i.e., a form of SELECT operation.

FIGURE 4
SOME DEFINED PSEUDO-PREDICATES

APL statements can be imbedded in the Horn clauses with the use of a pseudo-predicate named APL. The APL predicate, or goal, is true unless it fail to perform the expressions specified is the expression list provided within parentheses. An example of an APL predicate is:

$APL(\Delta A \leftarrow 1\ 2\ 3\ ;\ \Delta B \leftarrow 10\ 20\ 30)$

5.   CONCLUSIONS

The problems presented by XIMM are not unusual. There is the need for adequate data bases, analytical facilities, a rule base, a flexible environment, and a short development time. Certainly there are "packages" that can be used to solve parts of the problem. The solution method, that of using a system that is a unified embodiment of several concepts, provides a flexible environment and operational capabilities early in the implementation effort.

A self-extensible language, such as APL, can be used to implement concepts that are not inherent in APL itself. When extending APL, the emphasis is

on implementing generalized concepts that have broad application and a
strong theoretical base. APLDOT extends APL into relational data base
concepts with some additional benefits gained from the use of formulas and
elements of the data base. The relational structure of the APLDOT data
sets, along with the requirement that all components of APLDOT models
reside in these data sets, facilitate the implementation of report writers
and editors.

However, the unified environment presented here is not complete. There are
several inefficiencies that must be addressed. For example, while APLDOT
is an implementation of a relational data base system, it is efficient only
for relatively small data bases. As the data bases grow they will be
transported to a relational data base system, such as DEC's Rdb, which
would be more efficient for larger data bases. Another problem is that of
the localized name lists in APL. The threaded workspaces described by
James Ryan [Ryan 1985] in this symposium would be helpful.

Early implementation and the flexibility to incorporate new rules and new
data bases are key concepts that are not linguistic but are directly
affected by the choice of the problem solving language. Key factors in
early implementation are :

      a.   The statements that are written are directed toward the
             problem itself and not toward building the environment for
             the problem;

      b.   The rules and formulas can be used, experimented with, and
             modified as they are written. It is possible to obtain a
             useful result while the rule being developed is still in the
             editor.

REFERENCES

Clark, K. L., and McCabe, F. G., "PROLOG, a language for implementing
    expert systems," MACHINE INTELLIGENCE 10, (Hayes, P. and Michie, D.,
    Eds), Edinburgh University Press, 1982

Date, C. J., AN INTRODUCTION TO DATABASE SYSTEMS, Addison-Wesley, 1975

Jernigan, R., "Logic Programming in APL", CONFERENCE PROCEEDING, APL84, The
    Association for Computing Machinery, New York, NY, 1984

Kruba, Stephan R., "APLDOT - An APL Programmer's Modelling Lnaguage", APL83
    CONFERENCE PROCEEDINGS, The Association for Computing Machinery, New
    York, NY, 1983

Parsaye, Kamram, "Login Programming and Relational Databases", DATABASE
    ENGINEERING, Vol. 6, No. 4, Dec. 1983

Popplestone, R. J., "Relational Programming", MACHINE INTELLIGENCE 9,
    (Hayes, J., Michie, D., and Mikulch, L., Eds), Halsted Press, 1979

Robinson, J. A., LOGIC: FORM AND FUNCTION, North Holland, New York, 1979

Ryan, James L, "Threaded Workspaces - An Application Development Environment for APL", THE ROLE OF LANGUAGE IN PROBLEM SOLVING - I, North Holland, Amsterdam, The Netherlands, 1985

QUESTION AND ANSWER PERIOD

BARSTOW
In your title you called it an expert system. In the description I didn't see, to me, the two characteristics of an expert system, one of which is a program structure - that is an inference engine and a set of rules - nor did I see the methodology that I think is an important part of expert systems, which is to interrogate an expert to understand what his or her expertise is. Is that a part that you didn't show us?

DESAI
Yes, you are right. For the purpose of this conference I wanted to keep it simple so I could communicate some of the things that we trying to achieve. Alternatively, I wanted to present the union and the combination of a different PROLOG kind of application which scans different data bases, which is the basic purpose. In terms of the inference, different SCRIPTS could be combined to infer different kinds of issues within the data base. For example, some of the findings that we have we did not expect. They are basically inferred by the expert system.

BARSTOW
Were there experts that did this task before XIMM came along?

DESAI
Yes. There are experts, there is basically a team which addressed the basic idle materials management group. Idle materials, by the way, pop up in different of ways. One is through capital being over-bought or products being obsolete. So you have excess capital out there, you also have insurance claims which come back in. You have different reasons, for example, trade shows. All this material keeps coming back and you have to operate on this. You need experts to try to figure out what to do with this material.

GOLDFINGER
What is the current status of APLLOG as a language? Is it fully developed or is it still underdeveloped? Is it developed to a stage in which it can be used as a general language by other people?

DESAI
It would be correct ot say that APLLOG and other forms of PROLOG are still being developed.

DESAI
Thank you.

The Role of Language in Problem Solving I
R. Jernigan, B.W. Hamill, and D.M. Weintraub (Editors)
© Elsevier Science Publishers B.V. (North-Holland), 1985

MIRROR: A LANGUAGE FOR REPRESENTING PROGRAMS FOR REASONING

Aaron Temin and Elaine A. Rich

Department of Computer Sciences
University of Texas at Austin
Austin, Texas 78712-1188

A program is a description of a solution to a problem.
This description is useful for several purposes, such as
being executable and verifiable, and these uses help
determine the properties of the description. In this
paper we consider how using the description for a new
purpose, as data for answering questions about the program, effects the properties of the description. We are
building a question answering system which is validating
our conclusions.

INTRODUCTION

Recent research on on-line help systems has concentrated on making systems
that are responsive to users. Genesereth's Macsyma Advisor [Genesereth,
1978] and Wilensky's Unix Consultant [Wilensky, 1982] recognize user plans
which allow them to help users recover from errors. Finin's Wizard [Finin,
1983] is investigating when to volunteer advice to users. These systems
provide great flexibility at the user interface, but are founded on very
incomplete knowledge of the system about which they are providing assistance (the **domain** system). One may argue that the knowledge bases used
could be expanded to encompass any degree of accuracy. But this is the
wrong argument.

What one would like is to use a representation for question answering that
is similar enough to the source language of the domain system so that one
could mechanically translate from one to the other. This would avoid the
immense time needed to encode the system twice. It would guarantee the
consistency of the help system's knowledge base with the domain system.
And it would insure that the help system had as much detailed information
as existed about the domain system.

What we outline in this paper are the attributes of a knowledge representation for the question answering module of an automated on-line help system,
that will allow a help system to have a conveniently constructed, detailed
knowledge base, that is compatible with its domain system.

OPERATIONS SUPPORTED BY THE KNOWLEDGE REPRESENTATION

There are four question types that need to be answered by the question
answering system:

1.   Cause and effect (e.g. why did the system exhibit a particular
     behavior?)

2.   Comparison (e.g. What are the differences between the system's behavior when one gives these two different commands?)

3.   Lookup (e.g. what is the definition of a particular macro?)

4.   Syntax (e.g. what is the syntax of a particular command)

In what follows we consider most carefully categories 1 and 2. There are questions in category 1 which refer to interactions in the system that the designers and documenters may never have planned on, so that a detailed analysis of the implemented solution is the only place to find these answers. Answers to questions from category 2 need to be constructed on the fly, because there may be many possible comparisons in any large system, and anticipating any but the most common and obvious is tedious at best and impossible at worst.

As an example, consider asking what caused a particular behavior. In general we assume that behaviors can be tied to particular routines. Starting from the appropriate routine, we chain back through the calling sequence to find the relevant conditions which need be true to trigger the routine. These gives us (ranges of) values for certain variables. By finding where these variables are assigned values, and tracing these back to user input routines, we can reconstruct the input that caused the routine to fire.

GENERAL REQUIREMENTS OF THE KNOWLEDGE REPRESENTATION

The following general properties must hold for a program representation that is to be used to answer questions about the program's behavior:

o   It must be possible to determine a great deal about a program's behavior from a static analysis of its code.

o   It must be easy to match code segments and the conditions under which those segments will be executed with important components of the questions to be answered.

o   There must be language constructs that correspond to the ways that people typically solve problems.

o   It must be possible to explain the behavior of an individual program segment in a single paragraph of reasonable length and intricacy.

o   All code in the representation should be relevant to the domain of the program and of potential interest to a user. The representation should include as little "housekeeping" code as possible.

SPECIFIC REQUIREMENTS OF THE KNOWLEDGE REPRESENTATION

These general goals translate into the following specific features that are important in the representation language:

o   Support for modularity that reflects a top-down decomposition of the program into manageable modules, from which we can infer what code to explain together, and what abstractions are reasonable.

o   Typing mechanisms that help to constrain the search for particular
    objects of interest.

o   Abstract data types, because routines connected to particular data
    types deserve different explanations than higher-level control
    code.

o   Static and dynamic binding of types and sizes of variables, **as
    appropriate to the solution** instead of the programming system, to
    eliminate "housekeeping" code such as system dependent bounds
    checking. We call this **logical binding time**.

o   Global variables and absolute minimization of aliasing, so that
    the use of program objects can be determined from a static
    analysis of specific code segments, regardless of how deeply
    nested the call to that code may be.

o   Global variables, local variables, parameter passing by reference
    and by value, so that lifetimes of program objects can correspond
    naturally to lifetimes of problem-domain objects.

o   Flattened (unnested) conditionals, so that the conditions under
    which a particular piece of code will be executed can be
    determined with as little search as possible.

o   Separation of observing procedures (which do nothing but return
    values) from acting procedures (which may modify system
    structures) so that the set of situations in which side effects
    need to be considered is as restricted as possible.

o   Ability to return an arbitrary number of values from a function so
    that these values can be easily recognized when they are logically
    necessary.

o   Control structures, such as loops, whose definitions correspond to
    the types that are used so that logical control flow is as
    explicit as possible.

o   Informative procedure headings, because these headings summarize
    the function of the procedure and can be used without an
    examination of the procedure body.

o   Annotations that describe each identifier so that mappings from
    question to code and code to answer can be done.

THE CURRENT IMPLEMENTATION

We have defined a representation language adhering to these requirements.
This language is called Mirror (for more information see [Rich, 1984]). We
have a translator that compiles Mirror into Interlisp, which allows us to
test that solutions encoded in Mirror are valid. We also have a portion of
the text formatter Scribe and a prototype question answering system encoded
in Mirror.

The general structure of a Mirror routine is shown in Figure 1. Three
sample Mirror routines appear in Figure 2. These routines are related to

right-justifying a line of text in our version of Scribe. While several
features are not shown, they should give the reader an idea of the
implementation.


```
            (routine-name (routine-type
    (formal-arguments-with-types)
    (local-variables-with-types)
    (return-variables)
    (body-of-condition-action-pairs)))
```

FIGURE 1
MIRROR ROUTINE STRUCTURE


```
(ADD-CURRENT-LINE-TO-PAGE (ACTING-ROUTINE
   NIL NIL NIL
    (( T (POSSIBLY-SHIFT-CURRENT-LINE)
        (APPEND-CURRENT-LINE-TO-PAGE)
        (INITIALIZE-NEW-CURRENT-LINE))))))

(POSSIBLY-SHIFT-CURRENT-LINE (ACTING-ROUTINE
   NIL NIL NIL
    (((FLUSHLEFTP) (NILL))
     ((FLUSHRIGHTP) (FLUSHRIGHT-CURRENT-LINE))
     ((CENTERP) (CENTER-CURRENT-LINE))
     ((JUSTIFYP) (JUSTIFY-CURRENT-LINE))))))

(JUSTIFYP (OBSERVING-ROUTINE
   NIL  ((ANSWER BOOLEAN)) (ANSWER)
    (((AND (EQUAL
          INITIAL-STATEVECTOR:INHERIT:SVJUST
          T)
         (EQUAL
          CURRENT-STATEVECTOR:INHERIT:SVJUST
          T))
      (ASSIGNQ ANSWER (TRUE)))
     ((NOT
      (AND
       (EQUAL
        INITIAL-STATEVECTOR:INHERIT:SVJUST
        T)
       (EQUAL
        CURRENT-STATEVECTOR:INHERIT:SVJUST
        T)))
      (ASSIGNQ ANSWER (NILL))))))
```

FIGURE 2
SAMPLE MIRROR ROUTINES

There are two routine-types, ACTING-ROUTINEs (which have side effects) and
OBSERVING-ROUTINEs (which don't). Routine headers contain information
about all parameters, local variables and return variables. This information is all that is needed in some cases. The executable part of the
routine is a set of condition-action pairs. All conditions are OBSERVING-

ROUTINEs, and conditions between pairs are all mutually exclusive (so that order of evaluation isn't important). By having all the conditions for execution of some actions in one place, (all except for loop-termination conditions, which appear as part of the looping constructs in the actions) it is easy to chain around routines and collect the conditions relevant to some actions' execution.

These routines also demonstrate the three types of routine bodies Mirror has. ADD-CURRENT-LINE-TO-PAGE typifies a "decomposition routine", with a single branch that is always taken to trigger a series of actions. POSSIBLY-SHIFT-CURRENT-LINE is a dispatching routine, which routes the computation but does no computation of its own. JUSTIFYP is typical of a computation routine.

CONCLUSION

Computer science, more than any other discipline, is the science of problem solving. We begin with a problem, we derive a general solution, and we apply this to solve particular problems. While the most direct use of the solution is to obtain an answer, there are other uses for the solution (see Figure 3). And while each use has a preferred representation, in order to guarantee consistency we would like to have one representation of the solution which can be verifiably transformed into representations convenient to other uses.

Task to be Done	Representation into Which the Program is to be Transformed
Executed directly	Machine code
Interpreted	Symbolic program tree + symbol table
Examined and modified	Indented format + auxiliary information such as name reference charts
Proved correct	Verification conditions
Used to answer questions about itself	?

FIGURE 3
USES OF A PROGRAM AND THE REPRESENTATIONS REQUIRED

In this paper we have presented an approach to filling-in the '?' in Figure 3. We have presented the underlying qualities of a source representation, such that we can conveniently translate that representation into one that is easy to answer questions about.

As we suggested above, the language in which a program is written can affect the ease with which various activities involving the program may be performed. In general, many activities may be performed with any one

program. Thus, although different activities favor somewhat different
language features, it is necessary, from a practical standpoint, to try to
find one language design that does a reasonable job at supporting all of
the desired activities. When we began this project, we focused just on the
help-system process and ignored the other things. As the design progressed, we were encouraged to discover that the language characteristics
that support a help system are very similar to language characteristics
that are desirable for other reasons as well.

Despite all this, we have ignored some important questions. For instance,
we have ignored questions about the design of the program, and about design
decisions made during its construction. We have also ignored questions
about a program's purpose, as opposed to its capabilities. This is because
we don't have available the knowledge to answer such questions. It is
"compiled-out" in the programming process, and so is not explicitly present
in the program itself. An interesting open question is. If such knowledge
were available in a formalized way, as part of the source code, or as
specifications for that code, how would one go about answering questions
about it? Could we use the algorithms we are developing now, and maintain
the same requirements for the representation, or would we have new
constraints?

ACKNOWLEDGMENTS

We would like to thank Chris Lengauer and Kim Korner for their help with
this paper and the design of Mirror.

REFERENCES

Finin, Timothy W., Providing Help and Advice in Task Oriented Systems, In
     Proceedings of the Eighth International Joint Conference on Artificial
     Intelligence, pages 174-176, IJCAI, August 1983.

Genesereth, Michael, Automated Consulation for Complex Computer Systems,
     PhD Thesis, Harvard, 1978.

Rich, Elaine A. and Temin, Aaron, Representing Programs for Reasoning,
     Technical Report TR-84-02, The University of Texas at Austin, February
     1984.

Wilensky, R., Talking to UNIX in English: An Overview of UC, In Proc. AAAI
     2, 1982.

QUESTION AND ANSWER PERIOD

BARSTOW
I have been trying to figure out why you can get away doing this?

TEMIN
Why I can get away with it?

BARSTOW

Why you can get away doing this. Let me hypothesize an answer. One of the things that I think is true about source code is that there are probably thousands of decisions that were made in the course of writing SCRIBE, maybe millions of decisions, most of which Brian Reed didn't even realize he was making, and of those that he knew that he was making, very few of which have any explicit representation in the code. And recovering those design decisions is enormously hard to do, and perhaps impossible. So why should you be able to answer questions about the code when those decisions are unrecoverable? I think that the answer is that you have carefully restricted the kinds of questions you are trying to answer. You aren't trying to answer why questions - why is the code this way, as opposed to something else? Instead you are happy enough to know that this code reflects something in the domain. And that is the second part about why you can get away with this, and that is you are assuming, I think, that the structure of the code reflects the structure of the domain. Does that sound right?

TEMIN

Yes. Thank you very much. That is very nicely put. No, no, I really do appreciate that, that is exactly the kind of answer I want to give. Maybe I have done a little more thinking on this than you, so let me expand a little bit on that. Yes, we cannot recover any information that was "compiled out" in the design process. When we get information out of the program (for instance, if we are going to make any generalizations, or we are going to try to present information at different levels of granularity), we are depending heavily upon the hierarchical decomposition presented by the programmer. So we are going to assume that if function F, calls G, H, and I as sub-procedures, then if we want to talk about, in general, what G, H, and I are doing, we can just talk about F. That is because the programmer gives it to us. If we had a way to capture the information at previous steps in the design process, I think one of two things is going to be true. Either we are going to be able to use the same question answering mechanisms on different information that may be represented slightly differently but still have the properties we've described for Mirror, or we are going to need parallel question answerers in each level of the design stage. As things get translated down, you are going to want to ask information at different levels of detail, and about different stages in the design process and so we will need different question answerers. But I still think that as long as people use programs that are executed and if that representation exists, then you are going to want to ask questions about it, because things happen in the compilation that you are going to want to know about at some level. So I think its going to be useful, but it is not at all a panacea. Does that help?

CARBONELL

First of all I would just like to remark that I approve the domain, having become a SCRIBE hacker, not having been by choice but by necessity. I sure wish I had one of these things.

TEMIN

So do I.

CARBONELL

The serious question has to do with the only part of this that I perceive as a possible weak link, and that is the annotation part in there. Rather than having a sort of ad hoc collection of key words to provide the bridge, have you thought of trying to structure a semantic hierarchy for the domain that encodes all of the actions (and perhaps all of the state changes that these actions bring about in the system), so that you can index things by the desired state change, and you can also index things by the types of actions that the user knows? The user may suggest an action at a different level of detail than that at which it is implemented. For example, you may have a right justify and a left justify, but you may want to ask how do I get text justified, period? And that would be a higher node in the hierarchy, or it could be more than just one immediate node above as well.

TEMIN
I hope I was very careful to differentiate between the domain problem solver and the system problem solver. A lot of the reasoning that you are mentioning goes on in the domain problem solver, so that, for instance, the domain problem solver will look at justification, realize SCRIBE does not know about capital "J" Justification, but knows about left justification and right justification, and therefore would query the system problem solver about each of those individually, and then come back. It is true that you need that kind of knowledge around in order to have a reasonable help system. It is also I think true that you can use that kind of knowledge over very many document formatters. You can use it for NROFF. You can use it for TROFF, you can use it for TEX, some of it without modification at all, whereas the system problem solver itself depends heavily on the code and therefore changes with each program. It is not a real satisfactory answer. The other problem is that when make these domain models, we throw in a lot more knowledge, you start saying, well, heck, let us throw away the code altogether. That, for instance, is what Bob Wilensky is trying to do. He has got his own plans of what UNIX does, he doesn't look at the UNIX code at all, and it does some real nice things, but it is not necessarily talking about UNIX. It is talking about what Dr. Wilensky may think UNIX does.

CARBONELL
It also get amazingly complicated very fast, because he is implementing UNIX in the meta-level in his system. I was just talking about providing that bridge, not necessarily at the ground instance level, perhaps at one or two or three levels of abstraction above as well.

TEMIN
That sounds a lot nicer and I'll definitely give it some thought. Thank you.

R. RICH
I would like to go back one question, and I thought that was a good way of putting it. The user who is going to use this in practice, day in and day out, is really only interested in what? In other words, what must I do to cause the effect I want, or what is causing this effect that I don't want? We frequently, once we get that answer, mutter to ourselves, why did the knot-headed spindle-brain decide to do it this way, but there is nothing immediate that I could do even with that information, really, and so I think that your approach to answering what, has two justifications: the first one that was discussed earlier, that that is all you can do, and the other one is, that is really all that is helpful to the user.

TEMIN
Thank you. Yes, it is interesting to think of the possible scenarios in
which one might use such information, and you can use it to solve an
immediate problem, while sitting right at your terminal, or you could just
be trying to learn how to use the system and you may try to query about its
capabilities. Or I would like to think it is possible for another automated system, to go ahead and somehow phrase a question and check directly
with the question answerer. So if the operating system wants to know, for
instance, How long are you going to take to run? Are you a big program?
Do I really want to load you into core now? I am going to tell this person
to go away, or do I want to give you low priority or something? Then, in
fact, it could be used at that point. There are a lot of different levels.
It is unclear how many levels of information we can actually get out of the
code without having somehow to get at that decompiled information that Dr.
Barstow was talking about.

GUIER
I think this may be a dumb question, but I going to ask it anyway. Have
you tried putting MIRROR through MIRROR and see what you can come out with?

TEMIN
Have we tried answering questions about the question answerer? Not yet.
Do I want to? Very much. Are we going to get anything useful out of it?
Who knows? But since it is written in Mirror, we certainly can try it.
Part of the problem is figuring out what questions you really want to ask
of the question answerer that it can answer, as opposed to the decompiled-
out knowledge which you want to ask me. The other thing is, while we hope
that the question answering algorithms we have are general, and don't just
refer to document formatters, and don't just refer to systems with large
global state variables. We do want to figure out how to ask and answer
questions about systems that have an internal structure different from
Scribe. The question answerer is a little bit different. A numerical
program might be significantly different. The answer is we haven't done it
yet, but we plan to.

GUIER
I have a second question, which is motivated by a lot of the kinds of code
that I write, namely quick and dirty code to see if something works, then I
kind of forget the details about it, and then I would like either me or
somebody to go off and write a good bullet-proof, clean, efficiently running code to replace that. Am I correct if he had some guidance on why you
were doing it, and some of the things that that literally couldn't answer;
wouldn't that end up being a pretty good tool to rewrite it in well-
structured code, and more efficiently running code, and so on? I'm clearly
talking about not just translation, but improvement for production and
maintenance and so on.

TEMIN
I don't know. It certainly tries to embody some of the things we think of
as good structure, and insofar as you could use that to enforce good structure, I suppose....The structure is not enforced by us now, and we hope our
question-answering algorithms will still work somewhat on code that doesn't
necessarily adhere to all our conventions. For instance, you can write
code in MIRROR syntax that has very long procedure bodies. You can have 20
statements (we don't enforce a four or five statement limit, we just think
that is reasonable for succinct explanations) So you could write a piece

of code that was really long, for instance, without the modularity, because
it was just a nuisance to generate the module. Then you go and ask a question about it, and get back a very long-winded explanation, because the
question answerer would assume that a procedure body is a reasonable unit
of explanation. I don't see immediately how you could modify it to in fact
take advantage of the constraints it wants and go ahead and operate with
those. I don't think it is that good.

FALKOFF
If you produce representations of programs that are as perspicuous as they
must be in order to answer these questions, why shouldn't you write
programs in that representation to start with?

TEMIN
Well, let us see. First of all, there are too many parentheses. Second of
all, sometimes I wouldn't want to write the conditions that we have to
write. If you put three or four conditions and then you want to iterate
through all possible combinations of those and have a different right-hand-
side action for each, I'd rather have that done mechanically. It doesn't
look all that different from more conventional languages in many ways. The
constituents are not all that different from a lot of languages people do
write in, but I would find it a nuisance to actually put down the actual
syntactic pieces, and I'd much rather have a programming environment on top
of it, at least that had a little bit of smarts - to throw in types when it
knew that the function was called with variables of certain types, it could
fill in the formals for me, and that it would take care of all the parentheses and it would prompt me for things. But you can. We are working on
such an environment now. I write lots of code in Mirror now, but I'm not
going to ask anyone else to. However, one could write code in a similar
style in C, for instance, and (except for syntactic variation) the question
answerer would still work.

BARSTOW
Several of the comments have referred to the structure of code, and I think
it is important for us to keep it in mind why structure is there. The reason the structure is there is because of the incompetence of human readers
of the code, and the need to understand what the code is doing. It is not
a characteristic that is absolutely required of the code for the sake of
the machine. It is required for the sake of the people who are looking at
the code. So there are other ways for humans to get the same information,
and if we have some documentation of the development process then it does
not matter whether or not the code is structured, because the information
about why the code is as it is, is encoded in that development structure.
So I think it is important to remember why the structure is there. It has
to do with the readability of the code and no other reason.

TEMIN
I would like to take a little bit of issue with that; since I am down here
I guess I am allowed to. At some level of dealing with the problem and
figuring out the answer, a human being has to do it. Let us say the human
being has to come up with the initial pieces of the solution. The human
being has to understand it, he may have to justify it, and he has to talk
to other people, so at some point, he has to represent the problem, and
therefore, he wants to represent it modularly. Now I don't care whether
that is in predicate logic, or whether that is in me just talking in the
air to you, or what, at some point it has got to happen. Now whether it

happens in something that we call high level languages (or very high level
languages, or ultra high level languages), or programming environments, or
abstract description languages, I don't care. At some point, yes, I am
willing to believe that the machine can take over, and then who cares? The
machine can understand in other ways. Clearly, when we just give it a
string of bits which we call a machine language program, as far as we are
concerned, there ain't no structure at all. I'm talking about answering
questions at a level "before" the solution process can be entirely
automated.

BARSTOW
No, I agree with you completely. The structure has to be there, but it
does not necessarily have to be reflected in the textual level of the code.
That was the only point I was trying to make, and we sometimes accept that
as being an absolute necessity for the textual level, and that is not a
necessity. And so, I agree with you. I agree with you completely.

TEMIN   -   OK.

UNIDENTIFIED MAN
I was about to set up to organize your text to organize your thinking. Is
there a positive advantage of organizing your text so it helps you think?

TEMIN
If you are using text, you have to look at pieces of paper, but I am sure
that in the project that Dr. Barstow is thinking about, you don't even need
it. You may be thinking in geometrical figures which are analyzed in what
you are doing, or you may be thinking in terms of three-dimensional representations. I agree with you, that if you have to use text, then it is
nice to have it organized, but I would question whether you really have to
use text.

GUIER
I'd like to take exception to this last comment that you don't really need
structured code, if you have some other way to document it. My observation
is that most of the really fine programmers automatically write well structured code (in some reasonable definition of that). I think a lot of the
structuredness that ends up in programs reflects the programmer's understanding of what he is doing, as well as his attempt to give understanding
to the next person who is going to look at it.

TEMIN
But that is irrelevant to the code level. That is relevant to the problem-
solving process. The problem-solving process in many cases happens to go
directly to what we call the source code level. It needn't be that way,
and therefore one could organize one's thoughts modularly at some other
level, and what might be called code could be anything, as long as human
beings didn't have to look at it. I think it is an artifact of the fact
that people have to write code that talks to machines, and what we really
want to deal with is the problem solving process itself.

GUIER
I don't think that is true. Never mind.

TEMIN
Thank you all very much.

The Role of Language in Problem Solving I
R. Jernigan, B.W. Hamill, and D.M. Weintraub (Editors)
© Elsevier Science Publishers B.V. (North-Holland), 1985

APL: A PICTORIAL LANGUAGE

R. J. Bettinger

The MITRE Corporation
METREK Division
1820 Dolley Madison Blvd
McLean, Virginia 22102

Current research on the human brain has revealed that
there are predominantly two modes of thinking: verbal/
analytic/linear left-brain thinking, and nonverbal/
gestalt/nonlinear right-brain thinking.

Most computer programming languages depend heavily on
left-brain thinking, and thus do not utilize the right-
hemisphere capabilities available.

APL, on the other hand, incorporates both left-and right-
brain modalities of thought, and is thus a superior
vehicle with which to formulate problems for digital
computer solution.

INTRODUCTION

I read a book called Drawing on the Right Side of the Brain, by Betty
Edwards (1). A prominent theme of the book is that human perception can be
verbal/analytic/linear/left-brain, or it can be nonverbal/gestalt/non-
linear/right-brain.

Left-brain thinking is discrete, proceeding from premise to conclusion in
well-ordered increments of logic. Right-brain problem solving is more
intuitive, holistic, pattern-oriented, involving flashes of insight that
are best described as "Ah-ha!."

I became especially interested in the notion of right-brain/left-brain
modes of thinking when Dr. Edwards cited a paper written by Dr. J. William
Bergquist, a mathematician and APL specialist at IBM. His paper, "The Use
of Computers in Educating Both Halves of the Brain" (2), makes reference to
the left hemisphere of the brain as a digital computer, and the right
hemisphere as an analog computer. Bergquist suggests that the "ultimate
computer" will have APL built into it. This APL "will include primitives
for handling graphics, text editing, digital simulation, analog simulation,
numerical analysis, symbolic manipulation, and the language to link the
analog and digital functions." Hence, this "ultimate APL" will unify the
left and right brains and enhance creativity in problem solving.

Bergquist's and Edwards' respective works stimulated me to think about APL
in terms of left-brain and right-brain functions. I realized why APL is so
exciting - there are strong visual (right brain) modes of thinking embedded
in problem-solving with APL. APL is A Pictorial Language!

VISUAL CONCEPTS IN APL

Visual concepts are present throughout the language. For instance, data
objects have geometric interpretations:

APL data object	Rank	Geometric equivalent
scalar	null	point, has no dimensions
vector	1	line, has 1 dimension
matrix	2	plane, has 2 dimensions
3-D array	3	solid, has 3 dimensions

Rectangular arrays of rank ≥ 3 may be visualized as sets of two-dimensional
matrices, e.g.,


```
       ☐←2 3 4ρι24
  1   2   3   4
  5   6   7   8
  9  10  11  12

 13  14  15  16
 17  18  19  20
 21  22  23  24
```


Pictorially, the two planes may be enclosed in boxes, which are themselves
enclosed:

```
        -------------------
        | ------------- |
        | | 1  2  3  4 | |
        | | 5  6  7  8 | |
        | | 9 10 11 12 | |
        | ------------- |
        |               |
        | ------------- |
        | | 13 14 15 16 | |
        | | 17 18 19 20 | |
        | | 21 22 23 24 | |
        | ------------- |
        -------------------
```

This nesting principle extends in regular fashion to arrays of any rank.
For example, the nested display of the array 2 2 2 2 ρι16 is

```
-------------
  ---------
 |  -----  |
 | | 1 2 | |
 | | 3 4 | |
 |  -----  |
 |         |
 |  -----  |
 | | 5 6 | |
 | | 7 8 | |
 |  -----  |
  ---------
  ---------
 |  -----  |
 | | 9 10| |
 | |11 12| |
 |  -----  |
 |         |
 |  -----  |
 | |13 14| |
 | |15 16| |
 |  -----  |
  ---------
-------------
```

A VISUAL PRIMITIVE

"Visual thinking" is useful in understanding the workings of many primitive
operations. The outer product operator, $\circ.f$, where f is any primitive
scalar dyadic function, is commonly called table generator. This
description is apparent when the arguments are vectors, because for two
vectors α and ω the result is familiar: a matrix (table) of shape
$(\rho\alpha),\rho\omega$. For example,

```
      α←10 20 30
      ω←1 2 3 4
      □←α∘.+ω
11 12 13 14
21 22 23 24
31 32 33 34
```

How is it done pictorially? We draw the following diagram:

```
       +  |    1       2       3       4
      --- | ---------------------------------
      10  |  (10+1)  (10+2)  (10+3)  (10+4)
      20  |  (20+1)  (20+2)  (20+3)  (20+4)
      30  |  (30+1)  (30+2)  (30+3)  (30+4)
```

What about higher-rank arguments? Let

```
      α←100 200
      ω←10 20 30 ∘.+ 1 2 3 4
```

Then

```
      α∘.+ω
```

becomes

```
    100 200 °.+ 11 12 13 14
                21 22 23 24
                31 32 33 34
```

or

```
    +     | 11 12 13 14
          | 21 22 23 24
          | 31 32 33 34
    ----  | ------------
    100   |
    200   |
```

to yield

```
    111  112  113  114
    121  122  123  124
    131  132  133  134

    211  212  213  214
    221  222  223  224
    231  232  233  234
```

Returning to the "table generator" interpretation of °.f, we see that α°.+ω can be formulated as

$$((\rho\alpha),\rho\omega)\rho(,\alpha)\circ.+,\omega$$

with the tabular representation

```
              +  | 11 12 13 14 21 22 23 24 31 32 33 34
                 -------------------------------------
    (2,3 4)   ρ  100 | (to work out the result is left as
                 200 |  an exercise for the reader)
```

By using the visually-motivated "table generator" interpretation, the outer-product computation is reduced to an easily-manageable process.

MORE VISUAL PRIMITIVES

Another set of visual primitives consists of the three rotation functions: ⊖, ⌽, and ⌽. Each one is a compound character: a circle (signifying rotation) overstruck by a line segment (signifying the axis about which the rotation is to be performed).

Let us assume that we want to perform rotations on a matrix.

```
        □←M←10 20 30 °.+ 1 2 3 4 5
    11 12 13 14 15
    21 22 23 24 25
    31 32 33 34 35
```

We can indicate the effect of the rotation primitives over their respective axis domains by the following diagram:

```
\
 \
  \         ----- ⊖ ------
  |          11 12 13 14 15
  ⌽          21 22 23 24 25
  |          31 32 33 34 35
```

So, ⊖ affects columns (first coordinate axis), ⌽ affects columns and rows, and ⍉, affects rows (last coordinate axis).

We may use visual thinking about rotation in the following example. A common way to find the first occurrence of a substring of characters embedded in a larger sequence of same is to use ∘.= followed by ⌽ followed by ∧/ followed by dyadic iota. Let's use the "glass box" approach to the problem.

Let the embedded substring $\alpha \leftarrow$ '*IN*' be contained within the larger string $\omega \leftarrow$ '*SUBSTRING WITHIN STRING*'. Then, to find the starting location of within α within ω, we would specify

```
    ⎕←T1←α∘.=ω
0 0 0 0 0 0 1 0 0 0 0 1 0 0 1 0 0 0 0 0 1 0 0
0 0 0 0 0 0 1 0 0 0 0 0 0 0 1 0 0 0 0 0 1 0 0
```

The next step is to rotate *T1* somehow so that each successive row after the first is shifted left one column more than its predecessor, from top-to-bottom. This rotation will line up all succeeding columns below each other and set the stage for the reduction that follows. Then, ∧/ on the resulting matrix will return a boolean vector with bits set if α is embedded within ω, and reset otherwise. We indicate the rotation vector in the following way, using the vertical bar through the circle of the rotate function, ⌽ to indicate the axis about which rotation is to occur.

```
 ---      ------------------------------------------------
| 0 |    | 0 0 0 0 0 0 1 0 0 0 0 1 0 0 1 0 0 0 0 0 1 0 0 |
|   |  ⌽ |                                                |
| 1 |    | 0 0 0 0 0 0 1 0 0 0 0 0 0 0 1 0 0 0 0 0 1 0 0 |
 ---      ------------------------------------------------
```

to yield

```
      ------------------------------------------------
     | 0 0 0 0 0 0 1 0 0 0 0 1 0 0 1 0 0 0 0 0 1 0 0 |
     | 0 0 0 0 0 0 1 0 0 0 0 0 0 0 1 0 0 0 0 0 1 0 0 |
      ------------------------------------------------
```

Completing the process, we have

```
    ⎕←T2←∧/0 1 ⌽T1
0 0 0 0 0 0 1 0 0 0 0 0 0 0 1 0 0 0 0 0 1 0 0
```

And

```
    ⎕←T3←T2⍳1
```

7

Thus, the first occurrence of the substring α begins in position 7 of ω.

Generalizing the use of the rotate function to higher-rank arrays is straightforward. Write the array in "nested-matrix" display:

```
                        ---------------------
                       |   ---------------   |
                       |  |  1  2  3  4  |   |
                       |  |  5  6  7  8  |   |
                       |   ---------------   |
        T1←            |                     |
                       |   ---------------   |
                       |  |  9 10 11 12  |   |
                       |  | 13 14 15 16  |   |
                       |   ---------------   |
                        ---------------------
```

To rotate about the second axis (affecting column within-plane) using ⊖, put each rotation factor above the plane of the array to be rotated, e.g.,

```
            0    1    0    ¯1
           -----------------
          |  1    2    3    4 |
          |  5    6    7    8 |
           -----------------

            1    0    1    0
           -----------------
          |  9   10   11   12 |
          | 13   14   15   16 |
           -----------------
```

Then, collect each rotation vector into a matrix and apply the rotation matrix to the array along the second axis:

```
    T2←2 4ρ0 1 0 ¯1 1 0 1 0
    ⎕←T2 ⊖[2] T1
```

```
                        ---------------------
                       |   ---------------   |
                       |  |  1  6  3  8  |   |
                       |  |  5  2  7  4  |   |
                       |   ---------------   |
                       |                     |
                       |   ---------------   |
                       |  | 13 10 15 12  |   |
                       |  |  9 14 11 16  |   |
                       |   ---------------   |
                        ---------------------
```

The same technique will work for rotation about the last axis, using ⌽. The rotation factor would be written in front of the rows to be row-tated (sorry!).

The shape of the rotation factor for ⊖ and ϕ is determined by the following rule: You must use a scalar or one-element vector to rotate a vector, a vector to rotate a matrix, a matrix to rotate an array, and, in general, the rank of the rotation factor must be one less than the rank of the data object to be rotated.

The remaining rotation function, ⍉, is visualizable only for matrices, or in its dyadic form for arrays of rank ≥ 3. For a matrix, say M←10 20 ∘.+ 1 2 3, ⍉M is the familiar matrix transpose function that exchanges rows and columns. Thus,

```
      □←M
11  12  13
21  22  23
      □←ρM
2  3
      □← ⍉M
11  21
12  22
13  23
      □←ρ⍉M
3  2
```

The rotation takes place in the plane of the page. For higher-rank arrays, the dyadic form of ⍉ must be used. So, if A←2 2 3ρι12, then the within-plane transpose that can be visualized as transposes on each matrix of the array is

```
      □←1 3 2 ⍉A
1   4
2   5
3   6

7  10
8  11
9  12
      □←ρ⍉A
2  3  2
```

Specification of a data object by dyadic transpose is an advanced topic and is properly a separate discussion.

The take and drop functions, ↑ and ↓, can be visualized as "cutting planes" perpendicular to coordinate axes. For example,

```
      3 ↑ 1 2 3 4 5
 1 2 3
```

can be recast visually as

```
      ←------
      1 2 3 | 4 5
      ←------
```

and 1 ‾3 ↑ 2 5ρι10 becomes

```
                    | --------->
                ↑   |           ↑
                | 1  2 | 3  4  5 |
                ------- -----------
                   6  7 | 8  9  10
                    | --------->
```

so that, in each case, the enclosed elements are selected.

VISUALIZING PARALLELISM IN APL

Another frequently-encountered APL idiom is to remove redundant blanks from a character vector. Here is the problem: if

    α←'A B C   D   E'

then how do we produce the result

  A B C D E

where the excess blanks have been squeezed out? There are various formulations of the line of code to accomplish this task:

```
        (( ωv1φω        )/ω←α≠' ')/α
        (( v≠ω,[.5]1φω  )/ω←α≠' ')/α
        (( ωv1↓ω,1      )/ω←α≠' ')/α
```

Each solution has one element in common: the establishment of a boolean selection vector that is applied to α Finding the boolean vector is performed differently, but a visual interpretation demonstrates the underlying unity of each technique.

For motivation, consider
        □←α,[.5]1 φα
    A B C   D   E
    B C   D   EA

If each column containing at least one non-blank character were kept, and the remaining all-blank columns were omitted, we would have the solution to our task! Now, we must convert the visual expression into a boolean selection vector. Look at

        □←T1←ω←α≠' '
  1 0 1 0 0 1 0 0 0 1 0 0 0 0 1
        □←T2←1 φω
  0 1 0 0 1 0 0 0 1 0 0 0 0 1 1

Now, placing ω above 1φω, we have

        □←T3←T1,[.5]T2
  1 0 1 0 0 1 0 0 0 1 0 0 0 0 1
  0 1 0 0 1 0 0 0 1 0 0 0 0 1 1

Applying or-reduction along the first axis ("down the columns"), we have found our bashful boolean, viz.,

```
    □←T4←∨≠T3
 1 1 0 1 1 0 0 1 1 0 0 0 1 1
    □←T4/α
 A  B C D E
```

Quod erat demonstrandum by using visual thinking to restructure the problem into an APL-oriented non-looping solution.

SUMMARY

APL has both left-brain digital/analytic components and right-brain analog/holistic modalities of thought embedded in its primitive functions and data structures.

Problem-solving using APL can frequently be facilitated through the use of "visual imagineering." Many APL functions are inherently visual in nature, and by combining their properties with properly-depicted problems and sub-problems, elegant solutions can be formed.

Even in tasks where there is no immediate picture or visual image to be considered, it may be possible to portray the solution as a sequence of data transformations in which one or more steps has a visual component.

The combination of left- and right-brain modes of perception make APL an extremely effective tool for the expression of computer solutions to real-world problems.

REFERENCES

(1) Edwards, Betty, <u>Drawing on the Right Side of the Brain</u>, J. P. Tarcher, Inc., Los Angeles, 1979.

(2) Bergquist, J. William, Proceedings: Eighth Annual Seminar for Directors of Academic Computational Services, August, 1977, P. O. Box 1036, La Canada, CA.

QUESTION AND ANSWER PERIOD

SAMMET
I find this very interesting and, I think, very relevant to the purpose of this conference. Let me ask you the following question. There have been a number of graphics languages developed over a period of time, where sometimes they use primitives, like points and lines, sometimes they use other kinds of primitives to essentially move a cursor. Have you looked at any of those and then tried using those as the baseline to create the more algebraic aspect of this? That is one question, I then have a second one.

BETTINGER
Unfortunately I have not looked at graphics languages per se, unless you mean a library-like DISSPLA a subroutine library, and in that case you are still using text function codes. I'm interested in putting actual visual, or say abstractions, even, of visual characters into the line of code.

SAMMET
That is a partial answer to my second question which I have not asked you
yet. Let me try and state it more explicitly and maybe get a little more
detail from you. My question is, how would you use this? That is to say,
given the fact that you have written the APL code or intend to, and you
have this concept of the pictorial representation, is it your intent that
people ought to write programs including the pictures? It was not clear to
me what it was that you really thought people would do as a practical
matter if this were developed to its fullest. I think that this is very
interesting and I am trying to get a better understanding of what you would
do.

BETTINGER
You correctly anticipate my desire. Yes, I would like to see picture
representations of data processing included in programs, and they may or
may not be embedded in the line of text. Alternatively, you may have a
string of pictures that enters first with actual textual symbols, where the
picture is interpreted in some way to represent the processing to be done
on some object. Then that result is passed along for further, either
pictorial processing, or say, linear, processing.

WEINTRAUB
At the risk of eliminating what I am planning on saying later this evening,
I think that part of the dual brain thinking that comes from APL is because
APL as was taught to me originally by my third APL teacher, but first real
APL teacher, Waldo Renich, is a language where one can think simultaneously
bottom up and top down. And if one is getting stuck at some point because
of operations and working through, one can say, well what do I want? What
is that going to look like? Is it going to be N by M by I or whatever?
Then work backwards through transpose and reshape and indexing and all
those wonderful fantastic primitives we have in APL and especially since,
and this was one comment in a previous talk, APL does very strongly have
dynamically assigned arrays (APL2 is even stronger that way), one can work
backwards and simultaneously work at the problem from both ends and,
because of the interactive features of APL, see what is going on. You did
not point it out there, but monadic reshape, monadic rho, which is the
shape of an object, gives you a pictorial presentation. True, it does not
show lines etc., but it says this thing is now 2 x 30 or 50 x 7, and if you
use the following operations you can sit there and test, and it will be. I
think the dual hemisphere aspect of APL comes from this ability to see the
results, work backwards from a given result, and at any point, if you are
clever in your test data and just use shapes, you will be OK, you will
really learn something from this. And I think one thing we can learn as
APL'ers is how to teach this technique to potential APL'ers. On the other
hand, what other languages can learn from APL - and this is what I am going
to hopefully hit on a little bit later today - is the question of what it
is, what we are doing when we use these structures of APL to solve entire
systems, instead of sitting there and playing with individual data items.

BETTINGER
Thank you.

GOLDFINGER
I wonder if there might be an intermediate position in between standard APL
and something which is really pictorial? The idea comes from a frustration
that I had and I think maybe a lot of people have when they first learn

APL. And they say, gee, why can a function only have two variables? Because I want to put in a lot more variables. Then you end up packing them into the variables on either side. And the answer why a function can only have two variables, is because standard APL is basically written as a string, and there are two positions relative to the function, to the left and to the right of the function, before and after it, and that is the source of that limitation. But now-a-days, we are all used to using full screen editors and full screen systems, and there are some other positions that become available, other than left and right, such as above and below, or upper left and lower right, and so on. I wonder whether there may be a room for extending APL into, not a string oriented language, but a two-dimensionally oriented language, in which lines can proceed both from left to right, right to left, up and down and so on, and whether one might come up with some very different and perhaps concurrent ways of writing express-ions when you can sort of go this way and this way and meet some place else. That might be a somewhat intermediate position between a standard language and going to a fully pictorial one.

BETTINGER
What I would like to see would be, in some way, embedding a picture in the actual source code, and that becoming meaningfully interpreted.

GOLDFINGER
The direction I was thinking of actually came out of an application that we had in mind. We were, I guess about a year or two ago, working on some ideas of using the APL language to design the types of structures that would occur on the silicon chip. In the process of coming up with some of these ideas, we didn't draw pictures, there was no idea of using pictures as input. What we wrote looked like APL, but it did not just go from left to right, it went in all sorts of directions, it sort of mimicked the way in which things would be laid out on a chip, much like a block diagram is. But again, not with pictures, just with regular text. Sort of exploring the spatial aspects of a symbolically written language rather than going all the way to pictures.

BETTINGER
Yes. Thank you.

FALKOFF
There are several remarks I would like to make. First of all there is an old pictorial art, if you like, which has been lost. It used to be that really good machine operators could watch the lights blinking and know what was going on, and that doesn't happen anymore, I don't think.

BETTINGER
The lights blink too quickly now. The machine operators are just as good.

FALKOFF
They all have the VDTs now, and it is all mostly written in English. Sort of starting backwards, the last remarks, the matter of having multiple arguments for functions, that was solved in the very original implementa-tion of APL, when we introduced the brackets and semicolons. But prior to that, in the Iverson notation, we were able, in fact, to put four sub-scripts and superscripts around the name of a variable, but we found that there was that geometrical limitation. Now if you want to go to three-dimensions, I guess you could get eight, or something like that. I think

it is better to make that thing linear and, in fact, I have written a
paper, which I presented, in which I strongly advocate that we adopt that
semicolon/brackets as an alternative form of syntax for functions because,
in fact, mathematically, the indexing of a matrix, for example, is a functional mapping from the indices to the values of the matrix so the notation
is already there. It is a matter of persuading enough implementers to
adopt it. On the question of showing pictures, both APL2 and the I. P.
SHARP implementation, possibly STSC, I just don't know, when they went to
general arrays, actually nested arrays, felt that it was necessary to
provide the user with a tool for actually seeing some of these things sometimes, because heretofor it is relatively easy to visualize a matrix or a
three-dimensional array of a regular nature, but when you start enclosing
it, and have various depths it may become more complex, and does. So both
of those systems provide, not primitive functions, but functions that come
with the systems which actually display the shapes of the arrays with the
values in some formalization that they have invented for the purpose.

BETTINGER
Thank you. Are there any other questions?

BURNE
More of a comment then a question. I was really interested in the way you
are using pictorial representations as a means of expressing the symbolic
global attributes of your problem. I don't know if you are meaning to use
this pictorial representation as a means of helping the user, input the
data, input the aspect of the problem, or to resolve the problem. Maybe
you could comment on that.

BETTINGER
When I teach APL I do in a sense two columns on the board. One is and APL
code, one is a visual interpretation of it. I try to get them to visualize
the problem that they are working on. This does not always work. If you
are going to write a program to play the game of NIM, which is one of my
assignments, there is not terribly much to visualize there. But another
assignment that I give them dealing with matrices and vectors, then it is
very helpful for them to visualize what there are doing. In my own coding,
when I do programming, if I can not get an answer immediately, I will think
about, I will write a picture down, then try to do little doodles like that
and say what am I really doing and then translate that into the algebraic
notation. Also the APL environment, being interactive, is very helpful
there, because I can display something right in front of my eyes. It
frequently happens when I am really heavily involved in coding, I will go
into what I will call a different state of mind, I won't be thinking
verbally, you won't hear the voice in your head saying, OK, now what do I
do, you know step by step, because you want a holistic sense, a Gestalt
sense. If someone is going to talk to me, it will take me the snap of the
fingers to change my mind-set back to verbal communicative, I'm busy
visualizing. Maybe you have had that experience yourself, I don't know.

BURNE
So if I can understand you correctly then, this pictorial language could be
used as a means of expressing more of the global characteristics of your
problem in the sense of a more symbolic abstraction of your problem.

BETTINGER
I would say so. I would hope that the pictorial notation would be advanced
enough to be able to do that.

UNIDENTIFIED MAN
I just wanted to remark, that there was a very nice paper in the first
conference on LISP and Functional Programming on a pictorial representation
of LISP programs.

BETTINGER
Thank you.

The Role of Language in Problem Solving I
R. Jernigan, B.W. Hamill, and D.M. Weintraub (Editors)
© Elsevier Science Publishers B.V. (North-Holland), 1985

THREADED WORKSPACES - AN APPLICATION DEVELOPMENT ENVIRONMENT FOR APL

James L. Ryan

Director of Software Systems
Advanced Development Group
Analogic Corporation
Wakefield, Massachusetts 01880

The advantages of APL, relative to other programming languages, as an efficient means of algorithmic expression have long been recognized, and will not be dealt with here. Instead, the notions presented in this brief paper addresses a principle weakness of current APL systems, that is, the ability to partition a large scale application package into an appropriate number of easily programmed, understood, controlled, and integrated modules. The approach taken is that of evolution, not revolution. All of the notions presented here are simple amplifications of ones that have long been part of APL tradition.

In addition to its powerful and concise operators and functions, an integral part of APL has been its presentation in an interactive, workspace oriented, environment. The traditional APL workspace is composed of a collection of named objects which may be either operators, functions, variables, constants, or undefined. In addition, a "state indicator" describes the threading of execution through the various operators and functions. The state indicator also controls the duration of existance of names which are local to particular operators and functions. An APL workspace is dynamic in that the contents of the workspace are possibly altered as a byproduct of execution within that workspace.

The fundamental notion introduced here is that execution can thread, in an orderly and controlled fashion, through a collection of workspaces. This collection constitutes the application package, and each of the individual workspaces is a module within the package. Each of the modules has well defined interfaces which control the manner in which the modules can be linked together. Some additional conventions are introduced which establish the conditions under which the effects of execution become permanent.

Threading occurs when execution crosses workspace boundaries. It is the inter-workspace analog of function invocation. Threading has several attributes beyond those of simple invocation. It produces an instance of the workspace threaded into. All side-effects of a threading are then made provisionally to the instance copy of the workspace. If the threading terminates normally those side effects are materialized in the base

workspace. Abnormal termination discards the instanced side-effects and
the base workspace is undisturbed.

One of the other effects of threading is contention control. If the
workspace being threaded is attached to multiple processes, only one
instance of threading will be allowed at a time. Attempts to thread an
already threaded workspace are delayed until the current thread completes.
Threadable workspaces are modules with defined interfaces. For a
particular workspace its interface definitions specify the named entities
within the workspace which are available for threading from other workspaces. They also define the threading connections which may be made from
the workspace to ther workspaces. Thus each workspace can be conditioned
as a producer of services for other workspaces and as a consumer of
services made available by other workspaces.

Taken together with the current state of the workspace and its attachments
to other workspaces, interface definitions define the meanings of names.
All invocations occur through name reference. If a particular name is
defined completely within the workspace, evaluation proceeds in the traditional way. If the name is defined through an interface with an attached
workspace, a thread is constructed.

All of the actions which directly affect workspace threading are taken with
eight system commands. Note that these commands can be entered directly
from the keyboard or from an executing program. An abbreviated description
of each of these commands follows.

)ATTACH workspace

)DETACH workspace

Establishes or disestablishes the state of attachment between the consumer
workspace in which the command is executed and the designated producer
workspace.

)IMPORT name [surrogate] [workspace]

)EXPORT name [surrogate] [workspace]

Declares the intent to import or export a name. The exporter always
controls the name. An importer acquires access to the name. If a
workspace name is supplied, it represents a producer workspace from which
service will be imported by means of the supplied import name. If no
workspace name is given, the import name represents an access connection to
be made from the current workspace to a consumer workspace when it is
acting as a producer of services. A name may be exported directly or
through a surrogate name. The set of active import and export definitions
constitute the interface for the workspace.

```
)EMBARGO name [workspace]
```

Revokes, for the workspace in which the command is executed, any prior import/export declaration for the designated name relative to the designated workspace or to any consumer workspace (when no workspace name is supplied).

```
)CLONE duplicate-workspace original-workspace
```

Creates the designated duplicate workspace as an exact copy of the designated original workspace.

```
)ENTER workspace
```

Advances the execution thread into the designated wrkspace in immediate evaluation mode.

```
)EXIT
```

Terminates the threading through the current workspace and returns to the consumer workspace.

A couple of examples will illustrate the use of threaded workspaces. The form of the system commands used in the examples is that of Analogic APL. The use of field specifiers should be self evident.

For the first example, assume that you wish to use a shared data base system that has been written and placed into the library. Assume that the interface declarations for the cnsumer of this system are known. From your workspace you world execute the following commands.

```
)ATTACH W=PUBLIB.DATABASE

)IMPORT W=PUBLIB.DATABASE N=READ S=GETDATA

)IMPORT W=PUBLIB.DATABASE N=WRITE S=PUTDATA
```

You can now freely use the functions READ and WRITE as though they were no different than any other defined functions in your workspace. Each invocation of READ and WRITE threads into the workspace PUBLIB.DATABASE and invokes the function GETDATA or PUTDATA as appropriate.

So that you could enable the threading into PUBLIB.DATABASE, its author
would have had to have executed the following commands.

)EXPORT N=OUTPUT S=GETDATA

)EXPORT N=INPUT S=PUTDATA

Notice that the surrogate names can be used by either the consumer or
producer, if desired.

For the second example, assume that you wish to use a plotting system that
has been written and placed into the library. In this case, the plotting
system root work space will be used to contain state information unique to
the particular consumer. Assume that the interface declarations for the
consumer of this system are known. From your workspace, you would execute
the following commands.

)CLONE D=MYLIB.PLOT O=PUBLIB.PLOT

)ATTACH W=MYLIB.PLOT

)IMPORT W=MYLIB.PLOT N=SQUARE

)IMPORT W=MYLIB.PLOT N=CIRCLE

)IMPORT W=MYLIB.PLOT N=LINE

The functions SQUARE, CIRCLE, and LINE can now be used as though they were
an integral part of your workspace. When invoked, each of them will cause
a thread to be made into the workspace MYLIB.PLOT. These functions are
capable of modifying the contents of MYLIB.PLOT, for example, by saving
state information in global variables.

The specification of threading can be through multiple workspaces. The
author of PUBLIB.PLOT, for instance, could have specified, taking the role
now of consumer, attachments to one or more producer workspaces.

In this paper, only the underlying notions of threaded workspaces, and the
minimal facilities with which to effect them have been described. The
actual facilities used in Analogic APL are far more comprehensive than
those described here.

QUESTION AND ANSWER PERIOD

BARSTOW

How much experience have people had with using this? The reason I ask is
it looks to me like it includes many of the notions of interconnections
among concurrent procedures that look like they cut down on the modularity
that one might like to have. It seems to me that it might be very
difficult to evolve code that has been written using these kinds of
formalisms, so what experience have you had?

RYAN

The experience we have had is limited, currently, to working in a emulated
environment of this, because we are still in the process of implementing
the actual environment. I can only comment that the concerns you have
don't seem to have been problems to us, but I will also be fair and say
maybe we are being naive and we are not trying to create structures that
you perhaps you are envisaging. I would be glad to sit down and spend more
time with you and talk about these things.

The Role of Language in Problem Solving I
R. Jernigan, B.W. Hamill, and D.M. Weintraub (Editors)
© Elsevier Science Publishers B.V. (North-Holland), 1985

PROLOG APPLICATION IN SOFTWARE COMPONENTS REUSE

Naomich Sueda, Shinichi Honiden, Yoichi Kusui and Kazuo Mikame

Toshiba Corporation
1-1 Shibaura 1-Chome Minato-ku
Tokyo, Japan

In this paper, a method using PROLOG, a language suitable
for artificial intelligence description, in software
components reuse is proposed and its effectiveness is
discussed. In developing the image processing logic
simulator, which improves efficiency in algorithm checking
for image processing, the software components reuse concept
has been incorporated. In order to make it easier to reuse
software components,they must be easily searched, combined,
and quickly executed. Although the simulator was designed
using FORTRAN at the beginning, no good method could be
found for automatic combination of individual software
components. Then, the authors switched to PROLOG, where
parameter attributes for each component are treated as
knowledge and PROLOG infers the relationship between
parameters, which is a big advance toward automatic
combination of software components.

As shown above, an appropriate combination of the
traditional procedure oriented language and the artificial
intelligence oriented language helps widen the area in
which various problems can be solved, using superior points
of each language.

1.   PREFACE

In recent years, the software production volume has greatly increased.
Based on such a situation, it is a key factor for each company to improve
software productivity to maintain a competitive position with reference to
competitors. To solve this situation, methods to design each software
module as a component have been extensively studied. If software can be
treated as components, the components can be reused, easily modified,
designed with better reliability and maintainability, and easily understood. Many problems, however, must be solved to realize the idea. In
order to stock and reuse software components, internal and external
reference specifications for each component must be clearly defined. Also,
environments where the components can be easily utilized must be provided.

From the environmental standpoint, the software component concept has been
introduced into development of an Image Processing Logic Simulator, by
which a logical algorithm for image processing is checked. The authors
defined the environment in which each component can be easily used as
satisfying the following criteria: (1) search for each component should be
easy, (2) components searched for and found should be easily combined, and

(3) the resultant combination should be immediately executable. To solve
these problems, the authors chose FORTRAN and designed the first version.
As a result, points (1) and (3) could be solved, but point (2) was left
unsolved, that is, a lot of user interactions were needed and the result
was far from satisfactory. Thus, this approach reached its limits.

Then, the authors designed the second version, using PROLOG, an artificial
intelligence oriented language. It proved to be fairly effective for
solving the problems listed above.

In this paper, methods used to solve the problems are described and
discussions are given about their effectiveness.

2.   COMPONENTS REUSE

Before starting the discussion regarding PROLOG effectiveness, an outline
of the image processing logic simulator is briefly described.

2.1   DEVELOPMENT BACKGROUND

When the technology used for image processing is reviewed from the standpoint of pre-processing in two-dimensional image processing, established
methods have been readily available in many areas. Many software packages
for that purpose have been developed and used. Among them is SPIDER
(Subroutine Package for Image Data Enhancement and Recognition), developed
and edited by the Electrotechnical Lab of the Agency of Industrial Science
and Technology under the Ministry of International Trade and Industry in
Japan.

SPIDER consists of approximately 350 subroutine packages, which cover
various processing areas, such as orthogonal transformation, positioning
and emphasis and smoothing. They are described in FORTRAN.

Engineers engaged in image processing have currently chosen appropriate
software components among the group of subroutines (components) and
developed programs to combine such components. To escape from such a
troublesome job, a capability has been requested which would automatically
combine, execute, and verify necessary components without programming and
which would generate and register the resultant combination as a new
component.

In short, requirements are:

   *   Each component should be used without programming.

   *   A combination of components should be automatically set up.

To satisfy these requirements, the Image Processing Logic Simulator was
studied and developed.

2.2   IMAGE PROCESSING LOGIC SIMULATOR OUTLINE

The basic simulator idea is described here, based on the first version.
Its associated problems are pointed out.

2.2.1 CONCEPT

The concept of this simulator is shown in Figure 1. The Component Structure Interpreter refers to the group of components and the component attributes, and processes the image data after obtaining necessary information through a conversation with the user.

The component structure interpreter architecture is shown below.

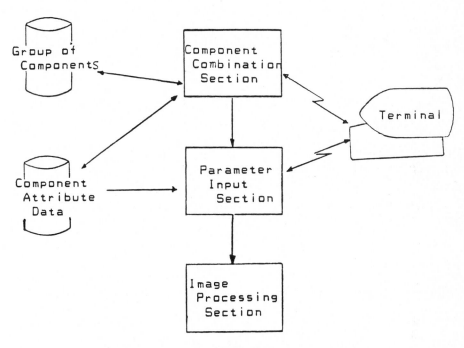

FIGURE 1
CONCEPT OF IMAGE PROCESSING LOGIC SIMULATOR

(1)　　Component Combination Section
　　　　This section selects components needed by the user, relates parameters for each component, generates a new component, and registers it into the component group and the component attribute data. (See Figure 2.).

(2)　　Parameter Input Section
　　　　Based on the component attribute data, validity of the input data of components (input value range, discrepancy between input parameters, etc.) is checked and input processing is performed.

(3)　　Image Processing Section
　　　　Based on the parameters input into the parameter input section, specified software components are executed. Relating operation among parameters in the component combination

section, which is the main part of this paper, is discussed in detail in the following sections.

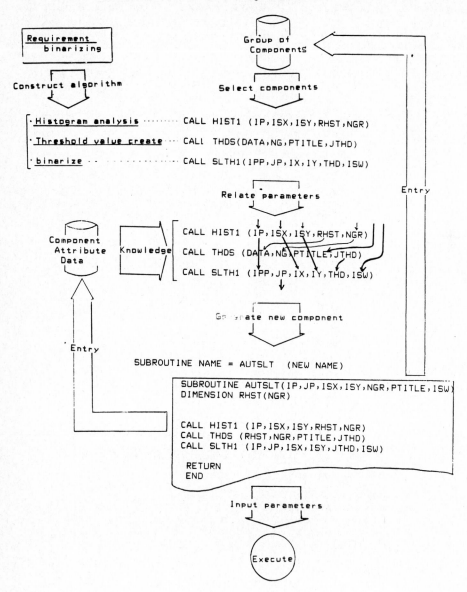

FIGURE 2
OUTLINE OF COMPONENT COMBINATION SECTION

2.2.2 COMPONENTS COMBINATION

The basic idea for the method to combine the components used in the first version of the simulator described in FORTRAN was that a user manually relates the parameters of each component. That is, the user specifies certain information into the simulator, wherein this parameter of this component has a relationship with that parameter of that component. The simulator selects several candidates by the following algorithm, and calls for a correct choice by the user.

Major component parameter attributes

*   Input/output made: Input, output, input/output,...
*   Dimension
*   Element size of dimension: Constant, depending on other parameters,...
*   Input type: Keyboard input, constant, operation between other parameters,...
*   Input value range

              .
              .
              .

<Combination algorithm>

    (i)   Image data parameters are not combined with general parameters.

    (ii)   Input, input/output, and input/working parameters must be combined with other parameters.

    (iii)   All parameters are able to combine.

    (iv)   Dimension agrees with each other.

     (v)   Element size of dimension agrees with each other.

    (vi)   Input value range for the parameter to be combined includes input value range for the parameter to combine.

Note: Relationship between the parameter to combine (on which another parameter is combined) and the parameter to be combined is as follows:

        CALL A (..., a, ...)   a: parameter to combine
        CALL B (..., b, ...)   b: parameter to be combined

2.2.3 PROBLEMS

When the method described above is used, appropriateness of the candidate chosen by the simulator (probability of right choice) is more than 90%, due to specification for names of components to be related on the parameters concerned. However, specification for the component names to be related for each parameter is a troublesome job for each simulator user. Therefore, it would be ideal for each user for the combination among components to be performed without any user assistance.

To determine the relationship between components, the following information
is generally used:

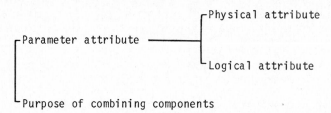

In this method, relationship validity is checked using only physical
relationships of parameter attributes. Several candidates to be related
are listed. This is the reason automatic determination cannot be expected.

The logical attribute shown above means the attribute humanly determined,
i.e., some intention such as "Let's assign such a role to this parameter"
is kept in mind.

(Example)

                    CALL HISTI (IP, ISX, ISY, RHST, NGR) ... Histogram
                                                              analysis
                                                              component
          RHST Attribute
          (Major physical attribute)
          *   Output parameter
          *   Real type
          *   Dimension is determined by RHST (NGR).
                  NGR-1
          *   Σ   RHST (i) = 1.0
                   i=0
          (Major logical attribute)
          *   Histogram data

"Purpose of Combining Components" means the reason why components A, B, C
and D are combined. This point is again discussed later on.

The authors considered that the probability of right choice by automatic
combination could be improved by introduction of the new concept, called
the logical attribute. Assume that the combination result can be checked
by the user and that correction can be performed when necessary.

3.   PROLOG APPLICATION

The first version written in FORTRAN was evaluated as described in the
previous section. For the component combination section of the image
processing logic simulator, PROLOG was applied to improve the weakness
pointed out.

3.1   WHY PROLOG?

Even if the logical attribute concept is introduced and as long as one to
one correspondence between one logical attribute specified by a parameter
and the other logical attribute by another parameter is retained, the

character matching method using FORTRAN is satisfactory and no need of PROLOG arises.

However, meanings of parameters do not correspond one to one. The example shown below indicates this situation:

(Example)

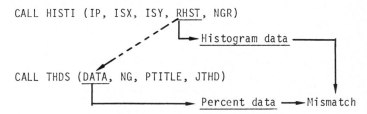

In the example above, the logical attribute for parameter "RHST" of the component "HISTI" is defined as a "histogram data" and that for parameter "DATA" of "THDS" is defined as "percent data". If a simple character matching is applied to this particular example, they do not match and cannot be combined. To solve this problem, each attribute has to be defined and a pattern matching process based on the defined attribute is needed. This mechanism is the so-called inference mechanism. In such an operation, FORTRAN is not suitable and the authors decided that the artificial intelligence oriented language PROLOG is appropriate.

3.2   COMPONENT ATTRIBUTE KNOWLEDGE

The attribute structure for each component is organized as shown in Figure 3 and is expressed as shown in Figure 4. An individual component and its parameters makes up an individual frame. Each logical attribute definition also takes the form of the frame. Each frame inherits necessary information from upper-level concept to lower level concept.

Organization of Component Attribute Knowledge

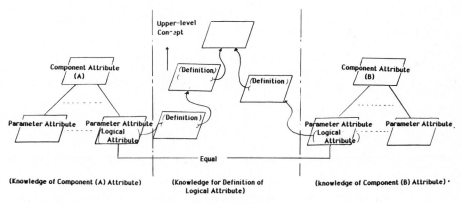

FIGURE 3
COMPONENT ATTRIBUTE KNOWLEDGE

```
(class hist1).
    parameter(hist1,ip,isx,isx,rhst,ngr).

(class ip).
    ako(ip,hist1).
    mode(ip,input).
    dimension(ip,2,para,isx,para,isy).

(class isx).
    ako(isx,isx.d).

(class isy).
    ako(isy,isy.d).

(class rhst).
    ako(rhst,hist1).
    ako(rhst,histogram-data).
    mode(rhst,output).
```

FIGURE 4
EXAMPLE OF KNOWLEDGE EXPRESSION

3.3   RULES FOR COMBINATION

An assumption that this parameter for this component relates to that parameter for that component is the basis of component combination inference. This is the so-called Backward Inference Method and it is a non-deterministic method. This mechanism can be roughly expressed in a process flow diagram, as shown in Figure 5 for convenience. Note that inference procedure is difficult to express in a flow diagram in general.

(1)   Rule for Selection
The rule for selection by the physical attribute is basically identical with that described in Section 2.2.2.

On the other hand, selection based on the logical attribute is inferred by interpreting the meaning of each parameter using the upper-level concepts.

However, if a structure of knowledge to define the logical attribute includes extremely abstract upper-level concepts, matching of some components at some level does not have practical sense any more.

For example, it is easily understood that a structure such as (histogram data) (percent data)...(numerical values) does not have any practical sense, even if components are related to each other, based on the numerical values concept. To avoid such a case, care has been taken to determine a way to give knowledge regarding the definition of logical attribute and the concept of the logical distance (a function of the number of interpretations) has been incorporated to quantify the discussion.

<u>Process for combination</u>

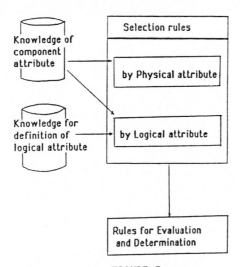

FIGURE 5
RULES FOR COMBINATION

```
s-rule1(*x)  :- mode(*x,input).
s-rule1(*x)  :- mode(*x,input-output).
s-rule2(*x,*y)  :- front(*x,*y).
s-rule3(*x)  :- dimension(*x,*a),
                dimension(*y,*b),
                equal(*a,*b).

s-rule25(*x,*y)  :- semantics(*x,*a),
                    semantics(*y,*b),
                    equal(*a,*b).

relation(*A,*B)  :- s-rule1(*A),
                    s-rule2(*A,*B),
                    s-rule3(*B),
                    s-rule25(*A,*B).
```

= = = = = = = = = == = = = = = == = == = = == = = = = = = = = = = = = = = == = = = = = = = == = = = = = = = = =:

```
If
    Input/output mode of *A is input or input-output
        AND
    *B is set in front of *A
        AND
    Dimension of *A equal dimension of *B
        AND
    Logical attribute of *A equal logical attribute of *B
Then
    *A is related to *B
```

FIGURE 6
EXAMPLE OF RULE EXPRESSION (SELECTION RULES)

Thus, the difference in logical distance can be checked after the logical attribute between parameters is matched.

(2)   Rule for Evaluation and Determination

Candidates selected using the rule for selection are finally determined by applying the following rules:

(i) For the parameters to be related, the physical distance is calculated based on the positional relationship (order of candidates in the coded program) of the candidates to be related.

(ii) Evaluation function
The index wherein parameter "a" will be related to the parameter "b" (different from probability) is calculated.

$f(a, b) = \alpha_1$ (logical distance) + α_2 (physical distance)

α_1 and α_2 are weighting coefficients.

Determination (a, x) :- comparison $(f(a, x) \geq f(a, y))$

```
function(*A,*B,*f)  :-  logical-length(*A,*B),
                        phisical-length(*B,*b),
                        multiply(*a,alph1,*x),
                        multiply(*b,alph2;*y),
                        add(*x,*y,*f).

decision(*A,*X)  :-  function(*A,*X,*XF),
                     function(*A,*Y,*YF),
                     greater(*XF,*YF).
```

FIGURE 7
EXAMPLE OF RULES EXPRESSION (EVALUATION AND DETERMINATION)

3.4   EFFECTIVENESS

The image processing logic simulator by PROLOG is being evaluated, using approximately several components. As a result, the right combination probability proves to be at the high level. Thanks to this performance, loads on each user in the parameter combination operations seems to be reasonably reduced.

4.   PROBLEMS TO BE SOLVED IN THE FUTURE

This image processing logic simulator has the following two unsolved problems.

(i)   Automatic combination processing has not been realized yet.

(ii)   It is expected that the time required for inference will increase in proportion to the complexity of the inference procedure, when the number of components is increased and the associated quantity of the knowledge is increased.

For point (i), this is because the information regarding "Purpose of combining components" which related components to each other as described in (2.2.3), has not been incorporated yet in this simulator. This can be shown in the following example.

CALL STAS (IP, ISX, ISY, <u>RMEN</u>, <u>STDEV</u>, <u>MAX</u>, <u>MIN</u>)

CALL DIVCIR (AP, BP, ISX, ISY, <u>C</u>, JERR)

This component combination algorithm is that the average value (RMEN), standard deviation (STDEV), maximum value (MAX), and minimum value (MIN) for the density of the image concerned are obtained by "CALL STAS" first. Then, the value for each pixel in the original image is divided by a constant value (C) by "CALL DIVCIR". At this time, the value of C can be one of RMEN, STDEV, MAX, or MIN, and only the user knows the right selection. Only enhanced knowledge of the component attribute cannot solve this problem.

The authors are currently engaged in research and development of the image processing expert system. Although no detailed explanation of the image processing expert system is given here, its general idea is that, when a user requests processing of an image, the system selects appropriate image processing components and furnishes a consultation to the user. At this time, the image processing expert system considers the combination of the image processing components. The image processing expert system is the user, from the standpoint of the image processing logic simulator. If this holds true, a step toward automatic combination seems to be obtained by inputting the purpose of combining components, intended by the image processing expert system, into the image processing logic simulator.

Furthermore, the inference mechanism currently utilized, in which all combinations are checked, will not be needed any more because the component combination purpose is clarified. Thus, the inference mechanism, in which components to be combined are predetermined, can be employed, i.e., it can be much simplified and problem (ii) is solved at the same time. The authors will continue research and development activities to combine both the image processing logic simulator and the image processing expert system.

5.   CONCLUSION

Throughout the research and development activities for the image processing logic simulator, the authors first constructed a system based on traditional languages, such as FORTRAN. Its limitations were reached which prevented research from going further. However, once the authors met languages suitable for artificial intelligence, such as PROLOG, it was possible to take a new turn in the research. It is considered that a variety of problems can be solved, if the traditional process oriented languages and the artificial intelligence oriented languages are appropriately chosen and used, considering their merits and demerits.

N. Sueda et al.

ACKNOWLEDGMENTS

The authors would like to thank Mr. A. Ito, and Dr. M. Arai of the Systems and Software Engineering Division in Toshiba Corporation, who furnished the chance to write this paper.

BIBLIOGRAPHY

Azriel Rosenfeld, Avinash C. Kak: Digital Picture Processing.

Avron Barr, Edward A. Feigenbaum: The handbook of artificial intelligence.

Nils J. Nilsson: Principles of Artificial Intelligence.

W. F. Clocksin, C. S. Mellish: Programming in Prolog.

Hideyuki Tamura and five others: SPIDER USER'S MANUAL.

Nakashima Hideyuki: The Language and Interpreters, Information Processing Society of Japan, VOL. 23, NO. 11.

Takashi Matsuyama: Knowledge Organization in Image Understanding, Information Processing Society of Japan, VOL. 24, NO. 12

Hideyuki Tamura, Katsuhiko Sakave: Three Kinds of Knowledge for Building Digital-Image-Analysis Expert System.

Shigeoki Hirai: Knowledge-Based System Tool, The Society of Instrument and Control Engineers, VOL. 22, NO. 9.

Suwa Motoi: The Current State and Future Trend of Knowledge Engineering, The Society of Instrument and Control Engineers, VOL. 22, NO. 9.

The Role of Language in Problem Solving I
R. Jernigan, B.W. Hamill, and D.M. Weintraub (Editors)
© Elsevier Science Publishers B.V. (North-Holland), 1985

SOLVING GRAPH PROBLEMS USING LOGRAPH (*)

P. T. Cox and T. Pietrzykowski

School of Computer Science
Technical University of Nova Scotia Canada
P. O. Box 1000
Halifax, Nova Scotia, Canada B3J 2X4

It is common practice to use graphical representations as
an aid to problem solving. From the earliest days in the
development of artificial intelligence, graphical representations have been commonly used any many attempts have
been made to systematise them. This interest was two-
fold: graphical representation as the visual aid, as well
as graphs, in the mathematical sense, as the formal
underlying model. The severe hardware limitations of
early computers restricted this approach reducing its
role to that of a secondary heuristic aid. As a result,
textual descriptions of problems and algorithms have
become increasingly complex in an attempt to attain the
representational power inherent in pictures. LOGRAPH, a
language that subsumes the expressive power of LISP and
PROLOG, is presented. In LOGRAPH, the graphical form of
a problem can be directly represented, not merely as a
heuristic aid, but as pictures on a screen that are
directly executed.

1. INTRODUCTION

It is common practice to use graphical representations as an aid to problem
solving. From the earliest days in the development of artificial intelligence, graphical representations have been commonly used and many attempts
have been made to systematise them (1,5,9,13). This interest was two-
fold: graphical representation as a visual aid, as well as graphs, in the
mathematical sense, as the formal underlying model. The severe hardware
limitations of early computers restricted this approach reducing its role
to that of a secondary heuristic aid. As a result, textual descriptions of
problems and algorithms have become increasingly complex in an attempt to
attain the representational power inherent in pictures. For example, early
lucid methods of developing programs and proving their correctness via
flowcharts have been vilified and replaced by textual methods (8). Nevertheless flowcharts, dataflow graphs and other similar graphical devices
have for a long time played an important role in both development and
documentation of computer programs (10,11).

The idea of using a graphical method of expression for predicate logic was
originally investigated by Peirce (14), and recently have been further
developed by Sowa (16).

(*)   This research was supported by NSERC grants A0286 and A0304.

The development of dataflow architecture (17) has forced the use of graphical means for expressing algorithms, giving rise to recently developed
languages such as GPL (7) and PROGRAPH (15). The inadequacy of Algol-like
languages for handling complex data structures (2), particularly graphs
(common in problem solving), gradually lead to their replacement for such
tasks by functional and logic programming languages such as LISP (12) and
PROLOG (3). The latter in particular is becoming increasingly popular
because of its natural suitability for problems requiring search. These
languages still suffer from two serious shortcomings. First, they still
use test to represent structures and processes that are inherently
graphical. This deficiency is even more obvious in these languages than in
traditional Algol-like languages which were primarily designed for processing textual data. Second, these languages directly handle only tree-like
data structures, and in order to process graphs, require various ad hoc
patches, like property lists in LISP and ground clauses in PROLOG, which
have to be modified during execution. LOGRAPH (6), on the other hand, has
several advantages. First, the graphical form of a problem can be directly
represented, not merely as a heuristic aid, but as pictures on a computer
screen which are directly executed. Second, data and programs are homo-
geneously expressed using exactly the same elementary objects. Third, the
expressive power subsumes that of PROLOG and LISP, allowing arbitrary
graphs to be directly represented and processed. Fourth, the possibility
of parallel execution is explicit and natural, while sequentiality can be
enforced if necessary. Finally, the graphical rather than textual nature
of the language provides the aid to human thought that similar devices have
always provided, as mentioned above.

2. A GRAPH PROCESSING PROBLEM

To illustrate the language LOGRAPH we will present a problem and show how
it can be conveniently represented and solved. The problem involves
finding the length of a longest path from some starting node in a directed
acyclic graph. This application in the critical path method for scheduling
tasks in some process, where nodes of the graph represent states of the
process, and arcs represent tasks that transform the process from state to
state (4). Each arc is assigned an integer label giving the time necessary
for performing the task. The solution to the problem is node to which
there is a path of maximum length from the start node. The length of the
associated path obviously gives a lower bound on the expected completion
time of the entire process. A crude method for finding such a node is to
search all possible paths from the start node, labelling every node on
every path with its maximum distance from the start. This can obviously be
improved by noting the length and terminal node of the longest path found
so far to a node with no successors, and updating this information only
when such nodes are found.

3. LOGRAPH SOLUTION FOR CRITICAL PATH PROBLEM

In Figure 1 is a simple directed acyclic graph with the lengths of the arcs
written as labels.

This graph can be represented in LOGRAPH as shown in Figure 2. The reader
should note that this Figure is a printout of the structure that the user
draws on the screen using the LOGRAPH editor. In this structure the nodes
and arcs of the graph are represented by "icons" some having the name node

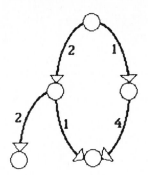

FIGURE 1

and others having integer names. The icons with integer names correspond
to the arcs of the graph and the integers specify their lengths.

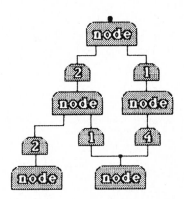

FIGURE 2

All the icons in this structure are examples of <u>functional cells</u> and are
connected by lines called <u>simple wires</u>, which are attached to the perimeter
of cells at points called <u>terminals</u>. Each functional cell has one distinguished terminal called the <u>root</u>, situated on its curved face. The other
terminals, all of which are on the flat face, are called <u>argument</u>
<u>terminals</u>. In our example, every integer constant cell has either one or
zero argument terminals, while the number of argument terminals of a **NODE**
cell is equal to the number of outarcs in the graph. Integer functional
cells with no arguments are represented simply as bold-face integers.

In Figure 3 is shown a set of LOGRAPH frames defining an algorithm for
solving the critical path problem. Again note that this Figure is a printout of diagrams that the user constructs on the screen with the editor.
The reader is encouraged to locate in this Figure the elements of LOGRAPH
which we will now introduce.

A frame consists of a set of compartments and a name. In Figure 3 there are three frames named **start, search** and **update.** Each compartment contains a network of cells connected by wires and cables. The cells in a compartment may be functional cells or literal cells. In the compartment of the frame **search** there is a variable functional cell which is distinguished by having no name. Each literal cell has a single curved face to which wires are attached at terminals, all of which are argument terminals. The order of argument terminals is significant for both functional and literal cells. In the former they are ordered from left to right, and in the latter are ordered clockwise starting from the arrow marker on the perimeter of the cell. The points to which wires are attached to the perimeter of a compartment are

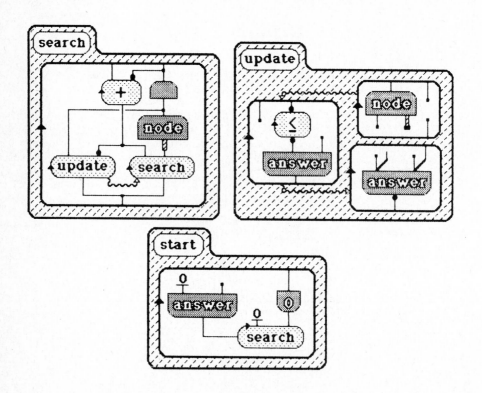

FIGURE 3

also called terminals, and are ordered in the same way as the terminals of literal cells. A wire is a set of terminals graphically represented as a maximal tree whose nodes are terminals and junctions (small black blocks) and edges are single wires. A wire consisting of exactly one terminal is represented by a simple wire terminated by a junction and is called a free wire. Obviously the graphical representation of a wire is not unique and can be chosen for visual clarity. A cable represents an unspecified number of wires (possibly none) and is always directly attached to the perimeter of some compartment or cell following all the specified wires. **Every** cell

and compartment has a cable, but for convenience empty cables are omitted
in the graphical representation. A cable may also be attached to a
terminal of a literal cell via a single wire; such a terminal is called a
for-all terminal. A cable, like a wire, can be free. In Figure 3 there
are two examples of cables; the one occurring in the compartment of the
search frame is attached at one end via a for-all terminal, while the
other, in **update**, is free. Compartments in a frame and cells in a compartment are partially ordered as indicated by wavy arrows called sequence
arcs. Any terminal of a cell may be marked by a priority marker, which is
a solid angled line meeting the cell at the marked terminal. A priority
marker indicates that this terminal belongs to two distinct wires, one of
which contains the angled line and is called a priority wire. A terminal
of a cell marked with a black spot is called a guarded terminal. The
reader is invited to identify examples of priority markers and guarded
terminals in the example.

SEMANTICS OF LOGRAPH

Before we discuss in detail the execution of the algorithm given in Figure
3 on the structure in Figure 2, we give a general description of the three
LOGRAPH execution rules. An execution rule can be applied to a cell with a
guarded terminal only if the wire attached to this terminal is attached to
the root terminal of a functional cell.

Merge:
If two functional cells have their root terminals attached to the same
wire, have matching names and matching terminals, they are replaced by
a new functional cell. Two cells have matching names if they both have
the same name or one cell is a variable; the cell resulting from the
merge is a variable if both merged cells are variables, and otherwise
has the same name as non-variable merged cell. Two cells have matching
terminals if the cable of one of them can be replaced by new wires and
a matching cable such that the two cells have an equal number of
terminals. The wire attached to a terminal of the new cell is
constructed by taking the union of the wires attached to corresponding
terminals of the merged cells, removing the terminals of those cells
and adding the corresponding terminal of the new cell. In the case
when a terminal of one of the merged cells has a priority marker, the
wire attached to the corresponding terminal of the new cell is further
modified by removing from it every root terminal which was not attached
to the priority wires. These terminals form a new wire which is added
to the compartment. In the case when an argument terminal of one of
the merged cells is attached to a cable which is also attached to a
for-all terminal of some literal cell P, there are several possible
cases which are presented in detail in a forthcoming publication. In
our example, the only case that arises is when the cable is matched
with a fixed number n of wires. In this case the literal cell P is
replaced by n copies of P each terminal of which is attached to the
same wire as the corresponding terminal of the original P, except that
corresponding to the for-all terminal. The n wires replacing the cable
are each attached to a different copy of P at the terminal corresponding to the for-all terminal of the original P.

Replacement:
If a literal cell and a frame have the same name and matching
terminals, then the cell is replaced by a copy of the interior of some

compartment of the frame. Each pair of wires, one attached to a terminal of the cell, the other to the corresponding terminal of the compartment, is replaced by the wire which consists of the union of these two wires minus the terminals of the cell and compartment.

Deletion:
If a functional cell has a free root wire then it is deleted. Terminals of this cell are also removed from the respective wires.

In order to execute an algorithm such as the one presented in Figure 3 on a structure such as that in Figure 2, it is necessary to provide a query which is a network of cells and wires (similar to the interior of a compartment). The query usually contains an initial structure (for example, that in Figure 2), connected to literal cells which share the names of some frames in the definition of an algorithm (such as that in Figure 3). An example of a query is shown in Figure 4.

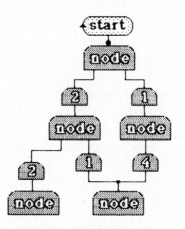

FIGURE 4

A lograph is a query together with a collection of frames; for example, the query in Figure 4 together with the frames in Figure 3. A lograph is executed by transforming the query using the rules described above. Note that these transformations may be performed in parallel where possible. Otherwise the order of execution is specified by sequence arcs and guarded terminals. An execution is said to be successful if it terminates in a query in which there are no literal cells and no conflicts between functional cells.

We now present a sequence of figures which illustrate a successful execution of the lograph consisting of the query and frames from Figure 3 and 4.

To obtain 5(a) the replacement rule is applied to the start literal in Figure 4 using the frame **start**.

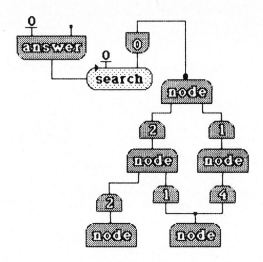

FIGURE 5(a)

Next we apply the replacement rule to the **search** literal in Figure 5(a) to obtain 5(b) using the frame **search.**

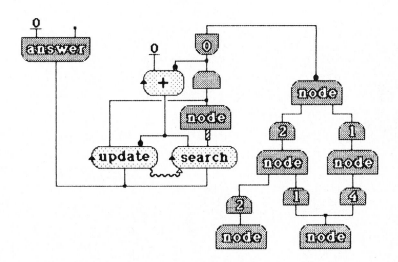

FIGURE 5(b)

Between 5(b) and 5(c), two transformations are performed in parallel: the **0** cell is merged with the nameless cell and the literal cell + is replaced. This cell as well as ≤ cell in the frame **update** correspond to system-defined frames. Note that the literal **update** could not be executed because its second terminal is guarded and the wire is not attached to any root terminal. Clearly the literal **search** could not be executed either because the **update** must be executed first.

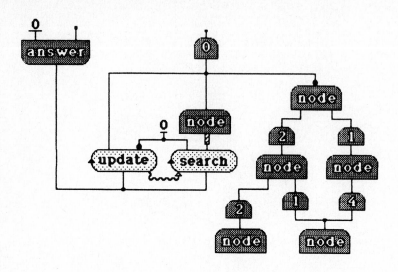

FIGURE 5(c)

In order to obtain 5(d), two transformation are performed in 5(c): the literal **update** is replaced and the cell **0** is deleted. The replacement of **update** uses the first compartment of the **update** frame because of the sequence arcs. This choice leads to a successful termination, however, if

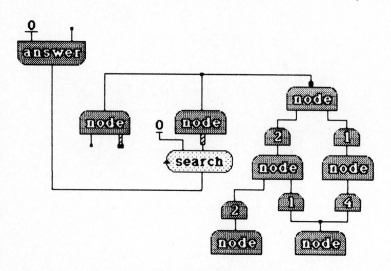

FIGURE 5(d)

this were not the case the other compartments would be tried in order until
one is found that 'leads to success. The first compartment succeeds if the
corresponding node in the graph has successors and is therefore not at the
end of the critical path. Note that the two top **NODE** cells could be merged
as well, but we postpone this action for the sake of clarity of presentation.

To obtain 5(e) from 5(d), the three top **NODE** cells are merged. Note that
in this case one of the merged cells has a cable which is attached at its
other end to a for-all terminal. As a result the **search** literal is
replicated.

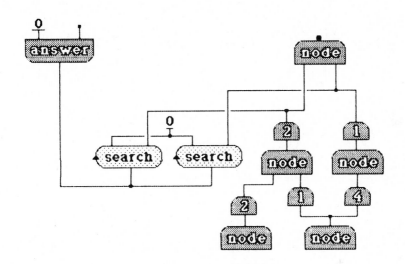

FIGURE 5(e)

Between 5(e) and 5(f) the leftmost **search** literal is executed by replacement. Note that both **search** literals could be executed in parallel, but
for clarity we execute only one in this step.

Figure 5(g) is obtained from 5(f) by a sequence of transformations strictly
analogous to those leading from 5(b) to 5(f). Note that the top **NODE** of
the graph cannot be deleted because its root terminal is guarded.

5(g) is transformed into 5(h) by actions analogous to those occurring
between 5(b) and 5(c).

Between 5(h) and 5(i) the **NODE** with the cable is merged with the lower left
NODE of the graph which has no argument terminals. As a result, the **search**
literal with the for-all terminal to which the cable is attached
disappears. Also in this step the **update** cell is replaced by the third
compartment of the **update** frame. Note that before this occurs, the first
and second compartments of the frame are tried by immediately lead to
failure.

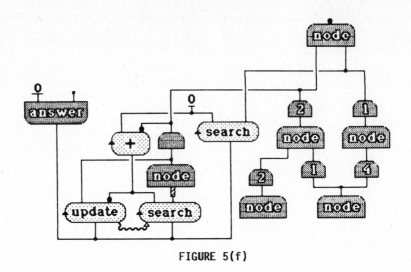

FIGURE 5(f)

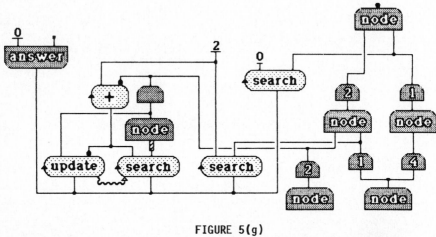

FIGURE 5(g)

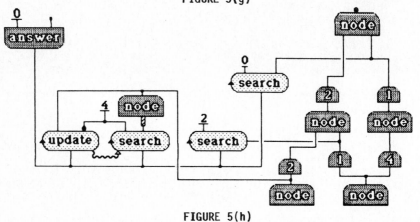

FIGURE 5(h)

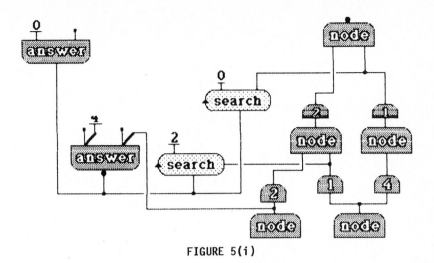

FIGURE 5(i)

The transformation of 5(i) to 5(j) is accomplished by merging the two
ANSWER cells. Note that the resulting **ANSWER** cell inherits the guarded
terminal, and has its argument terminals attached to 4 and the lower left
NODE cell of the graph. These values indicate the length and end of the
longest path found so far.

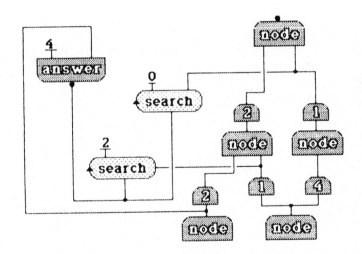

FIGURE 5(j)

We will omit further intermediate stages of execution and present in Figure
6 the structure remaining on successful termination. Note that this final
structure consists of functional cells only. The argument terminals of the
ANSWER cell are attached to 5 and the lower right **NODE** which respectively
represent the length and the end of the longest path from the top **NODE**.

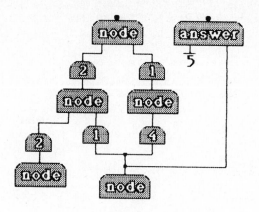

FIGURE 6

5. CONCLUSIONS

We have presented the language LOGRAPH in its preliminary form. Our intention is to develop an experimental implementation of LOGRAPH to test these
ideas by exposing them to interested users. We expect that it will be
particularly useful in applications to problem-solving, expert systems,
graph and database processing. This implementation will be in Prolog and
will draw on experience gained from the existing implementation of
PROGRAPH.

Following this pilot implementation, we intend to produce an efficient
version. One of the most interesting features of LOGRAPH, however, is its
potential for parallel execution, the consideration of which leads to a
variety of important issues. Some of these are as follows:

Communication between processes, using information about failure of one
process to cause termination of other processes which must eventually
fail.

Management of computing resources, particularly in the situation where
the demand for resources exceeds those available, leading to a control
strategy.

Structure sharing. Many processes can share parts of the structure;
the issue of how to share data structure between processes is critical
since explosive growth in the size of the structure limits parallelism.

Architecture. The most important issue is the design of an architecture suited to the specific needs of LOGRAPH. It seems that neither
the standard von Neumann or dataflow architectures are appropriate.

REFERENCES

(1) Amarel, S., On Representations of Problems of Reasoning about
     Actions, in **Machine Intelligence 3**, D. Michie (ed.), American
     Elsevier, N.Y. (1968), 131-171.

(2)   Backus, J., Can programming be liberated from the von Neumann style?
      Comm. ACM.

(3)   Clocksin, W., Mellish, C., **Introduction to PROLOG,** Springer-Verlag
      (1980).

(4)   Coffman, E.G. (ed.), **Computer and Job-Shop Scheduling Theory,** Wiley
      (1976).

(5)   Cox, P.T., Pietrzykowski, T., Deduction Plans: a basis for
      intelligent backtracking, **IEEE Trans. PAMI,** v.3 no.1 (1981), 50-65.

(6)   Cox, P.T., Pietrzykowski, T., **LOGRAPH: a graphical logic programming
      language,** Research Report 8404, School of Computer Science, Acadia
      University, Canada.

(7)   Davis, A. E. et al., **GPL Programming Manual,** University of Utah
      (1982).

(8)   Dijkstra, E.W., **A Discipline of Programming,** Prentice-Hall (1976).

(9)   Ernst, G.W., Newell, A., Some Issues of Representation in a General
      Problem Solver, **Proc. AFIPS,** vol. 30 (1967), 583-600.

(10) Floyd, R., Assigning Meaning to Programs, **Proc. Symp. Appl. Math.,**
      Am. Math. Soc., v.19 (1967), 19-32.

(11) McCarthy, J., Recursive Functions of Symbolic Expressions and their
      Computation by Machine, **Comm. ACM,** 3 (4), (1960), 184-195.

(12) McCarthy, J. et al., **LISP 1.5 Programmer's Manual,** MIT Press (1962).

(13) Nilsson, N. J., **Problem solving methods in Artificial Intelligence,**
      McGraw-Hill (1971).

(14) Pierce, C. S., **Collected papers of Charles Sanders Pierce,** C.
      Hartshorne and P. Weiss (eds.), Harvard University Press, v.3 (1933).

(15) Pietrzykowski, T., Matwin, S., PROGRAPH: a preliminary report, to
      appear in **Programming Languages.**

(16) Sowa, J. F., **Conceptual structures: information processing in mind
      and machine,** Addison-Wesley (1984).

(17) Treleaven, P.C. et al., Data-Driven and Demand-Driven Computer
      Architecture, **ACM Computer Surveys,** v.5 (1982).

QUESTION AND ANSWER PERIOD

JERNIGAN
We have some questions from the floor.

BOUDREAUX
This idea reminds very much of a notion that an American logician, Charles
Sanders Peirce, first developed about 1905. He called it the existential
graphs, in which he used diagrams, which look remarkably like the diagrams
you have just drawn for us, to express notions of first order logic. I
wonder if you are familiar with that.

PIETRZYKOWSKI
No, I am not. I appreciate ....

BOUDREAUX
It is in the collected works of Peirce in Volume III. A more modern source
for the same ideas, which I think are really very interesting, is a book
that was written by John Sowa of IBM, was published I think about three or
four months ago. It is called Conceptual Structures, and it deals with
things, that I think you might find of interest because, one, it has an
implicit or underlying graphic organization, graphic structure, and two,
the set of inference rules which he uses to operate on the graphs are a bit
richer, I think, and probably provably stronger than the ones that you are
using; the operations are simply merging your lines to identify the
separate variables. I think that is the second source which I can't really
recommend too strongly to you.

PIETRZYKOWSKI
Thank you very much. Speaking about the merging lines, of course, for the
people familiar with PROLOG, they see what is going on, the variables are
matched in this moment here.

R. RICH
This is probably a trivial question. Each of your inner boxes had orientation arrows on them. I thought I heard you say that if you forgot to put
an arrow on, the editor will put it in automatically because everyone has
it. If everyone has it, why does any of them have it?

PIETRZYKOWSKI
No, it will refuse to finish, it will ask me, where the arrows should be
put.

R. RICH
But if every box has something, then why does any box have to have it? Why
can't you just ....

PIETRZYKOWSKI
That orders the terminals, that is the beginning, that is the 12 on your
clock, you have to put the 12 on your clock somewhere.

R. RICH
Oh! It is not just a direction? It is also a location?

PIETRZYKOWSKI
That's all. Actually it will have a deeper meaning when we will do it for
full predicate logic. The left ones will be the positive, the right ones
will be the negative. But at this moment, it just orders them.

BARSTOW
In some of the experiments that we have done with graphical programming
languages, we have found that you can put less on a given area of the
screen with graphics than you can with text. We have sometimes felt that
although the graphics give clarity, we often want to see more than we can
conveniently draw in that location, and there is actually a fixed size to
the screen. Have you encountered any problems like this?

PIETRZYKOWSKI
That is the starting point, we are aware. We have a fairly good windowing
and zooming down, so we can, for the conceptual reason, put lots of them,
and when you want detail you zoom it up the thing. Thank you. That is a
very proper question. It has to be addressed.

BOUDREAUX
I might comment to David's point there, one way that Peirce solved the
problem, and I think it was 1906, was to use different levels, that is, he
would have pictures on several different levels; one he would call fur, and
another metal, and another paper, so he would name the layers that would be
visible by clicking through a graphic screen if there were such things as
graphic screens available at that particular time. That too might be an
interesting way instead of highlighting or blinking or zooming in magnifi-
cation. If you could have levels or layers in the system, that might keep
busyness down.

PIETRZYKOWSKI
This idea actually is interesting, but we wanted to have layers for educa-
tional things, like when you have a recursion, every call will stack them
and will make an impression in a three-dimensional thing. You have one
frame behind another, it is like divisible, and using little MacIntosh-like
tricks, you can have little pictures and you can pull one and examine what
is on it. Graphics, you have so much freedom for all sorts of your dreams
and desires, particularly if you implement in PROLOG. It is not much, not
a big deal to have this or that. By the way, I should mention that, of
course, you will say, so what, you will have the thing in PROLOG. It will
run forever because PROLOG is slow, as we well know. The point is that
this all this environment is to make the program LOGRAPH work - also all
sorts of data type compatibility checking, etc., - so it just has to be as
fast so as not to make the user bored to wait for the response. Once it is
all cleaned, there will be a compiled version which should run faster - of
course it will not be interpreted by PROLOG - but for designing the whole,
thing PROLOG is ideal.

The Role of Language in Problem Solving I
R. Jernigan, B.W. Hamill, and D.M. Weintraub (Editors)
© Elsevier Science Publishers B.V. (North-Holland), 1985

LANGUAGE, PROBLEM SOLVING AND SYSTEM DEVELOPMENT

Bruce I. Blum

The Johns Hopkins University
Applied Physics Laboratory
Laurel, Maryland 20707

This essay considers the role of problem oriented (or
implementation independent) languages for use in system
development. After a brief introduction to the problem
domain, a specific application class is identified:
Interactive Information Systems. A review of the
approaches to problem solving follows. One language and
development paradigm is considered in further detail. It
is briefly evaluated with respect to how well it - as a
problem solving language - models the application domain
and thereby facilitates problem solution.

LANGUAGES FOR SOFTWARE

We have been using high order languages (HOL) to implement programs for
almost three decades. Three of the earliest languages, FORTRAN, COBOL and
LISP, continue as the most widely used in their respective fields: scientific programs, business applications and symbolic computation. The
success of these languages is obvious; the revolution in computers would
have floundered without this mechanism for translating a problem statement
into a solution.

Language has always provided the mechanism for interfacing between the computational device and its user. Consider Babbage's observations of 1826:[1]

I soon felt that the forms of ordinary language were far too diffuse
... I was not long in deciding that the most favorable path to pursue
was to have recourse to the language of signs. It then became necessary to contrive a notation which ought, if possible, to be at once
simple and expressive, easily understood at the commencement, and
capable of being readily retained in the memory.

This language of signs was a natural (if not obvious) medium for reasoning
based upon abstraction.

When in the 1940s the first digital computational devices were being
developed, there was a tendency to retain the familiar language of decimal
notation. The Mark I, ENIAC and UNIVAC all operated on decimal digits.
Von Neumann's design of the EDVAC combined the concepts of the stored
program with binary arithmetic. A different pattern of communication
emerged; new languages now concentrated on the viewpoint of the machine.
The symbols progressed from ones and zeros to mnemonic codes and then
symbolic assembly languages. The new discipline of programming focused on
how to translate thoughts into processes the machine could perform.

Higher order languages (HOL) offered the promise of realigning that orientation so that the programmer could think of the problem to be solved. The
successful implementation of FORTRAN "provided an existence proof for both
the feasibility and reliability of higher level languages"[2] One can sense
the power of the new tool in the following report by Backus at the 1957
Western Joint Computer Conference.[3]

> A brief case history of one job done with a system seldom gives a good
> measure of its usefulness, particularly when the selection is made by
> the authors of the system. Nevertheless, here are the facts about a
> rather simple but sizable job. The programmer attended a one-day
> course on FORTRAN and spent some more time referring to the manual. He
> then programmed the job in four hours, using 47 FORTRAN statements.
> These were compiled by the 704 in six minutes, producing about 1000
> instructions. He ran the program and found the output incorrect. He
> studied the output and was able to localize his error in a FORTRAN
> statement he had written. He rewrote the offending statement, recom-
> piled, and found that the resulting program was correct. He estimated
> that it might have taken three days to code this job by hand, plus an
> unknown time to debug it, and that no appreciable increase in speed of
> execution would have been achieved thereby.

In this example, the problem statement was reduced from 1,000 lines to 47
lines. The new language allowed the user to describe the application in
the problem domain rather than the implementation domain. For simple problems, the FORTRAN statements were adequate. As the problems became more
complex, however, the programs grew in size. As a result, the orientation
again turned from the problem statement to implementation considerations.

Soon, applications consisting of millions of lines of HOL statements were
developed. Failures in the effective management of the development process
led to the NATO Conferences in Software Engineering (1968, 1969) and an
examination of the software process. The focus now was on the production of
system solutions to stated needs: a very complex form of problem solving.

After fifteen years of studying software engineering, some characteristics
of the system development process are well established:

-    Productivity is a function of project size. Individual produc-
     tivity declines as project size increases. Jones has illustrated
     this in commercial and computer systems applications.[4] Boehm has
     computed the relationship between effort (E in man months) and
     lines of code (L) as

$$E = 2.8 \ L^{1.2}$$

     for complex embedded applications.[5] Moreover, individual produc-
     tivity as measured in lines of code seems to remain constant
     independent of the language used.[6]

-    Most errors in large software systems are made during the early
     stages of development. Further, the majority of errors involve
     design and not programming. In 1976 Boehm reported that the ratio
     of design to coding errors generally exceeds 60:40.[7] A more
     recent analysis of errors in a large project reported the
     following distribution:[8]

36% -Functional specification incorrect or misinterpreted
16% -Mistake in control logic or computation of an expression
     component
12% -Requirements incorrect or misinterpreted
12% -Clerical error
24% -All other errors

- Systems are not static. Lehman has studied how large systems
  change and has formulated laws for program evolution.[9] Lientz and
  Swanson have studied commercial systems and have found that two
  thirds the life cycle cost is for maintenance.[10] Furthermore,
  over half the maintenance cost is for enhancement, i.e., solutions
  to previously unidentified problems.

If we select as our problem solving domain the meta problem of system
development, then how can the experience of the last decade aid us in
defining a language to deal with this problem? Two obvious solutions to the
problem of developing software are (a) to make the product smaller (thereby
reducing the number of people required and increasing the individual
productivity of those remaining) and (b) to find better ways to describe
the target application thereby (i) reducing the early design errors and
(ii) facilitating product evolution. Both of these solutions are difficult
to implement in the context of the traditional large scale system life
cycle. New languages - such as Ada - do not modify the development model,
reduce complexity, nor offer a clearer problem statement. Early detection
of errors continues to rely upon reviews, walkthroughs, and other examinations of descriptive materials: that is, human translation from a cognitive
model of needs to/from a formal statement of requirements.

Yet, there is also a reexamination of the life cycle[11,12] and alternate
development paradigms are being examined.[13,14] Winograd, in a 1979 paper
entitled "Beyond Programming Languages," offered a personal statement.[15]

I believe that the problem lies in an obsolete view of programming and
programming languages. A widely accepted view can be paraphrased: The
programmer's job is to design an algorithm (or a class of computations)
for carrying out a task, and to write it down as a complete and precise
set of instructions for a computer to follow. High level programming
languages simplify the writing of these instructions by providing basic
building blocks for stating instructions (both control and data structures) that are at a higher level of the logical structure of the
algorithm than those of the basic machine.

This view has guided the development of many programming languages and
systems. It served well in the early days of computing, but in today's
computational environment, it is misleading and stultifying. It focuses
attention on the wrong issues and gives the most important aspects of
programming a second-class status. It is irrelevant in the same sense that
binary arithmetic is irrelevant - the things it deals with are a necessary
part of computing, but should play a subsidiary rather than central role in
our understanding.

... We need to shift our attention away from the detailed
specification of algorithms, towards the description of the
properties of the packages and objects with which we build.

Winograd's solution was a new generation of tools where the emphasis would be placed upon "expressing and manipulating descriptions of computational processes and the objects on which they are carried out." Examples of this approach are executable specifications which use an applicative language to formally define the system behavior (see References 13 and 14 for examples). This approach requires very precise specifications which define what the system is to do. Unfortunately, both the behavior specifications of operational development and the procedural specifications of the conventional approach require a translation from the problem statement language to the implementation language. Thus, without debating the relative merits of the two development methodologies, it should be recognized that neither provides a concise "natural" language for problem statement and understanding except for a narrow class of problems.

THE SOFTWARE PROCESS

If we are to address the issue of system development, we must first understand what we mean by the software process. Our view of the process will - to a large extent - define the vocabulary which we use in controlling or directing the process. The most commonly cited view of the software process represents it as a cascading sequence of discrete activities (with feedback and iteration.) This is the so-called waterfall diagram. It implies a specific implementation approach which was derived from experience with equipment development. Figure 1 displays a model of the software process which is implementation independent.

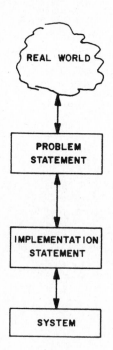

FIGURE 1
THE SOFTWARE PROCESS

Three basic transformations are identified. The first maps a set of needs
in the real world onto a set of requirements called the problem statement.
The second maps the problem statement into a complete and translatable
implementation statement. Note that this transformation assumes the
presence of permissive specifications.[16] That is, not all of the behavior
of the system is defined by the problem statement; thus the mapping is
into, and many mappings are available. The final transformation produces
the system from the implementation statement. This part of the process -
called compilation, assembly, etc. - is well understood and fully automated. Once the system exists, it is imbedded in the real world, thereby
modifying the real world environment (and perhaps the validity of the
problem statement). Thus, the figure represents only one iteration in the
software process.

Figure 2 adds a second dimension to this model. On the left are two concepts which measure the utility of the system. Correspondence measures how
well the system satisfies the needs of the real world environment. The
metrics are nonparametric and can be evaluated only after the system has
been installed. Correctness, on the other hand, measures how well the
implementation statement satisfies the problem statement. Parametric
measures exist, but there is no single measure for correctness beyond the
binary formalism. Moreover, many different systems can be equally correct
with respect to a given problem statement. Correctness and correspondence
are independent of each other; however, management of the software process
links them together.

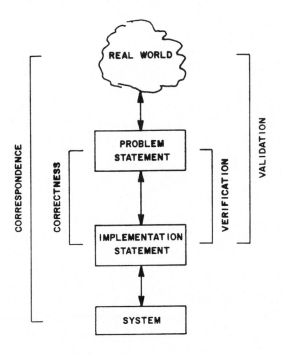

FIGURE 2
THE SOFTWARE PROCESS WITH QUALITY EVALUATION

On the right side of Figure 2 are the two processes which control correctness and correspondence. Verification provides the measure of correctness.
It is initiated with the first formal specification; its goal is to assure
that all derived products are consistent with respect to the higher level
statements. Validation serves as an estimator for correspondence. It
begins before there are any formal statements and continues until the
implementation statement is complete. Validation is the process by which
(a) agreement is reached on the problem statement and (b) the permissible
additions to the implementation statement are accepted. Stated another
way, verification removes objective errors and validation eliminates
subjective errors. Clearly, only verification can be automated. However,
automated tools can be used to assist in the management of validation.

Each of the three transformations shown in these two figures represents a
problem solving activity. The first, problem statement, relies upon an
understanding of both the application domain (in the real world) and the
potential capability of a system. The process is called analysis and the
activity is essentially a cognitive process. The major danger is that the
problem statement does not match the real world needs. We call this
application risk - that the correct system will not correspond. Risk is
lowered by modeling, simulation, prototyping and study.

The second transformation, that which produces the implementation statement, entails most of the development life cycle activity: design, sequence
of development steps with the output of each step at another linguistic
level,[17] e.g., module description, detailed design, PDL, code. Much of the
process can be supported by software tools. The elements of risk here are
that noncorresponding behavior will be introduced (application risk) or
that because of technical, management, financial, and/or schedule constraints it will not be possible to produce the desired product (implementation risk). The final transformation, production of the system, is fully
automated. (A broader, and more accurate, definition would include issues
of training, installation, documentation, etc. which this model avoids).

Figure 3 displays the state-of-the-art in terms of the two elements of
risk. Projects high in both dimensions of risk are beyond the state-of-
the-art; simplifying assumptions must be made. When viewed from a problem
solving perspective, it can be seen that there are qualitative differences
between projects which are high in one dimension of risk. High application
risk projects involve acquisition of knowledge of the real world and the
structuring of that knowledge to guide the software process. This typically
involves considerable social interaction, and natural language is the key
media for communicating and formalizing knowledge. High implementation
risk projects, on the other hand, begin with well defined statements of the
problem. In this case, problem solving relies more upon technology knowledge and the language has greater reliance upon the abstractions,
symbology and jargon of the target technology domain.

There are significant differences between products at the two extreme
dimensions of risk. It would serve little purpose, therefore, to consider
the role of language in problem solving without distinguishing between the
two. Since projects of high implementation risk are deeply embedded in the
knowledge domain of their technology, the considerations of language and
problem solving will be closely bound to similar issues which are
restricted to that technology. Consequently, in what follows we shall

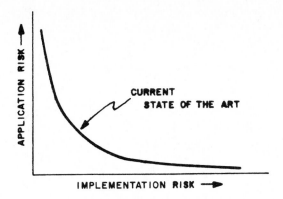

FIGURE 3
TWO DIMENSIONS OF RISK

limit our discussion to the development of projects in which application
risk is a significant factor.

An ideal environment for developing such systems would be one in which the
user could briefly state - in terms appropriate to the application environment - what was desired. That is, he would create the problem statement by
describing the real world needs. The system would identify inconsistencies
with respect to existing definitions. If correct, it would transform the
statement into some executable objects (a) to be tested as a prototype or
model, or (b) to be used as the final object. Iterations of this process
would assure correspondence. Such a language might use statements in a
traditional HOL block structure format, or it might use text, icons, and
alternate input/output media. The key requirement, however, is that the
language should represent a compact description of the problem to be
solved. Where the problem to be solved is highly procedural, e.g., an
interactive information system, one should expect the language to be
procedural. Conversely, where the problem to be solved is state dependent,
e.g., a message switching system, a nonprocedural, logically explicit
language is preferred. Thus, the problem statement language will depend
upon the vocabulary of the problem.

INTERACTIVE INFORMATION SYSTEMS

In order to provide a sharper focus for this discussion, we shall consider
only one class of application: interactive information systems (IIS).
These systems vary in scope from microcomputer applications to large
hospital information systems. There are many problem oriented languages
for this application class. Electronic spread sheets, word processors, and
similar work stations are all examples of problem oriented languages
designed for a single user. Another class of "programmerless language" is
the database query language. Systems such as QBE provide a very natural
method for specifying a potential solution to a problem.[18] OBE integrates
the functionality of the database query with text and graphics processing.[19] In each of these examples, the user is given a tool designed to
process statements which model the problem domain.

User interaction with larger systems requires tools of greater complexity. A class of tools called Fourth Generation Languages has been developed for commercial applications. The name implies that they are an order of magnitude more effective than the COBOL language they are intended to replace. Recall that COBOL was designed to be self documenting; that is, the code was intended to act as a problem oriented description. Its verbosity, the growth in application size, and complexity in data structures have limited COBOL's productivity. Thus, there was a ready market for a new class of languages which could suppress housekeeping and provide a more concise description of the problems to be solved. There are many FGLs; most are proprietary products. Typically, they trade off performance against productivity. In general, they do not have the flexibility of a general purpose programming language, although almost all can be interfaced with custom coded programs.

In each of the above examples, a tool is provided to the user which allows him to translate a problem oriented request for information into a statement processible by the tool. To apply these tools, some users need to develop new concepts such as a file or data type, but in most cases the concepts are presented in the context of their current schemata. Some of the tools are designed for the novice; they may use touch screens, icons, and cursor control devices to structure the language. Other tools, such as the DBMS query language, are designed for both casual users and computer professionals.

A LANGUAGE FOR INTERACTIVE INFORMATION SYSTEMS

The need to describe this application class has resulted in a diversion. This paper is concerned with the characteristics of a language used to implement systems. That is, in the current context, what are the characteristics of a problem oriented language for developing an IIS? This is quite different from the issue of developing a language to use an IIS. Yet one would expect both languages to have several common characteristics:

-   Designed for a specific user group: In this case, the users are professional analysts who understand the application domain and also are familiar with systems analysis, database structures, computer systems, etc.

-   Oriented to a specific problem class: In this case, the time sequenced activities of an IIS. This requires tools for the definition of data structures, the description of processes, and the identification of data-process interactions.

-   Embedded in a conceptual framework: In this case, there are many conflicting and overlapping methodologies for system design. The language designer must choose a philosophic orientation and strive for consistency. The choice is subjective and only can be evaluated pragmatically; consistency should be demonstrable.

As a first step in defining the requirements for an IIS definition language, it will be useful to introduce the concept of style. A style defines all processing conventions for a system: input, output, error processing, message formats, screen management, etc. For example, Martin, in his book Application Development Without Programming, shows how a three line QBE query can search a database and produce a report.[20]   He follows this example with the text of a 364 line COBOL program which will print the same report. QBE style allows the user to produce the report in the given format; to request alternate formats, one might require changes as extensive as the COBOL program. The word processing and automated spread sheet systems also have a style; one cannot use the tool outside the context of its established style.

Every language has its style. The style translates brief statements into more complex functions. The choice of style depends upon the application domain. For example, FORTRAN has a style which makes the printing of real numbers in scientific notation (E format) very easy; FORTRAN makes the listing of leading zeros for a number more cumbersome. COBOL has exactly the opposite characteristics. Pascal supports recursion; it is less modular (in a subroutine sense) than FORTRAN. Thus, a language style defines what functions it can perform with minimal housekeeping. Housekeeping represents the implementation specific instructions required to make a language support functions which are not embedded in its style. A word processor is an example of an environment with a great deal of functionality built into its style with little flexibility beyond the defined features. An assembly language represents the opposite extreme: the style implies very little, and neither does it inhibit functionality. Examinations of programs have shown that over 95% of the lines of code may be required for housekeeping.[21]

One goal in language design, therefore, should be to establish some style which reduces housekeeping with respect to the problem domain while also allowing sufficient freedom to solve problems which cannot be completely stated within the adopted style. With this as a fundamental principle, we now add some problem domain specific (philosophic) requirements.

1. The language must provide an easy to learn method for describing both the IIS structure and the functions of its implemented objects.

2. The language should be style-independent wherever possible. That is, lower level style conventions should be embedded in a knowledge base used by the language. That knowledge base, i.e., style, may be modified for different target environments independent of the language.

3. The language must contain statements in both human readable and machine readable form. Separate representations will be required to manage the descriptive text used in communication with the users (the problem statement) and the more specific statements used in implementation (the implementation statement). All language components should be managed in an integrated design data base (which preserves the links between the problem and implementation statements).

4. The statements which represent the implementation objects should be housekeeping free. Redundancy should be avoided; access to information already defined in the design database is assumed. All specifications should be minimal in the sense that they represent the most concise general description of the process to be implemented.

AN EXISTENCE PROOF

A problem oriented development environment, TEDIUM (TEDIUM is a trademark
of Tedious Enterprises, Inc.), follows this philosophy. It has been used
since 1980 for the implementation of sophisticated clinical information
systems and has been described in some detail elsewhere.[22,23,24] In what
follows, we outline some of its characteristics which might provide insight
into the requirements for a system development/problem solving language.
We offer it as an "existence proof" that such a problem solving environment
can be implemented with today's technology.

The unit for definition in TEDIUM is the application - not the program.
Within an application, there are both descriptive text and specifications
for implementation. The former are optional; generally they are completed
for use as part of the final documentation. (Since all design information
is maintained in an integrated database, the higher level text descriptions
are combined with the help messages to provide the bulk of the user documentation.) It is possible to create an application without using the
design and descriptive facilities of the language. An illustration of how
these tools are used has been documented.[25]

The implementation specifications are defined in the form of program specifications and a relational data model. The data dictionary of the data
model includes all validity tests and external descriptive materials, e.g.,
help messages. The program specifications are divided into two categories:

-    Generic specifications. These perform a single well defined
     function. Examples are a file management program which is fully
     specified when given the name of a table (i.e., relation). A menu
     program is fully specified when given a set of triples consisting
     of <text, input, action>.

-    Common specifications which are written in a housekeeping free
     procedural language.

All generic programs may be modified by replacing predefined subfunctions
with common commands.

Some examples will help. Consider a table called PATIENT

                        (ID,NAME,AGE,RACE,SEX)

The specification of a program to perform the file management functions for
that table which can add, edit, list and delete entries would be defined
as:

                EDIT_PATIENT ENTRY program for table PATIENT

The style defines the user interactions, the functions performed, etc. To
list the data in the table, another generic program (also defined within
the style) is available. To print a projection on PATIENT in a specific
format, one might write the following procedural specification (i.e.,
common program)

    HEAD   Page headings as desired
    For Each   ID in PATIENT

```
Get        the desired attributes from PATIENT
WRite      the line in desired format
END
```

(The format is not as shown; however, the TEDIUM command statements are in
one to one correspondence with this example.)

Generic programs may be modified by the inclusion of TEDIUM commands.
Thus, the generic report may be modified not to list the patient name, or
the generic entry program may be modified to prompt for the elements in a
different order. That is, the style and generic programs provide a convenient baseline. The command language allows the user to augment this
baseline in order to produce a more customized application. The cost to
produce a custom product is increased housekeeping. (The modification of a
generic program is similar to the concept of defining a subclass in
Smalltalk[26] except that there is only one level of inheritance.)

As noted, all procedural statements use the TEDIUM command language. Here,
too, housekeeping is held to a minimum. The statement to prompt for an
input of ID, perform all error tests defined in the data dictionary, and
respond to help messages is:

        Input ID

As a second example, consider the command to print the prompt

        Add Edit Delete List

and then continue processing only with a return of A, E, D, or L. The
command also replaces a null entry with the first option (A), and manages
help requests or a break in the flow. It is written:

        PRompt Add Edit Delete List.

Note that in each of these cases the statement describes the user perspective. Moreover, it is minimal in that no shorter statement could convey
the same information. (The lower case "nput" and "ompt" are noise added
for readability.)

Once a program has been specified, an executable program can be generated.
Generation uses the encoding of the style, the specification, and other
data in the design database. The specification language is implementation
independent, and alternate styles can produce different target languages.
In this way an executable object can be created and tested as soon as the
problem statement is complete for that object. The programs may be treated
as a prototype or model. Programs may be discarded or altered until they
satisfy the desired objectives. Once a program is accepted, it will be
complete in the sense of performing all functions defined as implicit in
the style; it will also be consistent with all other definitions maintained
in the design database.

From these brief examples it can be seen that TEDIUM satisfies the requirements previously outlined for a system development language.

  -     It is concise. Within the context of the style, all specifica-
        tions are minimal.

-    By use of modeling and prototype evaluation, the designer is given
     an early opportunity to evaluate the validity of the design.
     (Since the programming transformations are performed by the
     generator, verification is limited to those functions not already
     embedded in the style.)

-    The language describes entities which are natural to a designer.
     The availability of the design database provides online access to
     all definitions; the integration of text and data supports the
     preparation of documentation. The implementation statement is
     retained at a high level, thereby facilitating evaluation.

To this point the discussion of TEDIUM as a system development, problem
oriented language has been descriptive, embedded in philosophy, using
informal concepts such as style. Can this be evaluated? First, there are
some pragmatic measures: TEDIUM has been used to implement and maintain
several large sophisticated information systems (including TEDIUM).[27,28]
Performance of these products is considered to be comparable to custom
coded programs using the same target language. Second, TEDIUM has been
used by over one hundred individuals. Installations exist in four states,
the District of Columbia and Australia. These facts qualify it as a viable
prototype worthy of further study.

There have been several analyses of TEDIUM productivity. One study
compared "lines of code" for several projects and several languages and
found that TEDIUM was four times more compact than MUMPS and twenty times
more compact than COBOL.[29] Another benchmark was run against a well documented student exercise.[21,30] The application definition was about 20% of
the size of the student's HOL product and the difference in effort required
- after adjusting for differences in experience - was between 5:1 to 10:1.

Perhaps a better measure of the effectiveness of TEDIUM in describing an
application is the number of times a program must be modified before it
satisfies the user's needs. That is, how many times must the program be
generated (compiled) before it is correct. Further, once it has been
accepted as correct, how many times must the program be generated to effect
changes during evolution. Clearly, the lower the number of generations,
the better the language is at modeling the problem. In one study of a
production application (101 programs), it was found that an average of 2.7
generations (compilations) was required to test and debug the programs and
an average of 2.1 generations (compilations) was required to modify the
programs through the life of the application.[31] Forty-three of the
programs were generated only once and 47 were never modified. This clearly
indicates that the style embodies an effective representation of the
problem domain.

CONCLUSION

New computer interfaces are altering what we mean by computer languages.
At one end of the spectrum are the tools designed for end users. At the
other end are tools for developing systems. There are several different
paradigms currently under consideration. These include traditional high
order languages, such as Ada, operational development modes using executable specifications, comprehensive environments such as Smalltalk, and the
approach presented here: development of a specialized language for describing applications in a problem domain. Each of these paradigms presents a

different view of the implementation process, and each uses a language suited to that problem view. It is not clear that TEDIUM represents more than a simple extension of the user oriented tool paradigm. Nevertheless, it does provide a useful object for the study of how languages can be applied to the problem solving aspects of system design.

REFERENCES

[1]   Babbage, C., On a Method of Expressing by Signs the Action of Machinery, 1826, cited in B. Shneiderman, Software Psychology, Winthrop Publishers, 1980, p.65.

[2]   Wegner, P., Programming Languages – the First 25 Years, IEEE Trans. in Computers, (C-25,12) 1976, p.1209.

[3]   Cited in J. Bernstein, The Analytical Engine, Random House, NY 1966, p.74.

[4]   Jones, C., Program Quality and Programming Productivity, IBM Technical Report TR 02.764, 1977, pp.42-78.

[5]   Boehm, B. W., Software Engineering Economics, Prentice-Hall, Inc., Englewood Cliffs, NH, 1981, p.75.

[6]   Wasserman, A. I. and S. Gatz, The Future of Programming, Comm ACM (25,3) 1982, pp.181-206.

[7]   Boehm, B. W., Software Engineering, IEEE Trans. in Computers, (C-25, 12) 1976, p.1230.

[8]   Basili, V. R. and B. T. Perricone, Software Errors and Complexity: An Empirical Investigation, Comm. ACM (27,1) 1984, pp.42-52.

[9]   Lehman, M. M., Programs, Life Cycles and Laws of Program Evolution, Proc IEEE, (68,9) 1980, pp.1060-1076.

[10] Lientz, B. and E. Swanson, Software Maintenance Management, Addison-Wesley Publishing Co., 1980.

[11] McCracken, D. D. and M. A. Jackson, A Minority Dissenting Position. Systems Analysis and Design – A Foundation for the 1980's, W. W. Cotterman et. al. eds, Elsevier North-Holland, NY 1981, pp.551-553.

[12] Blum, B. I., The Life Cycle – A Debate Over Alternative Models, Sofw. Eng. Notes (7,4) 1982, pp.18-20.

[13] Balzer, R., T. E. Cheatham, Jr. and C. Green, Software Technology in the 1990's: Using a New Paradigm, Computer (16,11), pp.39-45.

[14] Zave, P., The Operational Versus the Conventional Approach to Software Development, Comm ACM (27,2) 1984, pp.104-118.

[15] Winograd, T., Beyond Programming Languages, Comm ACM (22,7) 1979, pp.391-401.

[16] Turski, W. M., On Permissiveness of Specifications, Software
      Process Workshop (in press).

[17] Lehman, M. M., Stenning, V., and Turski, W. M., Another Look at
      Software Design Methodology, Imperial College of Science and
      Technology, Research Report DoC 83/13, 1983.

[18] Zloof, M. M., Query-by-Example: A Data Base Language, IBM Systems J.
      (16,4) 1977, pp.324-343.

[19] Zloof, M. M., Office-by-Example: A Business Language that Unifies
      Data and Word Processing and Electronic Mail, IBM Systems J. (21,3)
      1982, pp.272-304.

[20] Martin, J., Application Development Without Programmers, Prentice-
      Hall, Englewood Cliffs, NJ, 1982, pp.31-37.

[21] Boehm, B. W., An Experiment in Small-Scale Applications Software
      Engineering, IEEE Trans. S.E. (SE-7,5) 1981, pp.482-493.

[22] Blum, B. I., A Tool for Developing Information Systems, H. O.
      Schneider and A. I. Wasserman (eds), Automated Tools for Information
      Systems Design, North-Holland, New York, NY, 1982, pp.215-235.

[23] Blum, B. I., An Information System for Developing Information
      Systems, National Computer Conference, 1983, pp.743-752.

[24] Blum, B. I., Three Paradigms for Developing Information Systems,
      Seventh Inter. Conf. Software Engineering, IEEE Computer Society
      Press, 1984, pp.534-543.

[25] Blum, B. I. and C. W. Brunn, Implementing an Appointment System with
      TEDIUM, Fifth Annual Symposium on Computer Applications in Medical
      Care, IEEE Computer Society Press, 1981, pp.172-181.

[26] Goldberg, A. and D. Robson, SMALLTALK-80: the Language and its
      Implementation, Addison-Wesley, Reading, MA, 1980.

[27] Lenhard, R. E., Jr., B. I. Blum, J. M. Sunderland, H. G. Braine, and
      R. Saral, The Johns Hopkins Oncology Clinical Information System, J.
      Med. Sys. (7,2) 1983, pp.174-174.

[28] McColligan, E. E. The Core Record System, B. I. Blum (ed), Informa-
      tion Systems for Patient Care, Springer-Verlag, NY, 1984, pp.260-269.

[29] Blum, B. I., MUMPS, TEDIUM and Productivity. First MEDCOMP, IEEE
      Computer Society Press, 1982, pp.200-209.

[30] Boehm, B. W., T. E. Gray and T. Seewoldt, Prototyping vs. Specifying:
      A Multi-Project Experiment, IEEE Trans. S.E. (SE-10,3) 1984, pp.290-
      303.

[31] Blum, B. I., System Sculpture, Correctness and Correspondence: A Case
      Study (draft paper).

QUESTION AND ANSWER PERIOD

BARSTOW
Do you have anything to say about the efficiency of the code that TEDIUM produces?

BLUM
Yes. It is essentially as efficient as custom code. The only difference that we can see in looking at it is that it is about 20% larger. But if you stop to realize that the largest partition we use is about 7K characters, that is not too much of a problem. People outside of Johns Hopkins also produce code that they perceive to be as efficient as custom code.

BARSTOW
Another way to ask the same question might be, could two programs which are in the TEDIUM language and have a very similar appearance end up being implemented in very different ways?

BLUM
You have several choices. There are generic programs. For example, there is a generic program that does the add, edit, delete: the standard file management functions. You can generate that simply by specifying "Entry program for this particular relation or table." That will create a reasonably large implementation with a lot of error checking that you may not necessarily need. You could also do exactly that same sort of thing if you knew that certain chunks of the entry program were not required for your application or if you wanted to do things in a slightly different way. In this case, you may find it easier to write and maintain it as a common (non-generic) program. The two implementations would have different specifications, but the same general functionality would be expressed. The guidelines I give to all the users are that they are going to spend more of their time maintaining the system than they will spend in writing it the first time. Thus, I suggest that they choose an approach which is appropriate for understanding what they are doing which will be meaningful to those who follow.

WEINTRAUB
That is a general rule.

BLUM
Yes. I can not tell all the people what to do, but I can sure tell those who work for me. They don't listen either. (laughter)

GUIER
I believe you said you implemented this in MUMPS. Do you have any estimate on how difficult it would be to implement it in something else, like PROLOG or APL?

BLUM
There are two things. One is the language that the system runs in and the second is the target language that it generates. The TEDIUM specification language itself has nothing to do with the target implementation language. This next calendar year I am going to rewrite the generator in TEDIUM. (Everything in TEDIUM now is written in TEDIUM except for the generator and a few utility programs.) When I do that, I will also build into it the ability to modify the style and the target language. At that point

virtually all of TEDIUM (except its utilities) will be is written in
itself. Then I will be able to regenerate TEDIUM in some other target
language.

GUIER
You mean almost like a SYSGEN or something of that sort?

BLUM
Yes, sort of like that. How well TEDIUM will work in some other target
language is hard to say at this time. However, it also turns out that for
under $10,000 one can get a multi-user MUMPS system. In that case, one can
simply treat the system itself as a processing environment or analyst work
station. It doesn't really matter what language the system is written in,
the only important factor is the target language in which one wishes to see
the output code.

The Role of Language in Problem Solving   I
R. Jernigan, B.W. Hamill, and D.M. Weintraub (Editors)
© Elsevier Science Publishers B.V. (North-Holland), 1985

INVITED ADDRESS:

OVERCOMING LIMITATIONS IMPOSED BY CURRENT PROGRAMMING LANGUAGES

Ben Shneiderman

University of Maryland
Department of Computer Science
College Park, Maryland 20742

My topic today (which was requested by the organizers of
this conference) is "Overcoming Limitations Imposed by
Current Programming Languages". I want to comment first
on some ways to study programming languages, and then to
pursue the main topic, first by looking at what is
attractive about current programming languages, and then
by looking at some of the problems and suggesting some
solutions. I'll close with a couple of provocative ideas
and the beginnings of something of a theory about how
people think about programs and how they create programs.

THE STUDY OF PROGRAMMING LANGUAGES

The idea of studying computer programmers experimentally goes back at least
15-20 years; to date, there have been a large number (perhaps hundreds) of
experiments which focus rather narrowly on specific issues within programming. I'm very pleased to see that there is a continuing growth of
interest in this (we are planning a conference in 1986 about empirical
studies of programmers). There are a lot of small issues and some major
issues which have been studied [1, 2]. Lots of papers have appeared
examining indentation strategies or control structures (for example, what
kind of control structures are winners and which ones are losers); more
recently, some interesting papers have appeared about data structures,
attempting to see what kinds of structures are comprehensible and which
ones are not. People have done experimental studies about design and
different phases of programming, using novice programmers, non-programmers,
and professionals. What is satisfying to me about this kind of work is
that it gets past the arguments about "my language is more 'natural' than
your language" or "it's 'better' than that language", and instead applies
traditional scientific methods to understanding programming behavior.

The list of research paradigms starts with what a psychologist might call
"introspection": you sit quietly in your office and invent a new language
construct or guideline, and that is where a lot of important good ideas
begin. I wanted to go further than that, go beyond the point of arguing
about whose idea is better, and instead try to apply techniques of
controlled experimentation and data collection, to try to find out what
really happens when people write programs. For example:

-    how often do they use certain control structures or data
     structures,

- are flowcharts or pseudocode helpful,

- what is the frequency of error when certain constructs are used,

- what happens to the frequency of errors when the length of modules changes?

In fact, rather surprising results have come out of experimental research which show that the folklore of programming is not always supported by the evidence!

What I would like to promote most strongly here is the notion of controlled experimentation. I favor the psychological flavor of those studies: You start from a lucid and testable hypothesis (something small, where you can be precise, and other people will agree as to what you are talking about). You alter a small number of independent variables and measure a small number of dependent variables. Each little experiment (and some of these seem like picking at small points) does, I believe, provide us with a clear cut result. Just like a tile in a mosaic; by itself, it doesn't "save the world", but many, many hundreds of these tiles add up to an emerging image of human performance. This image leads us not only to explanatory theories, but also (hopefully) to predictive theories. In the end, we should be able to take a new programming language and say, for example, that for this community of users for this kind of task, this language should lead to about 6.4 errors per thousand lines of code, but if you change the command structure thus and so, it might bring it down to 4.7 errors per thousand lines of code. We would like to get there - we're still far, but we would like to get closer. This is the traditional reductionist approach of the scientific method, and I find it is the path to rapid progress in trying to understand the nature of human effort in programming.

EVALUATING CONTEMPORARY PROGRAMMING LANGUAGES

The efforts that we and others have applied have led us to some understanding of the successes of contemporary programming languages, and also of some of the weaknesses, some of which we see how to overcome. I'll talk about the modest refinements, extensions and improvements that we might apply (or that we still need to test in some cases), and then I would like to gamble at offering some more important directions of change including my particular interest (an idea that I call direct manipulation) which has been referred to by several of the speakers in this conference already.

SUCCESSES OF CONTEMPORARY PROGRAMMING LANGUAGES

As I dealt with the (rather negative) topic of "the limitations" I thought I owed it to myself to think first positively and to ask what were the things I admired about contemporary programming languages. I came up with this list (you may be able to help me and add some more items or you may disagree with some of these):

- Precision
  The first thing that strikes me is precision, and that it is a great gift. We see other notations that have brought precision (such as music notation, chemistry notation, knitting instructions, mathematics). There is a great power in precision because

it helps us think, it helps because we can write down our ideas. When we can realize our ideas on paper, reflect on them and preserve them, we have great power.

- Effective Means of Human Communication
Contemporary languages provide an effective means of communicating among people. If we didn't have musical notation and I had to explain the symphony each time or play it for you each time, it would be a more difficult world.

- Power
It is remarkable what is accomplishable in contemporary programming languages, even the older ones such as FORTRAN or COBOL. There is great diversity of expression, there is great capacity to manipulate numbers, strings, elaborate data structures, and actually to influence devices in the physical world like printers and display screens. Those are only the beginning of what we're going to see happening in the way programs will influence the real world.

- Permits Modular Decomposition
How wonderful it is that each programming language (some better than others) supports the modular decomposition of programming. This enables several things:
 1. Individuals can work on large programs by breaking them down into smaller sequences of programs.
 2. Groups can work together (and that is quite remarkable to me).
 3. We can have style guidelines so that groups who compose independently will actually compose fairly similar structures.
 4. We can actually comprehend large programs written by other programmers.

- Extensible
Extensibility (some languages more than others) indicates to me a remarkable wisdom by the creators of languages. They enabled programmers not only to write programs, but to make it easier for others to write even larger programs. The simple notion of FORTRAN subroutines was a great idea, and the more elaborate structures that ADA supports are even more exciting. Some languages even permit the redefinitions or addition of syntax. Sometimes too much extensibility is the problem, since there is a danger of people extending in the wrong way, but still the notion that languages are extensible is to me a very important contribution.

These successes represent a great advance. These powers are implicit in all that we do in programming, but to understand the directions that we want to go, I think we have to focus on these successes.

WEAKNESSES OF CONTEMPORARY PROGRAMMING LANGUAGES

Contemporary programming languages have weaknesses (and very serious ones,) and this is what leads us to hear complaints about the software crisis and the problems in programming:

-   Error Prone
    First and foremost I think the error proneness of most programming
    languages is disturbing. Errors do occur, and they occur in
    places where, in spite of the best efforts of the brightest people
    to eliminate the errors, the errors remain. This is a serious
    problem; one of the questions we should be looking for is how to
    reduce the rate of errors in the use of programming language.

-   Sometimes Tedious
    I quibble with myself about the word "sometimes" (maybe it should
    be "sometimes not tedious"). More often than not, I find that
    it's a great effort to write a lot of syntax and to make sure I
    get my "BEGINs" and my "ENDs" together, and be to sure that my
    semicolons are there. There are a host of details that have to be
    attended to. There is an opportunity to reduce the tediousness
    while preserving the power and all the features that we want.

-   Requires Substantial Training
    The substantial training required is another problem of
    programming languages. Learning your first language, learning
    your second language (but then also learning the additional
    features within programming languages), takes a long time. You
    have to learn not only the programming language constructs, but
    the ways in which you can put them together. For example, an APL
    programmer may know all of the constructs but be a very poor
    programmer because he/she is not fluent in the "idioms" (as Perlis
    has called them) of APL, or the plans, templates, structures,
    schemata of programs. How can we make those plans, templates,
    schemata more obvious, less difficult to discover? Recall the APL
    example of this morning, "A Pictorial Language" of Ross
    Bettinger. The substring match was an example; I would never
    discover the way to do substrings in APL, yet it is one of the
    central features, one of the procedures or schemes that are used
    frequently in that kind of APL programming.

-   Insufficient Level Structuring
    In spite of the efforts to permit modularity and in spite of the
    efforts to allow higher level abstraction, I find languages
    extremely weak in the definition of level structures. For
    example, there is only one form of conditional, the "if" state-
    ment, and that is supposed to take care of very high level
    predicates as well as very low level details. I don't see
    sufficient features in a language that allow me to write a high
    level specification and a very different language for the low
    level detail.

-   Weak Facilities for Checking Correctness
    There are some efforts to include language features which are
    integrity constraints or predicates which are checked, but I still
    think that we could do a lot better in adding features to insure
    correctness.

-   Difficult to Maintain or Revise Program
    This comes back to issues that were mentioned many times at this
    conference, that the bulk of effort in programming is spent in

maintaining code. Can we find ways in which we will make it easier for people to read programs and then modify them?

MODEST REFINEMENTS, EXTENSIONS, AND IMPROVEMENTS

I'll start with my list of possible directions and suggest some modest improvements.

PURSUE STYLISTIC ISSUES

Stylistic issues I think represent small changes, having to do with the surface structure (the "cosmetic issues").

- Commenting
  Think about the commenting style of programs. What kind of comments do we put? In one study we found that high level block comments at the beginning of a module turned out to produce a higher level of comprehension than many comments sprinkled through the program. We traced the effect to the level of the comment, in terms of the knowledge conveyed to the user. I believe comments should be at one level of abstraction above the level of the code. For a knowledgable programmer it does very little good to repeat the function of the code in a comment.

  Another problem with too many comments is that they lengthen a program and lead to clutter, making it harder for the programmer to study the program. Increased page flips are a further impediment. This issue becomes a serious problem when working on a screen with only 24 lines and 18 of them are comments! You will find yourself scrolling back and forth. How nice it would be if we could issue a command to eliminate all the comments once we've read them and then just study the program. This is a simple notion, once you understand the problem that exists.

  In one of our research projects we've built a system with two screens: one screen has a program, the other screen has the comments. As you scroll the program, the comments scroll to keep up, but they are off on the side and you don't have to look at them. If you want a third screen that has flow charts, or pseudo code or whatever, and that scrolls to keep up as well, then go ahead and add it! It may be advantageous. That is the kind of study we're doing right now, to to see if multiple displays which are synchronously scrolled are an advantage; potentially they are also a distraction (every idea has positive and the negative aspects).

- Mnemonic Variable Names
  We (and others) have found that there are interesting patterns associated with mnemonic names. Meaningful names do facilitate comprehension. Too long names get to be a problem. Too short names can be confusing if you have more than 20 names in a module (for up to 15 names, people seem to make do with whatever bizarre names you supply). Single letter names can be an advantage in some cases.

- Indentation
  This is another cosmetic issue; it has a very modest effect (less
  than some of the advocates would have you believe). There are
  negative effects of indentation as well. Certainly, modest levels
  like two space indentations are winners, six spaces gets to be a
  problem.

- Use of Control and Data Structures
  I don't want to dwell too long on this area. Yes, it does matter
  if you use nicely formatted programs without "go-to's" or with
  one-in and one-out control structures. That does help, and
  choosing the right data structure for the problem does make a
  difference: A recent study showed that a pointer-linked tree
  structure turned out to be more comprehensible than an array
  structure, because the problem that was dealt with was more
  natural, more convenient in the pointer linked-tree structure than
  in an array situation [3].

- Modular Design and Parameter Passing
  There are some interesting results that show a tendency towards
  excess modularity in some programs. People are insisting that
  modules should be one page long. But by adding lots and lots of
  modules you do risk additional errors: the overhead involved with
  invoking the function, passing the arguments, returning from it,
  returning the arguments, etc. Some NASA and IBM data are among
  the mounting evidence that (for professional programmers) somewhat
  longer modules (on the order of 200 to 300 lines) have produced
  the lowest rate of errors per line of code. There are interesting
  questions here, because we still don't quite understand how people
  organize programs: long linear programs that are well organized
  turn out to be very good idea! But unfortunately, in many sites
  there is a rule that says "no more than 100 lines per module," so
  programmers arbitrarily break the program, and that disrupts
  comprehension.

  We found in a little study that adding blank lines enhanced
  comprehension of the program. We took a 70 line complex
  program. One group received that, and another group received the
  same 70 line program but 5 blank lines were inserted to show the
  module organization. The extra blank lines made a significant
  difference in the comprehension scores for that group. This
  strong result has been replicated by another student.

EXPLORE NOVEL CONSTRUCTS

Many novel constructs have been proposed:

- Enhanced Data Types
  Enhanced data types (and more elaborate structures) are an
  interesting idea, as long as those data types are appropriate to
  the application domain. There is a danger in allowing programmers
  to create very complex data types which are not naturally part of
  the application domain.

- Higher Level Control Structures
  I find myself limited by looking at the "if-then-else" and "do-while," and want to see bigger patterns that control the flow of programs. Proposals for elaborate CASE, SELECT, or SEARCH control structures might be worth exploring.

- Plans, Specifications, Integrity Constraints
  Here I would encourage you to read the work of Elliott Soloway of Yale University, who has been trying to make a catalog of the plans that people use in their understanding of programs as well as in their composition of programs [4].

DEVELOP NEW ASPECTS OF PROGRAMMING

I think there is an opportunity here to develop and make a real contribution by bringing programming to new application areas.

- Concurrent Programming
  New programming languages such as MODULA-2 and ADA offer interesting constructs for creating concurrent programs.

- Rule-Based Systems
  Novel rule-based systems such as OPS5 support the expression of complex decisions such as in expert diagnostic systems.

- Graphics Programming
  I think there are cases where there is an opportunity to do much better than trying to use the crude versions of BASIC, FORTRAN or other languages, and twist them around to do concurrent programming, rule-based programming, or graphics programming. I think we are yet to see good graphics programming languages or sound programming languages.

- Sound Programming
  I'm still disappointed every time I look at the proposal for programming languages to program sound: I see just a flurry of numbers on the screen, for example, "sound, parenthesis 3 comma 4 comma 19 and minus 1." That is not a language in my mind (it's not much better than reading it in binary, right?).

- Interactive Programming
  It is still a difficult chore to write a complex interactive system in traditional programming languages. There is an, opportunity to create new environments such as Tony Wassermans "Rapid/Use" system [5] which allows you to program interaction, or the work of Rex Hartson at VPI on dialogue management systems [6]. These seem to be attractive areas.

- Control of Devices in Three-Space
  I think a lot of the work in this decade will be making computers control the real world, whether it is controlling automobiles, devices in your house, or "robots", or rather "programmable manipulators" or "flexible manufacturing systems" (those phrases are not as jazzy as robots but I think they more accurately describe what is going on).

There are interesting opportunities. The IBM contribution of the AML language is an interesting situation, but I think there are more possibilities when thinking about programming in 3-space.

DESIGN ADVANCED PROGRAMMING TOOLS/ENVIRONMENTS

This is the next way that I think we can overcome some of the limitations: the environment approach, or "enhanced tools". These might include:

-   Syntax-Directed Editors
    Tim Teitelbaum's work at Cornell is now is probably the most well known example, but there are a lot of projects that will offer novice and professional programmers the chance to create programs in a more effective way with fewer errors and more rapid performance [7].

-   Correct and Complete Operations
    This is tied in with Syntax-Directed Editors, and was alluded to earlier in the talk about LOGRAPH, where it was described how it was not possible to make errors: the system would not allow that to happen. That seems to be a very good principle! Rather than spending time on designing parsers to handle errors and provide excellent error messages that are specific and constructive in positive tones, I think you want to look for the ways to substantially reduce the chance that the user will make an error [8].

    One example of this (the simplest case) I call "Correct Matching Pairs". In slides yesterday, there was an example of the unmatched left parenthesis problem. That is an example of an error that should never occur. How can that never occur? Well... It occurs because we still hang onto the old world of teletypes and we still think of paper, typing one character after another character.

    I think that when you type the left parenthesis, the right parenthesis should appear on the screen at the same time. Your cursor goes between (in "insert" mode), and you can type the contents. If you try to delete either, they both go away. After all, they're really one character, so why should you have to be forced to type them as two?

    I think you can see that that principle (it's a small one) in fact has many little applications, such as PL/1 comments, slash-asterisk, asterisk-slash, etc. How many times (I see some bowed heads here) have you forgotten to close the asterisk-slash so that the rest of the program becomes one great big comment? Well, it ought to be one idea, one indivisible keystroke. Similarly for any pair of correct matching pairs. You can see it in many more places, such as quoted strings. Most word processors have some notation for arranging for boldface (e.g., Control B, string, Control B) - you forget the second Control B, and the rest of the document is boldfaced. Anytime you have a matching pair of symbols, you ought to consider the single keystroke strategy.

Some may say that this is too harsh a rule. There are conditions
where you wouldn't want this to occur. Fine! There are softer
ways to arrange it: When you put the first one up, it blinks
until you put the closing one. Or there is a message on the
screen. Or there is some way that you do not allow that problem
to propagate through the system to a later point.

To repeat, the "correct matching pairs" is to me the simplest
example of a correct and complete operation. There are more
elaborate ones, such as "if-then-else" templates. You ought to be
able to use one uninterruptable operation to drop that in, so that
you don't forget the details.

You might want to do the same thing in air traffic control
systems, where (even today) the level of operation of the
controller is one little step at a time: To raise an aircraft
from 14 to 18 thousand feet, you raise it to 16 thousand feet, and
then you may get distracted and forget to move the plane up to 18
thousand feet. Typical error. You would like to have a way that
the operator can describe the multiple step intention!

Even if you can't complete an operation when you initiate it, you
still need to describe the full intention at the point that you
decide on that action. I think that there are great opportunities
to find ways to reduce the number of possibilities of error.

- Reusable Libraries of Code
  These still have a long way to go. I find the attempts at them
  still very simplistic, without an adequate understanding of what
  the user's position is when trying to reuse code. Just having
  code out there in some vast library does not ensure that the user
  will come along and use it! We need better ways of telling the
  user what the code is, and what it has to accomplish, and giving
  them some assurance that in fact it does work.

  I think the social mechanisms of reusability are not thoroughly
  explored. It is interesting (for example) that the Collected
  Algorithms of the ACM include certifications and other
  testimonies; the successful subroutine libraries are ones where
  you have some assurance that it has been tested, it has been used,
  and if something is wrong, there is someone to go to.

- Program Analysis Tools - Slicing
  This is another popular topic. I cite one example: "slicing" (by
  a colleague of mine at Maryland, Mark Weiser) [9]. This was a
  very clever idea, to be able to have a program analysis tool that
  takes a program and slices it. By "slicing" he means extracting
  from the thousands of lines of code the 30 lines of code that
  affect a particular variable. Wouldn't it be nice to do that
  automatically? The compiler has the information (because the data
  flow graph is in there), and if you can extract it and show it to
  the user, it may offer an interesting way of studying programs and
  finding the errors.

- Verification, Testing and Debugging Tools
  These are another category that many attempt.

- Maintenance Tools - Transformations
  This is a favorite of mine. You would like to, for example, take
  an if-then-else statement and turn it into a case statement with
  two cases, because you intend to add 2 or 3 more cases. Or take a
  do-while and turn it into a repeat-until. Or take some
  mathematical expression, split it, and insert a temporary
  variable.

  Now we get to case like: You see this piece of code? This reads a
  person's weight, does something to it, writes it, copies it, and
  prints it out. Now, let's do the same thing to the height. It
  should be able to generate the code that does the same thing to
  yet another variable. Or: You see this program. Instead of a
  single value for the height, why not maintain the last 12 values
  for the person's height? So turn it into a vector! APL supports
  this (in many cases) automatically, but wouldn't it be nice to be
  able to give an instruction which converts a scalar into a vector
  and maintains it in the proper way?

  There are other maintenance transformations that might be
  possible. The question is, can we find a compact list which is
  meaningful and effective, that covers the space of things we want
  to do. To compile such a list, we have to go back to try to
  understand what the patterns of maintenance operations are. What
  things do people do when they maintain programs? What kind of
  changes are made? You'd like to be able to say, "All right, every
  time a certain variable is used as a divisor, I would like to
  insert a guard statement, an 'if', in front of it." And the
  transformation system will do that in an automatic way. Or you'd
  like to say, "Well, this program does a loop. I would like to
  insert some sort of counter." It would then prepare the
  declaration, suggest places for the initialization, for the
  incrementation for the test, and so on. I think that these offer
  great possibilities to overcome the limitations of programming
  languages.

MAJOR CHANGES (INCLUDING DIRECT MANIPULATION)

The bigger changes are harder to see. Let me gamble on a few.

REPLACING PROGRAMMING WITH TOOLS (EDITORS, SPREADSHEETS)

Some might say that an editor or spread sheet is really a different kind of
programming language. Okay. I could accept that suggestion. But it seems
to me you're really transferring the locus of programming from one kind of
person, "the programmer", to another person that is more "the user". More
and more of what we call programming is being done with powerful, generally
available software tools.

ENABLE USER TO CREATE PROGRAMS

These may not involve complex programming languages. You want to reduce
the need for writing a messy program in an old fashioned language. It was
striking when one university converted from using COBOL on a daily basis to
using a fourth generation language; it just changed the nature of
administrative life at the university. The previous work style was this:

If you needed a new report from the database, you put in a request. The programmers would work on it, and the average time for preparation and test of the new program was 2 months. Unfortunately, there was a 3 year backlog. That was the problem. So when the administrative staff was offered a tool which they could use directly, instead of preparing requests for a few new programs per month, they were (within a few months) writing 1500 queries of their own per month! This changed radically the nature of their use of computers. I think there are opportunities still to do that in many fields.

IMPROVE TRAINING AND MOTIVATION

This is aligned with Barry Boehm's book about Software Engineering Economics [10]. If you remember, the cover of that book showed that the major factor in the development process is the quality and the ability of the people, the programmers, their motivation and their training.

Can we think creatively about motivating programmers in exciting new ways? Do we offer them clinical therapy to help sort out their personal problems? Do we give them management training courses? (Audience comment: Don't ever do that!)

All right. There are some radical ideas about peak experience in programming when people write a thousand lines without error. Can we study how it is that some people seem to accomplish these magical feats? Can we teach others to do the same thing [11]?

DIRECT MANIPULATION

My own interests now turn toward the notion of direct manipulation programming, which I summarize by saying "do it and redo it, rather than talking about doing it." It seems that the experience of programming is to talk about doing something, when it would be a much more natural process to go and do it! It is a lot easier, and the feedback is clearer.

Let me try to make some clarification about what I mean by that (there is a long description of this that I struggled over several years to write, and it appears in the August 1983 IEEE Computer) [12]. I tried to find out and characterize what it was that brought a great sense of enthusiasm about some computing systems. I saw this years ago, when display editors became more common, replacing line editors. My analysis of this suggests that one obvious advantage with a display editor is a full page display of the object of interest. The object of interest seems to be the central force here. You want to see what is happening! With a display editor, you see a full page (hopefully with the right fonts, boldface, underscores, page breaks, and centering), you see everything as it would appear if you pressed the print button. This is the "WYSIWYG" idea: What you see is what you get. The cursor is visible, replacing the implicit line pointer and the narrow cardboard tube view of the line editor. You deal with physical cursor action. Instead of having to remember some sort of command (you know, "down 3, next 4"), you just move the joystick, mouse, or cursor key in a way that seems natural. You seem to be "flying" or "driving" that cursor. It seems to be something that you can control and manipulate.

With display editors, insertion and deletion are done by keystrokes, not by command, not "CHANGE (slash) old string (slash) new string (slash)" (don't

forget the last slash). Instead, you move the cursor to the point you want
and you start typing and the text goes in. Or you press the delete and the
characters start shifting over. Its almost as much fun as Pac Man! The
characters just sort of zoom in from the right side, and hopefully the text
rewrites to keep it current and correct at every point.

A study we did during the summer with fifth and sixth grade remedial
reading kids (and learning-disabled kids) found that one of the central
problems they had was that the cursor's position gave them confusion: with
many text editors, the cursor is on one character and if you press delete,
it will delete the character to the left. This was a severe problem,
especially if their task was to delete the first character in the line
(they would bring the cursor over, press delete, and the editor would
delete the return from the previous line and bring the two lines together,
to the complete confusion of these kids). Even in microscopic details, you
have to try to make the screen show reality as much as possible. The extra
rule that says, "OK, the cursor is here but it is going to delete the
character to the left," is a source of confusion and many errors, in spite
of substantial training and practice.

This direct manipulation idea is applicable in even microscopic ways.
Cursor motion in some editors require "up", "down", "left", "right"
commands. I claim these are more difficult than showing arrow keys. I
favor the four-way arrows, placed appropriately. Jim Foley (my friend and
colleague at George Washington University) ran an experiment with some of
his students and found a 30% increase in errors with word command keys and
linearly-placed direction arrow keys, as opposed to appropriately arranged
directional arrow keys. So it did show up to be a reality (what surprised
me was that the first two categories were approximately the same; it wasn't
just the visual representation, but the physical placement as well.
There's a very strong "cue" to the physical placement).

A joystick is even more direct, and the touch panel ("reach out and touch"
the spot you want) seems to be more directly manipulating the objects of
interest.

There are lots of examples that I've tried to catalog, including:

-    "Form Fill-in"
     Showing on the screen a simulation of the form that is meant to be
     filled in.

-    Query-by-Example
     Moshe Zloof's database facility allows users to fill in relational
     tables with prototypical responses [13].

-    The Spatial Data Management System from the Computer Corporation
     of America, has this scenario: You see a globe of the earth on
     the screen in full color, you zoom in with your joystick on the
     Pacific seeing the Russian and American convoys and zoom in on the
     Russian ships which are red silhouettes. Then the names of the
     ships appear, and as you zoom in on one, the details about the
     ship appear. If you put your cursor on the Captain's name and
     press the button, a full color image of the Captain appears (you
     know, bearded face, etc.) with text below describing the

biography. If it is the wrong one you just pull back and zoom
around to another ship [14].

We've built similar environments with more modest cases, and they
turn out to be very satisfying.

-    Video Games
     I think video games are the best example you have of common and
     very successful applications of these principles. If Pac Man is
     on one side, you just move the joystick to the other and Pac Man
     goes zooming off. It seems clear, natural and predictable as to
     what is happening.

-    Visicalc
     Another one of the great ideas in programming (and a successful
     product, as well) has that flavor; it is not graphics, but it is a
     visual presentation of the world of action. The weakness of
     Visicalc is that the spread sheet model indicating the linkage
     among items is written in a traditional language: the language for
     describing the spreadsheet is not a direct manipulation
     language. The presentation to users has much more the feeling of
     direct manipulation (when I asked Dan Fylstra of Visicorp "Why is
     Visicalc so successful?", his comment was "It jumps!" That
     conveys the sense of the liveliness, the fact that you are
     manipulating this world of linked actions).

-    Car driving is my favorite example of direct manipulation. You're
     driving along and you want to go left, so you turn the wheel
     left. If you've gone too far, you just turn back a little.
     Imagine driving by typing "left 30 degrees". Then you have to
     type another command to see where you are. And then undo....

In the future, the nature of computer use (office automation and
programming) will be much more like driving a car or flying a
plane, much more fluid and continuous.

That raises some good questions. How do we allow higher speed?
How do you build a safe highway for high speed programs?

Some of the direct manipulation ideas are embedded in older
systems (and other areas like CAD/CAM systems), where you see the
object, and you sort of drill a hole and you can see the drilled
hole on it. This is also used in automobile design, architecture
or circuit layout.

Macintosh is, of course, a system which takes advantage of these
ideas. Macintosh does several things very cleverly, like the
rubber rectangle. Once you've selected and you put down the first
point, the rectangle expands and contracts as you move it. We
found that this is dramatically different from a system where you
put one point down and then the second point and then it draws the
square. The rubber rectangle has a much more satisfying feel to
it.

The rubber rectangle is one of a series of cases where the
continuity of action is extremely important. The disruptiveness

of screen actions is very harmful, even with things like jump
scrolling versus smooth scrolling. Smooth scrolling is a winner
in readability by substantial amounts, as opposed to jump
scrolling. It does require more hardware to do it, but the human
factor of the readability is much enhanced by smooth scrolling.

So, Macintosh is good in that it has:

- continuity
- selection rather than typing commands
- informative feedback, in that at every point when you do
  something you can tell that it has had an effect. You can
  see what happens. The check marks move on the screen in
  response to what you've done. You don't have to inquire what
  has happened. You don't have to issue a command. You can
  see everything that is happening in a very natural way.

So, this is how I characterize direct manipulation:

- Appropriate Cognitive Model
  It depends on finding a clever cognitive model, a way of
  representing reality in the application domain, that will elicit
  the user's analogical reasoning and tap into the user's full range
  of knowledge. This is the trick: To find some domain where the
  users come with all the knowledge, so that then you don't have to
  teach them so much, and the system is very much more in consonance
  with the world of actions. We have to look at areas where there
  are well-defined application domains independent of the computer
  domain.

- Rapid Incremental Reversible Operation
  - Rapid
    Operations should be quick, so you can see what is happening,
    and the penalty for an error is small, so you can undo it
    quickly.

  - Incremental
    Operations should be smooth and continuous. The jumps should
    be small, lots of small ones that can be done rapidly. So
    you appear to do big operations quickly by piling together a
    lot of little ones.

  - Reversible
    By this I don't mean just the "undo" command (that seems to
    be the first attempt at reversibility, which is useful in
    some cases). I think the right way to do it is to have
    natural inverses for every operation as part of the system.
    Again, in the car, you just turn the wheel and if its too
    far, you don't "undo" the last turn, but you bring the wheel
    back, turning it right.

    I think you can look in systems where pairs of complementary
    operations are available. They tend to be fairly easy to
    learn, and it's quite natural then to see what to do if you
    get into trouble; you just reverse it or apply the inverse
    operator.

I once designed a little system where I had just objects with
"plus" and "minus": "plus log" was logon and "minus log" was
logoff; "plus item" was add an item, "minus item" was delete
an item; "plus" and "minus" moved the cursor up and down;
"plus help" got you help, "minus help" allowed you to send
messages to the designer. Makes sense, right? You begin to
add these things together.

- Immediate observation
  Immediate observation of the effect of action means no hidden
  side effects or change in underlying states that are not
  visible.

- Physical action, doing things, pressing buttons or making
  selections rather than typing commands. The machine can
  display a thousand characters a second, when you can only
  type 1 or 2 a second; why not let it show you a large number
  of choices and then have you point at one? One implication
  of this is that, in programming, you should never have to
  type a variable name more than once (this is also part of the
  correct and complete). Once you've typed it, it's there.
  From then on you should only have to select it (sometimes
  that can be a hassle with a lot of names, but it may be very
  much more convenient than not having to retype each time. I
  think there are ways to make that happen conveniently).

  Since there is no (or less) command language to memorize,
  training is simplified, and there are few error messages,
  e.g., when the Pac Man hits against the right wall, and
  you're moving the stick to the right, you don't need a
  command to say "illegal move!" It seems clear enough that it
  blinks and stays there, because the world of action is very
  clearly represented to the user.

I think this is just a beginning stage, and perhaps you can offer
examples of other systems that have some of these features, and
there are lots of them (more and more of them every day): Lisa,
Macintosh and Xerox Star are often cited examples.

WRITING PROGRAMS IN A DIRECT MANIPULATION LANGUAGE

What is missing here (and what I'm interested in now) is how to write
programs in a direct manipulation language. All of the systems (like
Macintosh and Xerox Star) do not allow you to program in that language. To
create those visual facilities, they use some other language. They do not
offer the user a way to do a set of actions, record them, and then redo
them or play them back.

This presents an interesting challenge, which has to be met if we are
going see these systems spreading still further. If you issue a set of
commands, you can store that set of commands, give it a name, and then run
that name. There are lots of variations on that theme. What is harder is
introducing conditionals and looping structures. For simple looping
structures, you can usually make out (the way APL handles them) by
converting the logical process from a looping one to one where the whole
collection is handled at one time. But there are some times (even in APL)

when you need to have a predicate which stops the loop; these seem to be
hard to describe, as are conditionals. We're looking at ways around that.

There are others that are pursuing this, sometimes under the term of
"programming by example." For instance, Dan Halbert is doing a PhD at
Berkeley and working at Xerox PARC on something called "Small Star", which
is a modest version of programming in the Star environment. Harold
Thimbleby of York University, U.K., and Don Hatfield of IBM Cambridge
Scientific Center have developed advanced editor designs applying direct
manipulation.

SYNTACTIC/SEMANTIC MODEL

I have attempted to weave together individual results and guidelines into a
theory. This is an idea that began with Richard Mayer (a psychologist)
when we were both working at Indiana University. It first appeared in
1976, and I've applied this notion for interactive systems as well as for
programming [15]:

I claim that there are two kinds of knowledge that a person brings to using
a computer (see Figure 1). Let's take for the moment the person being a
programmer:

-    Language-Specific Syntactic Knowledge
     I claim there is an enormous amount of detail programmers have to
     know about a programming language. That is messy stuff. It is
     very hard to make a logical order out of it. There is no way you
     could predict that the assignment operator is "colon-equal" in
     Pascal, "equal" in FORTRAN, and "back-arrow" in APL. There is no
     logic to it. Nor any logic about semi-colons being terminators or
     separators, nor could you predict what the looping keyword would
     be in a language: "do", "for" or "loop". There are times when
     those syntactic details trip you up ("What does the function SQR
     do?" Well SQR does square root in Basic and square in Pascal, but
     in FORTRAN square root is SQRT). There is just a lot of clutter
     that you have to remember about each programming language, and
     this kind of syntactic knowledge is highly system-dependent. It
     must be rote-memorized (there is just no way around it). It is
     subject to forgetting, unless you frequently rehearse it. If you
     go away from the language for 6 months, when you come back, you
     can't remember the syntax - but if I gave you a program written in
     the language, you could probably read it, because you'll be
     prompted by the syntax, and the semantics are more stable in
     memory.

-    Semantic Knowledge
     This knowledge (I claim) is of a different kind, and I separate it
     into two categories.

     -    Problem domain
          The problem or application domain has been mentioned here (a
          number of people use that term). This has nothing to do with
          computers. You're going to program mathematical functions,
          stock trading, or medical applications. It may take you 20
          years to learn that.

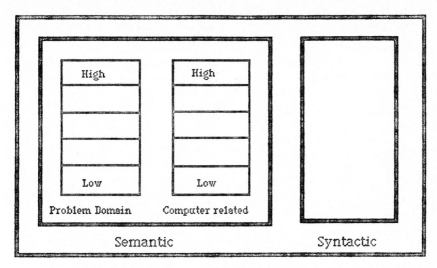

Knowledge in Long-term Memory

FIGURE 1

LONG-TERM MEMORY IS CONJECTURED TO CONTAIN SYNTACTIC AND SEMANTIC
KNOWLEDGE. SEMANTIC KNOWLEDGE IS LEVEL-STRUCTURED AND FURTHER DIVIDED
INTO PROBLEM DOMAIN AND COMPUTER-RELATED CONCEPTS.

- Computer knowledge
  There is a separate world of knowledge that is related to
  computers. Here you learn constructs about use of
  computers: there are directories, there are files, and files
  have a first record and a last record, and records are of
  fixed length, and each record contains atomic items which
  contain characters, and the character set is so and so, etc.

I claim that both of the semantic domains are structured in a more orderly
way (top down if you wish), from big concepts to lower level concepts.
Human thought processes dealing with complex issues tend to utilize a
level-structured, modular tree-structured strategy. In many worlds (many
application domains) you can hear people talking about these separate
levels. In orbital mechanics, people represent a low level of detail (by
the polynomial equations), and then a higher level of actions (that takes
place in three-space), and then another level (which would be the intent to
rendezvous two rockets in three-space). Thus, there are clear levels of
abstraction; one has to break problems down to lower levels of detail in
the problem domain.

If the person comes as a knowledgeable programmer, he or she may still have
to acquire the problem domain to be successful. If one comes as a
knowledgeable person in the problem domain, one may have to acquire the
computer-related domain or the syntactic domain. The person who knows one
computer-related domain (knows programming in Pascal and is trying to learn
Ada), well that person will have to (in many places) substitute the
syntactic details, or know when they are the same.

I should point out that the lines are very neatly drawn in this abstract
model of the way people organize their knowledge in their heads. In fact,
I realize that these boundaries are not so sharp, and that sometimes some
issues might be both syntactic and semantic. It can be very hard to
separate when something is in the application domain and when something is
in the computer-related domain. In spite of those problems, I do find it
useful to organize my thinking in terms of this model.

The model is also useful, for example, in deciding how to teach people
about programming. To a novice student, I start by a motivating example in
the problem domain, then I convey the semantics of the computer domain, and
only at the end do I show the syntactic details. If I have someone who is
a knowledgeable programmer, then I can just try to teach the problem domain
in which they need to do their work.

The model also helps me to organize and understand a little bit about the
kinds of errors that people make: Whether these are truly syntactic errors
(which would be minor flaws) or logic and semantic errors. Whether these
are errors that result from a misunderstanding from the application domain,
or are errors that result from some problem with the computer related
issues (e.g., some misunderstanding of what the semantics of a certain
control structure are).

I invite you to criticize this model (or perhaps refine it). Here, it is
only laid-out in outline. If you don't like this, I would encourage you to
think explicitly about alternatives. I think this field needs a deeper
understanding, some theory, some models about the way programmers work, the
way they learn, they way they design, the way they debug. And to that
challenge, I invite you all. Thank you.

REFERENCES

[1] Shneiderman, Ben, Software Psychology: Human Factors in Computer and
     Information Systems, Little, Brown and Co., Boston, Ma. (1980).

[2] Weiser, Mark and Shneiderman, Ben, Human factored software design and
     development, in Handbook of Human Factors, Salvendy, Gavriel
     (editor), John Wiley and Sons, Inc., New York (to appear 1985).

[3] Iyengar, S. S., Bastani, F. B. and Fuller, J. W., An experimental
     study of the logical complexity of data structures, Information
     Processing and Management, (to appear 1985).

[4] Soloway, Elliot, Ehrlich, Kate, Bonar, Jeffrey and Greenspan, Judith,
     What do novices know about programming?, in Badre, Al and
     Shneiderman, Ben (Editors), Directions in Human-Computer
     Interactions, Ablex Publishing Co., Norwood, N. J., (1982), 27-54.

[5] Wasserman, Anthony and Shewmake, David, The role of prototypes in the
     user software engineering methodology, in Hartson, Rex (Editor),
     Advances in Human Computer Interaction, Ablex Publishing Co.,
     Norwood, N. J. (to appear).

[6] Hartson, Rex and Johnson, Deborah H., Dialogue Management: New
     concepts in human-computer interface development, ACM Computing
     Surveys (to appear).

[7] Teitelbaum, T. and Reps, T., The Cornell program synthesizer: A syntax-directed programming environment, Communications of the ACM 24, 9, (September 1981), 563-573.

[8] Shneiderman, Ben, Correct, complete operations and other principles of interaction, in Salvendy, Gavriel, Human-Computer Interaction, Elsevier Science Publishers, New York, (1984), 135-146.

[9] Weiser, Mark, Communications of the ACM.

[10] Boehm, Barry, Software Engineering Economics, Prentice-Hall, Englewood Cliffs, N. J. (1981).

[11] Molzberger, Peter, Aesthetics and programming, Proc. ACM CHI '83: Human Factors in Computing Systems, available from ACM, P. O. Box 64145, Baltimore, Md. 21264 (1983), 247-250.

[12] Shneiderman, Ben, Direct Manipulation: A step beyond programming languages, IEEE Computer 16, 8, (August 1983), 57-69.

[13] Zloof, Moshe, Query-by-Example, Proc. of the National Computer Conference, AFIPS Press, Arlington, Va. (1975).

[14] Herot, Christopher F., Spatial management of data, ACM Transactions on Database Systems 5, 4, (December 1980), 493-513.

[15] Shneiderman, Ben and Mayer, Richard, Syntactic/semantic interactions in programmer behavior: A model and experimental results, International Journal of Computer and Information Science 8, 3, (1979), 219-239. Reprinted in Curtis, Bill, Human Factors in Software Development, IEEE Computer Society EHO 185-9, Piscataway, N. J . (1981), 9-23.

QUESTIONS AND ANSWERS PERIOD

BOUDREAUX

One question which does occur to me is, it seems to me that one of the underlying problems which you've identified is the problem of how one accurately goes about segmenting and breaking into meaningful chunks a variety of things, programs, problem domains and so forth. It seems to me that the thesis that you are working on is that if only we could get the chunks right, then we could make the job that we are doing in the field of computer science much easier. Do you think that there is one way in which we could chunk languages or do you think that is an example of something which is, in general, field-dependent?

SHNEIDERMAN

I think chunking is much harder than people give credit for. The creators of Ada abstractions believe that if they only offer the facilities, the users will come forward and produce wonderful abstractions that we can all share and borrow and build on. And I think that it's a much greater challenge to deduce a useful abstraction than is understood. I think we

can talk about information hiding, modular decomposition with low cohesion,
but I don't have a complete understanding of how to create successful
abstractions.

BOUDREAUX
Do you see any connection between that problem that occurs in this field,
that is the field of attempting to break the design of algorithms and
whatever into meaningful pieces or chunks, and a corresponding problem
which occurs in the field of pattern recognition for relatively unstruc-
tured visual domains - the analogy in the world of robotics being picking
and placing, where I'm dealing with an un-palletized system? There the
recognition problem can't be so easily....

SHNEIDERMAN
Yes, it is an interesting analogy, I think. In the robotics vision problem, the hard part is how to input the person's whole world of knowledge.
The person will recognize that if the Statue of Liberty doesn't have a
hand, there is something missing. But it is very hard to get computers to
recognize that. I think we have to pay more attention to what is in
people's heads to understand how to build better systems.

GUIER
I found your talk really fascinating. You started out by talking about
metrics and how we might do some really controlled testing and so on.
There's one dimension that I, at least, didn't see in your talking about
the testing, and I think did not see it later on in some of your examples
and details, and that is that different people are just inherently differently skilled in programming. We seem to have no trouble at all in
assuming that musicians are going to be good musicians and bad musicians,
and that there are the soloists and they're the ones that play in the top
orchestras and then teach music and so on. I think many of us fall into
the category of trying to call all programmers kind of dependent on the
amount of training or experience, but forget the lack of an innate skill.

SHNEIDERMAN
Oh, innate skill. Well, I don't know about innate skill. It is hard to
say if we can breed better programmers; I'm not sure that I want to go too
far with that. But, the studies do show at least a 10 to 1 ratio in
performance with programmers with the same job titles and salaries. I
think we don't have good ways of evaluating the quality of programs or the
quality of the programmers, that is, their productivity and the quality of
the code they write. Those are really important directions.

GUIER
I have sort of a personal bias, that the answers to some of your questions
are going to be very sensitive to that.

SHNEIDERMAN
Unquestionably.

GUIER
Some of these very cute ideas you have on transformations on whole blocks
of code, for example, may not be at the appropriate level for some sorts of
people.

SHNEIDERMAN
I fully agree. I think programming has matured enough that we have a rich
diversity of people and no one thing is going to be good for everyone. I
liken it to photography, in that 100% of the population take pictures and
100% of the population use computers in some way. Ten percent are serious
amateur photographers or serious programmers. One percent may be professional programmers. Now there's no one camera you're going to sell to
everybody, and there are going to be Hasselblads and 8x10 cameras for
special communities. So I think, yes, we're going to see a rich diversity
of programming ideas.

RADA
When you were talking about a search for models or cognitive models, you
presented a paradigm in your own research, I gather, where you would take a
domain and some users and do an experiment to test whether one model was
better than another. What is your feeling about the value of other
people's work? Take, for instance, Jaime Carbonell's talk yesterday where
he would discuss reasoning and propose that there are eight fundamental
ways in which people reason and this might cover all the reasoning types
that people can use. In general, if you look to other work in cognitive
science, what is your feeling about the value of using results from that
kind of work in software engineering?

SHNEIDERMAN
I'm intrigued by your question. I wish I had a clever, quick answer for
you. I thought the list that Jaime Carbonell presented was interesting. I
think the gap is very great between the kind of small experiments I'm
proposing and the kind of inventive programming languages for that domain.
So I would like to think about it more, but I don't have a quick answer.

RADA
I guess it seems to me that on the one hand you can appreciate that we
don't have much understanding of knowledge or cognition and that it seems
that what we need is a vast amount of experiment first. You take a particular well-defined problem, you do an experiment, you build knowledge. And
to the extent that we can start to make generalizations, that we have
enough experimentation to make generalizations, and that we can describe
the nature of human cognition within our lifetimes, then perhaps it's
already time to start to look at what some other people have done...

SHNEIDERMAN
One interesting point that comes from it is the notion of semantic knowledge being independent of syntax. I teach a course in Human Factors in
the Computer Science Department. I gave them all a small 20-line program
in PASCAL and asked them to study it for two minutes. It was a simple
program that added up and printed some values. Then everyone turned the
sheet over. Half of them I asked to rewrite the program from memory as
well as they could. All of those who were competent PASCAL programmers had
no problem. There were two students who were majors in other departments
and didn't know PASCAL; they had a very poor time with it. The other half
of the students I asked to rewrite that program in any other language that
they were fluent with, and they did equally well. And they had just two
minutes - a very short time to do it. I claim that people represent the
semantics of a program in their heads independently of the syntax. I use
that example to suggest how this theory about syntax and semantics and the
organization of knowledge is helpful in making some predictions that you

were asking about, and yet, we need to refine that more, but I think that's the direction I would like to go - a theory elucidated by an experimental result which leads to some refinement of the theory that offers, hopefully, better predictive powers.

WHITFORD
Dr. Rada, I think, took the words out of my mouth while I was trying to prepare or think of how to ask questions, so I think I'll try to restate it. I've done a little bit of reading recently in psycholinquisitics and epistemology, and I'm thinking in listening to you that there might be a body of knowledge or at least some theories about the relationship between concept formation within the mind and transferring that concept into meaning through, natural language; that maybe there are some findings even within that ground that might be relevant; that if we just focus on programming languages, we may only gradually move them to being more and more efficient. We may not make profound changes or arrive at any insight that we might get from looking at the more universal model of mapping concepts and languages.

SHNEIDERMAN
Let me just say, yes, to the psycholinguistic connection. I think that could be very fruitful and I feel honored to call myself 20% of an experimental psychologist, but there's a lot more that I need to know, and I think the infusion of knowledge from other disciplines will enrich programming substantially. It struck me that the developers of programming languages did not consult psycholinguistics, the learning, the memory, the problem solving literature. We're now reaching the point where the marriage between disciplines is coming about. So I'm hopeful and I encourage attempts in that direction.

WHITFORD
I guess that's what I was trying to say. It seems like computer languages sort of happened, and we don't look at them from outside computer languages. We only look at computer languages and we compare them to each other and then we fault them or we pat them on the back, so to speak, but we don't really look at them in a broader context.

UNIDENTIFIED
I thought that your distinction between syntactic and semantic, and then within semantic between the problem domain and programming, was just the right distinction to make. I would like to make one observation and then one suggestion. The observation is that we often think of that distinction as being one between specification and implementation. And, in fact, I think that one of the interesting things about the programming process is that the entire process is a mixture of knowledge from both parts, the problem and computation. It's not simply a question of defining the domain and then implementing from there, but in fact, the skills in programming involve both.

SHNEIDERMAN
I think it depends on what kind of programming. If you're working in an application domain that you're familiar with and you know the programming language fluently, then I think you can really make a clean specification and write that at a high level and then refine it. However, I'm sympathetic to the other kind of programming, where you implement a little bit and see what happens. Balzer characterized it nicely as the "inevitable

interweaving of specification and implementation." There are situations where you are in a novel world where you have a new programming tool and an application domain you're not familiar with and you can't see far into the distance and write it all down and then code it. It seems to me that if you're learning something new, other than the application domain, the computer-related, or the syntax of the language, you need to have the opportunity to play. When I talk with professional programmers, I try to encourage them to separate that part from the part where they're sitting down to write the real code, you know, the serious stuff.

SAME MAN
I'm not arguing against discipline. What I'm suggesting is that the decisions that the programmer makes during the programming process depend upon what he or she knows about the domain. I'll give some simple examples. He may find that the specification, we'll take a mathematical specification, turns out not to be tractable. So you have to make some kind of approximation to that, and the kind of approximation you make depends upon what you know about the domain.

SHNEIDERMAN
Absolutely. I agree. I think it interesting in the light of the theorem proving systems. There are heuristics that emerge out of the application domain. In geometry theorem proving, you get a whole set of heuristics from the visualization of the shapes. Then there are computer related things that you know about tree pruning algorithms that are independent of the application domain, and the capacity to define both of these makes for a successful system.

SAME MAN
Now, let me make this suggestion. I think that one of the things that we've learned in work on expert systems is that introspection is a terrible way to try to figure out what we know. A much better way is to try to implement it in some kind of computer program where there is some hope of observing what whatever it is that we've written down actually does. And so the sort of suggestion I would make would be to look at a variety of different domains and try to build computer programs that write programs in those domains as a way of understanding and codifying both the domain knowledge in that domain and the computer knowledge in that domain.

SHNEIDERMAN
I think that's interesting.

PANEL SESSION:
LANGUAGE REQUIREMENTS FOR
EFFECTIVE AND EFFICIENT
PROBLEM SOLVING

The Role of Language in Problem Solving I
R. Jernigan, B.W. Hamill, and D.M. Weintraub (Editors)
© Elsevier Science Publishers B.V. (North-Holland), 1985

PANEL SESSION:
LANGUAGE REQUIREMENTS FOR EFFECTIVE AND EFFICIENT PROBLEM SOLVING

Panel Chairman: David Barstow, Schlumberger-Doll Research
Panel Members: Jaime Carbonell, Carnegie-Mellon University
                  John A. Carlton-Foss, Human-Technical Systems, Inc.
                  Adin Falkoff, IBM Research
                  Andrew Goldfinger, Applied Physics Laboratory, JHU

BARSTOW INTRODUCTION

This is the panel session on Language Requirements for Effective and
Efficient Problem Solving. It was actually hard to tell how that was
different from the topic of the entire workshop. So I think that the ball
game here is wide open. Let me start off with some quick comments about
what the topic might mean. It seems to me that there are actually three
areas of effectiveness or efficiency that we might want to think about.
One is the effectiveness or efficiency with which the user can communicate
the problem - how easy is it to state a problem? The second is: how can
languages help us to solve a problem well? That is, what is the nature of
the solution process and what language features would help or hinder
that? The third is whether there is some measure of the quality of the
result of the solution process, what language features might help us find
the best or the worst of the alternative solutions. So those three areas
are ones that I think important for us to keep in mind.

Now, in the organization of the panel, what we are each going to do is make
a few minute presentation, and then once all the presentations are over, I
would like to see the discussions opened up to everybody. So let me start
off the first speaker who will be Andrew Goldfinger from APL.

GOLDFINGER PRESENTATION

A man was taking a tour of the United Nations and he was shown by his
guides to the cafeteria. The guide commented that it was interesting to
watch the people eat at the cafeteria because the different delegations sat
down and ate together, and by watching the behavior at each of the tables,
you could learn something about the personality of the people from that
country. "Look there, for example. You can see that the people from the
British delegation are seated very quietly and listening to each other and
talking in turn, and you can tell that the British people are fairly
reserved and quiet and calm. Now, if you look over there, you can see the
Italian delegation. They are talking with their hands and you can see that
they're fairly emotional people and very excitable." And the man said, "Oh
yeah, I see it. It's kind of interesting. Who is seated at that table
over there? He said, "That's the German delegation." "I can see the
Germans are very polite people because one person is talking and talking
and talking and everybody is just listening and waiting and listening and
waiting. They're just waiting for him to finish and are very polite." The
guide said "Well, I'm afraid you have that all wrong. You see, what is
happening is they're waiting for the verb."

Now, I see there are some people here who know German, and for those of you
who don't, the way the German language seems to operate is as you speak it
or if you write it (believe it or not, some people actually speak that way
in Germany), various nouns and clauses seem to queue up in the sentence and
then you end up with a big flurry of verbs that sort of unfold everything.
And really, if you stop to think of it, it's pretty much like a push-down
stack, and the language is kind of like a reverse Polish notation, in which
the verbs, which are sort of like operators, resolve the nouns, which are
sort of like data that occur in the sentence. And I would wonder how HP
calculators sell, let us say, relative to TI calculators in Germany. Do
people who speak German find themselves more comfortable with a reverse
Polish notation? And if I were developing a computer language to be used
in Germany, might I want to develop a stack-oriented language, because the
people may tend to think that way?

For any of you who have ever been to Tokyo, there's also a somewhat frustrating situation that arises when you try to find your way around and you
discover that in a city of 11 million people, they don't have street names,
they don't have addresses, and this is somewhat puzzling to those of us who
don't come from Japan; but to the Japanese, it's really quite natural. In
thinking that through, it occurred to me that that is somewhat similar to
what happens to their language. Those of us who speak any of the Western
languages are kind of appalled at the fact that a college educated person
in Japan has to know about 7,000 Kanji characters, especially since it is
possible to write the language in what they call "Ramaji", which is a
phonetic transliteration of it into English characters, or even into what
they call "Kana", which are their own phonetic characters. But the
Japanese don't seem bothered by having 7,000 characters, because after all,
the words are each unique and have different meanings, so why should they
be reduced by some sort of atomistic reduction. Similarly, if you write
down the address of a place in a city, you are taking a very unique
location and you are reducing it atomistically to a lot of equal qualities
to describe it. They don't seem to think that way. And one would wonder
(Ross Bettinger in his talk spoke about left versus right hemisphere thinking; there are some experiments that have been done on NMR patterns in the
brains of Westerners and Japanese in thinking, and the actual activity
within the brain in Japanese is more distributed between the hemispheres)
that maybe pictorial or holistic languages, languages that don't involve
reduction, would be much more comfortable and much more useful to Japanese.

One other example of language profoundly influencing the way people speak
is in the case of Biblical Hebrew, which, by the way, is a slightly different language than the modern Hebrew which is spoken in the state of Israel.
If you look at Biblical Hebrew, you'll find that the tense structure is
very different than we are used to in Western languages. You don't really
have past, present, and future. You have perfect and imperfect - complete
action versus incomplete action. And if you go and study the Bible, and if
you make the assumption that just because things that are mentioned in a
certain order in the text, that they are occurring in chronological order,
you will very soon get into trouble and find out why there are all sorts of
people who write comments on the Bible and can't quite understand what is
going on. But to the people who spoke that language, it was apparent that
the order in which things should be stated has nothing to do with chronological order, it has to do with conceptual order. And, if we were to find
one of these people from 4,000 years ago who had Biblical Hebrew as their

native language, maybe they would be much better programmers in concurrent programming than in sequential programming.

The reason that I bring all of this up is that I think there is an issue underlying all of this. That is not what is the best language for a given problem, although that is I think an important issue. But I think there is also another issue, and that is what is the best language for a particular person. There exist individual differences between people, both on a national basis, and on an intra-national basis - you versus your next door neighbor. And I think that there are some very important issues as to how we match the language, not only to the problem but to the person. We here at APL, at the Applied Physics Laboratory, are just beginning an informal experiment in this. I am associated with a graduate student who is in a counseling program, and there exist certain psychological instruments - a psychological instrument isn't something like an oscilloscope, it is what psychologists call a test; you know you, take the paper and you write it down and they call that an instrument, because they want to pretend they are hard scientists - but, there exist some fairly standard psychological instruments which measure certain aspects of personality, such as thinking, style, and the way that people handle information, the way they take in information, the way they make decisions. And through this graduate student, what we are going to be doing is giving these tests to a group of people here at APL, who are all professional technical people but some of whom, on their own, have chosen to use a language like PL/I while others have chosen to use a language like APL. Our hypothesis is that in some of these areas - the ways in which people scan information, take in information, and so on - we will find different populations, and that it is personality which led the person to choose the language. So, hopefully, at the next one of these conferences, we'll be able to report on that. But for now, I would just like to voice the hypothesis that we need to look not only at the problem, but also at the person.

QUESTION AND ANSWER PERIOD FOR GOLDFINGER PRESENTATION

QUESTION
What do you think comes first? Does the language influence the thinking or vice-versa?

GOLDFINGER
As far as I understand that's an ongoing debate in the field of psychology. But it's really kind of irrelevant to some extent to our purposes, because most programmers are adults who already have developed both their language and their thinking patterns. All we need to know is that the two are associated, we don't really need to know what the cause and effect are.

CARLTON-FOSS PRESENTATION

I found those comments very interesting, because when I took an intensive course in Chinese, I found there was a huge transition between learning the Chinese itself and reading about it in English - a very difficult transition to make.

I would like to take this opportunity to develop two ideas further than was possible in my original presentation, interplay among the phenomenal world which we perceive and interpret, the technologies we use to help us to

perceive and measure that world, and the person, the "we" in the system
(see Figure 1).   There is a tendency among one or another group to truncate
that "we" from this large system and claim that a certain style of perception, interpretation, and model building is "objective;" or to truncate the
assisting technologies, claiming they are "inhuman," and therefore irrelevant or undesirable.   I would argue that all three parts are essential for
successful perception, problem solving, and model building.   The images
from the "we" provide us, for example, with the material necessary for
understanding.   One must accept and use this part of the system if for no
other reason than that it permits us to recognize the proclivities we bring
to addressing a problem.   Appropriate languages can help us to keep our
image-based perceptions and models in order.   One must also accept the
importance of technologies.   In part, technologies are important because
they extend our human limits.   For example, a chunk of information can be
processed in a characteristic time of 25 - 200 msec by the human working
memory (Card, Moran, and Newell, ch. 2) [see reference list in Carlton-Foss
paper, this volume], whereas modern computer-based systems can process
information much more rapidly.   It is therefore very important to consider
the realities of all three parts of the large problem solving system.

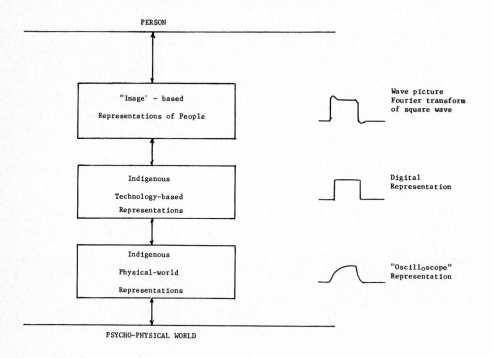

FIGURE 1

Second, I provide an example of a root image and the way that its application changes as one applies it and its associated language(s) to solving
problems by analogy at increasing distances from the domain of the original
problem solved.   I will choose the root image associated with the Newtonian
problem of a wave passing through a "slit".

Root Level. The dominant effect is that of the wave passing unimpeded through the "slit", such as the entrance to a harbor thoroughly protected by breakwaters. The wave emerging from the slit can be modeled as the superposition of waves that impinge on the slit and reflect from surfaces. Second order edge effects occur, with some multiple reflection of waves, some exogenous perturbations (e.g., surface ripples due to wind), and some absorption of wave energy by the breakwater. Arithmetic, trigonometric functions and algebra can be used to model the phenomenon.

Level 2. This image can also be applied to a classical treatment of the passage of light through a slit (or hole) of a dimension on the or wavelength of the incident light. In this case the edge effects are very significant in relation to the pass-through effects. Models still are based on the assumption of wave superposition, but here one must formulate and solve the problem using integral and differential calculus, and use Bessel Functions and/or graphical or integral functions (i.e., the Cornu Spiral) to express the solution.

Level 3. The image can again be applied in an elementary quantum mechanical treatment of diffraction of light through a slit. This requires a shift in mathematical language to functions of a real variable, to operators on a Hilbert Space, and using an underlying statistical concept (e.g., see Messiah, pp. 115-118) [see reference list in Carlton-Foss paper, this volume]. The root image has begun to break down, we have intrinsic uncertainty as part of any model to be developed, and there are undesirable side-effects of the shift in language. For example, even though we believe that a particle (wave packet) moves through free space continuously, it is impossible to demonstrate this rigorously when motion is defined by the Schrodinger Equation (or its variations). The distance norm for much of a quantum-level is $||R(t + dt) - R(t)||$, and this cannot be shown to approach zero as dt approaches zero.

Level 4. The image can also be applied to the passage of electrons through a slit. Because of the spin of electrons, this requires full quantum mechanic modeling of the phenomenon, and presents the scientist with well-known paradoxes. I view such paradoxes as, in part, the result of the breakdown of the root image in modeling the phenomenon. They represent the incursion of other root images into the experimental and theoretical picture. Even though problematic issues can be rationalized, any model necessarily contains new root images combined with the original diffraction root image.

Breakdown. The root image can be pushed beyond its full usefulness. For the wave-through-a-slit image, this breakdown occurs at the nuclear level in the "shadow scattering" model of the neutron-nucleus interaction (e.g., see Evans, pp. 94-95) [see reference list in Carlton-Foss paper, this volume]. The model is presented as an introduction and as a means of initially structuring the nuclear scattering data at the beginning of nuclear physics courses. It again requires only algebra and some simple calculus as mathematical languages. Thus, the power of the root image is dramatically reduced, the language required becomes simple again, and the model becomes less capable of representing the key phenomena in the domain of interest.

Thus one can see that, in extending a root metaphor by analogy from its
initial domain to new domains, it is extremely important to pay attention
to all aspects of the application. The metaphor may have greater or lesser
face validity in a new domain. But one must push far beyond similarities
and differences. In addition, a broader application may require modification of the metaphor, a shift in the language used for modeling, the
introduction of other root metaphors, and acceptance of models with greater
complexity. Finally, one must pay careful attention to signs that the root
metaphor is no longer applicable. These signs are at first very subtle,
and one must think very rigorously about what the metaphor actually brings
to a problem. Later, however, the deviation between experimental results
and model-based predictions becomes so great that there is no question
about the breakdown.

BARSTOW INTRODUCTION

Our next speaker is Adin Falkoff of the IBM T. J. Watson Research Center

FALKOFF PRESENTATION

There is much that I've heard here in the last two days that I've found
very interesting and very encouraging. Many of the matters discussed are
things that we've thought about in the past in our work in APL, and there
are many things that we're still giving thought to. For example, we have
found the idea of introducing more mathematical functions into a language
to be somewhat controversial. Many mathematicians think that you shouldn't
do that. Thus, in an early version of APL2 we introduced eigenvectors and
eigenvalues and polynomial zeroes as primitives, but before the program was
let out as a full fledged product the primitive calls were removed,
although the functions remain available in a secondary form.

BARSTOW COMMENT
Can you tell us why they were removed?

FALKOFF PRESENTATION

Because some respected mathematicians have held that it is not a good idea
to include functions for which different algorithms might be required for
different purposes, and the sophisticated user would probably want to write
his own. That doesn't really negate the idea that the unsophisticated
user, or even the sophisticated one, might still find it handy to have it
available. So it was compromised, and the functions were left encoded in
the assembly language, but made available under some alphabetic name rather
than a primitive symbol. So they're there, but not there.

On the question of language requirements for effective problem solving, I
think that I want to start out with an ad hominem argument. A language for
effective problem solving has been evolving over many centuries. It's
called mathematical notation. And I think that in one way or another
mathematicians have incorporated into their notation, into their way of
thought, things which enhance problem solving. We make the claim that as
far as programming languages go, APL has followed more closely in the
direction of mathematical notation evolution than other programming languages. Therefore, I assert that somehow or other, we are closer to the
right track and more effective for problem solving than we might otherwise

be. To support this I will now discuss several aspects of APL which I believe tend to influence the user's thinking about his problem; which I take to be the underlying theme of this conference.

Referring to Figure 1, which might have been better titled "Thought-Influencing Aspects of APL," we see the first item is arrays as primitive objects. I think that this is extremely valuable and very important. It

```
          INFLUENTIAL ASPECTS OF APL


     * Arrays as Primitive Objects

          - Thrust towards generality
          - Modelling in terms of patterns


     * Functional Notation

          - Modularization
          - Suppression of detail
          - "Non-procedural"


     * Operators

          - Function classes
          - Function transformations


     * Shared Variables

          - Cooperating processes


     * Encouragement of Experimentation

          - Interactive implementation
          - Interpretive with no declarations
          - Workspace organization


     * Primitive Notions Used as Models for
       User-Defined Objects

          - Functions with array arguments
          - Ambivalent functions
          - Defined operators
          - "Abstract" data types
```

FIGURE 1

does affect the way you think about a problem. People report that they
think differently when using APL, and become uncomfortable when forced back
into a scalar mode of thought. I've put down two specific effects here.
One is the thrust towards generality, and I think that's been touched upon
earlier in this conference, so I'll just remind you of one basic idea. In
the reduction of arrays, such as the sum or product over a vector, APL
defines the result of reducing an empty vector to be the identity for the
particular function. This is perhaps a small-seeming generality, but it
turns out that it often provides unexpected insights, and frequently
obviates the need for special-casing in complex contexts.

Another effect when you deal in terms of arrays is that you tend to see and
account for conditions that you might overlook when working element by
element. Still another aspect, which I haven't specifically noted in the
Figure, is that it tends to make you think in terms of transformations of
arrays, and so is conductive to a functional mode of thought.

I refer next to modeling in terms of patterns, which was touched on today
in one talk, but, if you'll permit me, I would like to give you an example.
The example has to do with the old warhorse of finding primes, and the
program, called PRIMTO, is shown at the top of Figure 2. The idea for this
program, as far as I know, is due to J. W. Bergquist, of IBM Scientific
Marketing in Los Angeles, who conveyed it to me some years back.

It's the familiar sieve of Eratosthenes, transformed to make use of the
array-structuring power of APL. I've worked out the example here for all
the primes in the first 20 integers. This is shown in the center part of
Figure 2, where five boolean matrices are displayed. I start with a matrix
that has a zero and 19 ones, and reshape it into a matrix whose width is
the first prime, producing a 10 by 2 matrix, as shown. Clearly, all the
numbers represented in the second column will be multiples of two, and
therefore composite. So the next step is to set to zero all the numbers
below the top one in that column.

I then look for the next prime by looking for the first one in the second
row. If you think about that, it turns out that there is a theorem - I
don't know the proof - that given a prime, N, there will always be another
prime before you get to 2 times N. So you're guaranteed to find a prime in
the next row. I look for it and find that it turns out to represent the
number three, and so I restructure the array to a 7 by 3, making sure that
none of the first 20 has been lost. And again, I set the last column to
zeroes, except the top one. The process ends here, because five, the next
prime, is greater than the square root of 20. I now ravel the last matrix,
and make the correspondence to the integers shown below. You can see that
it has selected the primes. So there is a way of thinking about doing the
problem that depends entirely on thinking in terms of an array, and on
knowing that you have the ability to manipulate that array as an array.

At the bottom of the Figure I've written an expression which I think
represents the fundamental definition of prime numbers; namely, if you
consider a complete multiplication table, the numbers that are missing are
primes. This very simple definition comes directly from thinking in terms
of arrays, in this case the APL outer product, which forms tables. Reading
the expression, we see that the vector of integers formed on the right is
tested for membership in the multiplication table it generates, and the
negation of the result is used to select the primes.

```
     Z←PRIMTO N;S;M;NP;R
[1]  S←N*0.5 ⍝                   set end criterion
[2]  M←(N,1)⍴~N↑1 ⍝              initialize boolean matrix, M
[3]  NP←1 ⍝                      initialize NP
[4]  L1:NP←NP+M[2;]⍳1 ⍝          find next prime (in second row)
[5]   →(NP>S)/L2 ⍝               test for termination
[6]  M←((R←⌈N÷NP),NP)⍴M ⍝        reshape M to width of new prime
[7]  M[1↓⍳R;NP]←0 ⍝              apply the sieve
[8]   →L1 ⍝                      repeat
[9]  L2:Z←(N⍴M)/⍳N ⍝             select numerical values
```

```
    0        0   1      0   1      0   1   1      0   1   1
    1        1   1      1   0      0   1   0      0   1   0
    1        1   1      1   0      1   0   1      1   0   0
    1        1   1      1   0      0   1   0      0   1   0
    1        1   1      1   0      1   0   1      1   0   0
    1        1   1      1   0      0   1   0      0   1   0
    1        1   1      1   0      1   0   0      1   0   0
    1        1   1      1   0
    1        1   1      1   0
    1        1   1      1   0
    1
    1
    1
    1
    1
    1
    1
    1
    1
    1

 0  1  1  0  1  0  1  0  0  0  1  0  1  0  0  0  1  0  1  0
 1  2  3  4  5  6  7  8  9 10 11 12 13 14 15 16 17 18 19 20
```

```
(~Q∊Q∘.×Q)/Q←1↓⍳N
```

FIGURE 2

Well, part of the title of this session has to do with efficiency, and this last expression, as a method of finding prime numbers, is grossly inefficient, but it certainly leads you to think about fundamentals. It's bad for computation, but it's good for thought. PRIMTO, on the other hand, turns out to be extremely efficient, and in a very few seconds on a reasonable

size machine, you can get boolean arrays for, say, the first 500,000
primes. It's really good. Well, so much for the influence of arrays.

Referring to Figure 1, the next aspect of APL which I consider to be
important and rather obvious, is the functional notation; and there is no
question here of the reliance upon prior history of mathematical notation.
Functional notation reinforces thinking in terms of transformations of
arrays, or transformations in general; it allows you to compose functions,
and it allows you to modularize in what I consider to be a very healthy
way.

You modularize by making functions for concepts. Any large design project
is really a matter of successive language design. You're designing a language at each level; that is, as you build up to the final (user) interface,
the objects of interest take on appropriate personalities (abstract data
types), and the functions defined on them reflect the significant transformations at that level. Functional notation allows you to do that in an
organic way. There is an obvious analogy to mechanical design, where you
have assemblies, and subassemblies, and so forth.

The next point is that functional notation, if you wish, can lead you to
"non-procedural" programs. Now, Iverson, has for some years espoused the
notion of what he calls "direct definition", in which he conveniently
assigns a function name to a simple APL statement or a special conditional
form. By standarizing the names of the function arguments, and introducing
some conventions for local and global names, it becomes feasible to represent meaningful concepts in a single line, and these one-line definitions
can be automatically compiled into APL functions. Now, if you do this, if
you carry it out as far as it can be carried out, you wind up with essentially a collection of statements like a set of mathematical equations. The
order in which you write them doesn't matter, because at the time of execution each one will call on some other function that it requires wherever it
may be. So all you have to know is the first one to call, and the rest
follow. The collection of functions, listed in arbitrary order, then
constitutes a non-sequential program.

Furthermore, execution aside, the collection can be studied like any set of
mathematical relations. That this should obtain is not surprising, since
Iverson developed this notational variation primarily because of his
interest in education and his desire to use APL in the exposition of mathematics. But it should be noted that functional notation has this property
generally, and more complex "sequential" programs can also be viewed as
both statements of relations and directions for execution.

The next aspect of APL noted on Figure 1 is the notion of operators, functions which transform and compose functions. They serve in many cases to
form function classes such as inner product, which transforms the class of
scalar functions in such a way as to generalize the narrowly defined matrix
product to include a very large number of useful variations. Transforms of
single functions include reduction, scan, and outer product - all of which
have been defined in APL - and others, such as function inverse, which have
not. A recent development of particular interest is the notion of duality:
a form of composition which results in one function being used to transform
the data - for example, from rectangular to polar coordinates - to be
followed by the application of the second function in the new domain and
the use of the inverse to the first function to restore the modified data

to the original domain. The power and potential of such an operator - and concept - for problem solving should be obvious.

Quickly running through the remaining items in Figure 1, another concept that we've explored in APL is the notion of shared variables. It rests on the observation that if two processes are to cooperate they must communicate, that they can communicate only if they have something in common, and what they have in common is necessarily in the nature of a variable. If one or more such variables are available to two different processes under some kind of protocol by which we can regulate their use, you can generate all forms of communication (simplex, half-duplex, etc.), and develop various disciplines of concurrency. Such a protocol has been in use on some APL systems for more than a decade. Shared variables are also very helpful in system design, as their use tends to force clarity in the boundaries between modules, resulting in clean designs.

Those, I think, are aspects of the language which significantly influence the user's way of thinking and help to solve problems. Another aspect of APL which depends to some extent on its implementations, but is characteristic of them all, is the encouragement of experimentation. I think this point has come up before. Certainly, if you're trying to solve a problem it helps to be able to try things. So we have interactive implementations, where you don't have to commit to whatever size or whatever type a variable might be, where the workspace organization allows you to save things which are related, come back to them later, even save the state of an incomplete operation, review it, take parts out, try them in a different context, and so forth. Experimentation is often an important part of problem solving, and these are things which greatly facilitate it.

Finally, I would simply say that our experience is that the primitive notions of APL have been used as models for user defined objects. APL is extensible to the extent that you can define things and make them part of the language, and the syntax is designed so as not to become more complex when you introduce user defined objects. So we find people defining functions with array arguments, defining functions with ambivalent syntax, like the APL primitives, (minus, for example, is both a dyadic and a monadic function). Recent APL systems have introduced the notion of defined operators, and use of abstract data types has long been fostered by APL systems. One man's abstract data type is another man's primitive. It depends upon your implementation and how deep you want to go underneath it. Any data type ultimately is determined by the functions that are defined on it. Some languages require you to make declarations of data types, and then check to see that they are used properly, and in APL that's not done. But almost every application necessarily defines some set of data types in terms of the functions that are defined on the objects of interest, as pointed out in the discussion of functional notation.

I think I'll stop right here. Thank you.

QUESTION AND ANSWER PERIOD FOR FALKOFF PRESENTATION

BARSTOW
Any short questions, or does anyone want to ask what his last comment was?

R. RICH

I think all of us would agree that APL has stuck as closely to the mathematical notation as a procedural language could, but in mathematics that
notation is specification, rather than the definition of a procedure. In
other words, seeing which of the first 500,000 numbers is prime is not a
mathematical approach to the discussion of prime numbers and there are
languages like PROLOG in which you can state the situation that you assume
obtains without having a procedure to generate all the instances of it. Is
it clear what I mean by the difference between a procedural definition and
a specification? Would you like to comment on whether you think that
introducing not only the programming notation, but also the mathematical
attitude towards that notation as let x^2 equal $y^2 + e^2$ and tell me what you
now know about it other than by generating instances of it, is this a
useful approach at all?

FALKOFF

Well, I always have difficulty with this notion of non-procedural languages
because it turns out that if you scratch the surface, there is procedural
underneath. You tell me, you know, let x be the square root of n or something like that, and I don't know what to do with that. If it's going to
be useful somewhere along the line, I don't see that that's any different
from saying x gets the square root of n. I haven't told you how to get the
square root of n in either case.

R. RICH

If I wanted to discuss prime numbers as a mathematician, there are certain
things I want to know about. I don't want to list the first n of them. I
want to know what their general properties are. I want to know if there is
an infinite number of them, I want to be able to prove the theorem you
alluded to, I want to know that they get sparser and that they get sparser
in a particular way. I want to do lots of things. I don't just want to
list them, and there are lots of things I can say about x equals the square
root of n that don't have to do with its application to a particular n and
a particular x. That's the order of ideas that I'm asking you.

FALKOFF

Except that I'm afraid that I don't quite get the point because you're
entirely free to write expressions about these prime numbers without necessarily computing any of them. And to say that computing the first 500,000
has nothing to do with the mathematics of it, I think is rather unfair.
There are a number of things like various conjectures and Fermat's Last
theorem and stuff like that where the mathematics of it has turned out to
be raising the bounds by computation.

GOLDFINGER

I would like to react to what you said about PROLOG not having the problem
of assignment or specification or, I think you used the word "procedure",
it wasn't procedural, but I think there's an analogous problem that exists
in PROLOG and that is that PROLOG resolves all of its cases by construction. It always finds a solution and, therefore, can't handle truly an
existential quantifier. It's always universally quantified. So that
really underlying PROLOG exists the same problem. That it's going to go
and that it's going to do something. And you can't really work with a
specification of what you mean, and I don't know of any language that can
execute on a computer that doesn't have that problem. I was just thinking
whether there were any, but I don't know any. Maybe some other people do.

BARSTOW
I'll give a simple example of one. A system of equations simply describes
a relationship and you don't use it to compute anything until you decide
which things are the dependent variables and which things are the independent variables. And it's only after identifying those that it turns into a
description that you could use to actually compute something.

GOLDFINGER
Some of the languages like MACSYMA and so on?

BARSTOW
Sure.

FALKOFF
Well, you seem to be saying that as long as you defer execution, then it's
a non-procedural language, and I agree with that.

BARSTOW
Well, I thought that I was saying that the non-procedural side is that you
define the relationship without defining how it's going to compute. But, I
think maybe we should go on to our next speaker. So, I'm going to have to
ask you, unless it's real short.

BOUDREAUX
One quick question. Do you regard the sparseness of a language as an
intrinsic virtue or an extrinsic virtue?

FALKOFF
You're asking me?

BOUDREAUX
Yes.

FALKOFF
You'll have to tell me what you mean by sparseness.

BOUDREAUX
One of the points that you made in favor of APL was that it was intrinsically sparse and you could eliminate as many operators as you.

FALKOFF
I'm sorry. I don't think I made that point. In fact, it's very rich in
functions, and I think that's good.

BOUDREAUX
Oh. Thank you.

BARSTOW
I forgot to mention in the beginning that Saul Amarel had to leave early so
if you counted the number of people up here and came up with the wrong
number, that was why.

GOLDFINGER
It was an error in index origin, I think. (laughter)

BARSTOW INTRODUCTION

The next speaker was introduced earlier, Jaime Carbonell from Carnegie-Mellon.

CARBONELL PRESENTATION

Talking about numbers, I'll throw out a couple of them. I'm going to be speaking for approximately 10 nano-millenia. If you can't quite figure out what that is, it's about the same thing as a micro-decade. I'll be finished speaking before you get there. I want to make two points. The first is that a programming environment is at least as important as the language in which a system is designed, especially for large R&D systems in which building a system entails part of the learning process itself. The second point that I'm going to be making has to do with (the reason I'm making two points is because I don't have enough time to make only one) the second point is that it is sometimes difficult to scope the knowledge that's required to solve a task, and as languages become more and more abstract and move more and more towards the specification rather than by the step-by-step execution, very often implicit knowledge must come in that must be contained somewhere within the system, and how the implicit knowledge is brought in and what implicit knowledge is brought in is a major problem. It is an instance of what in artificial intelligence is called a frame problem.

Let me come to the first point now about the importance of programming environments. In language that does not have the capability for rapid prototyping, a language that is not interactive, a language that does not allow fast modification, one in which one does not have tracing facilities, breaking into a code, stopping at a particular point, setting break points, for example, direct interpretation of the source code, one which is not extensible in itself, etc., is one that is highly impoverished; it doesn't matter how elegant the language itself might be, without these additional abilities, these additional facilities, it becomes very difficult for a person to utilize that language in an effective manner. Moreover, if we look towards building larger and larger systems, some notion of automatic help or version control for modular establishing modules and building up hierarchies of these modules for doing things that we don't quite do today, such as automatic checking of communication protocols that have been established a priori by the programmer between the modules, automatic documentation aids, automatic help facilities, that the system both provides you about the language and also allows you to embed within the system as it is built rather than after the fact (when the person is building the help facility sometimes he may have a misunderstanding of how the system is supposed to be used) many of these things, some of which exist today, and some of which don't quite exist, provide the tool chest of added on capabilities that go on top of the language itself that make it highly effective, that make it humanly possible to deal with these languages. I think that this aspect of it has not been dealt with all that much in this conference so far and I wanted to make sure it did get raised.

For the second point, the importance of background information and the fact that in writing systems at higher and higher levels of specifications it becomes more pronounced, I think I will go back to the missionaries and cannibals problem. Remember, this is the one in which there are 3 missionaries and 3 cannibals on one side of the river and they have this little boat that fits only 2 of them, and in the process of carrying them across the river, you never want to have more missionaries than cannibals on one side or the missionaries might convert the cannibals. (laughter) Well, let us think about what a completely non-procedural problem solving engine would look like, one in which the procedures are not specified to it, just the problem is specified. Let us assume, moreover, that it has additional knowledge about the world. For instance, it knows that there might be a bridge somewhere. One solution that it could come up with is, well, why don't you go down, find a bridge, all of you together, all six of you and walk over to the other side of the bank in the opposite location to where you were before. Well, one obvious counter to that is, but there is no bridge. The system can say, but you didn't tell me that there was no bridge.

Then we have to add another axiom to the system that says, if I don't tell you that something is there in the problem specification, you can assume that it is not there. Now go off and solve the problem again. This time the system comes up and says, but the problem is impossible. Well, why is the problem impossible? The boat can't go anywhere, it has no oars. Well, what do you mean, it has no oars? You didn't tell me that it had oars and you told me to assume that anything that you didn't tell me about wasn't there, such as the bridge. What gives the oars a privileged status over the bridge? Well, that's a serious question and until we can solve questions like that it is very difficult for a system to implicitly add in information which is not defined as part of the problem specification in order to arrive at the solution. I hope that these two topics get discussed after Dave has had his chance to make the last presentation.

BARSTOW PRESENTATION

Let me give one way to think about the first point that Jaime made about programming environments. I've sometimes asked myself what ADA would have been like had the designers started out with the idea of building a programming environment instead of building a programming language. I don't know what the answer is, but I often wish that they had done that. I'm going to be very focused here and talk about a particular kind of problem, a design problem, by which I mean a problem in which you are trying to determine some composition of primitives which satisfies some kind of specification. And if you want examples of this, you can think of hardware design or software design. One aspect of design problems is often that you are supposed to find the best such solution, where best might mean use of the fewest chips, or is the fastest running or something like that.

One way to think about design problems is in terms of a design space. (See Figure 1) The idea here is that each circle represents a problem or a subproblem, and the process of designing something to satisfy a specification is one of either deciding that your specification is so primitive that you can solve it immediately, or that you break it into some composition of subproblems or subspecifications. You can imagine a space that you are exploring, each node of which represents some composition of subproblems that, if you solved them all, would constitute a solution to the original problem. Now the question is, how can you explore such a space effectively? And here, there are really two issues: one is that you want to get down to the solution at the bottom that is the best, that is the most efficient, say in terms of memory or the fewest chips or something like

that; the other is that you want to get there as fast as you can. In other
words, two out of the three aspects of efficiency that I mentioned at the
beginning.

What I would like to suggest is that a wide spectrum language will help you
in both respects. What I mean by a wide spectrum language is that the com-
posers and the components that are composed may be either primitive (these
are the very base level things, from which you are supposed to construct
the thing you are designing) or they may be abstract. You can think of
them as being conceptual things that perhaps describe a whole space of
possibilities. Second, the abstractions, the abstract compositions, the
abstract components may have multiple implementations which are described
as compositions of other kinds of components. And if you have this kind of
language, it provides a way of exploring the space.

In fact, you get two kinds of help in the search process of exploring that
space. The first is that the abstractions provide islands in the design
process. They are indices that give you hooks into any specialized knowledge which you might have about solving particular kinds of problems. For
example, you're designing a chip and you know something about memories. If
you can decide that one of your subproblems is going to be to design a
memory, then you can bring everything to bear that you know about memories
to solving that problem, and thereby reduce the amount of search.

A second kind of help is that these abstractions provide constraints that
you can often use to determine constraints on other parts of the problem
rather than the one you were just focusing on. For example, again in the
digital hardware domain, if you decide you can break the system down into
two parts that have to be run in sequence, and let us say, you know the
total time to run those two pieces, say 50 nano-seconds. If you know that
the first one can be implemented with something that will take 30 nano-
seconds, then that gives you a constraint that you can apply to the second,
namely, that you have to reduce that one down to 20 and that will help you
prune away some of the possibilities you might be exploring.

So, all I'm trying to suggest is that wide spectrum languages in which
parts of the language can be, and deliberately are, defined in terms of
other parts of the language give you a way of simplifying the search
process for design problems.

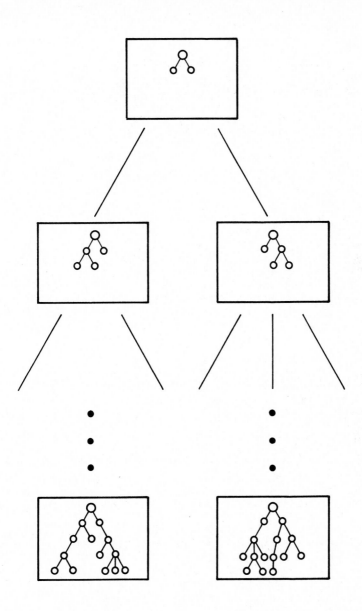

FIGURE 1

QUESTION AND ANSWER PERIOD - GENERAL PANEL DISCUSSION

BARSTOW
And now, let us open it up for general discussion.

E. RICH
I had a thought while listening to several of the people talking and I
wanted to see what you guys thought about it. It seems like something that
keeps coming up here is that there are things that we would like computers
to do for us, call it compilers or something else, that are very straightforward (everybody knows how to build a FORTRAN compiler that takes arithmetic expressions and produces machine code); there are other things that
we would like them to do for us that are much less straightforward. People
have talked about reducing non-procedural languages to procedural execu-
tions. Some examples that came up today were computing eigenvalues or
solving the missionaries and cannibals problem. One of the ideas that's
come forward in artificial intelligence is that what you want to do if you
want to make a black box - if you make an object that you, say, include as
part of a language and you give to people for free to use - is that it
ought to be possible to provide a formal definition for what the black box
does. In other words, it does exactly this in this fashion and it, well, I
don't care about the fashion, but you give it this input, this is what
you're going to get back as the answer. And if you have operations that
you can't do that for, because they contain a lot of knowledge or something, then they ought not be in black boxes. In artificial intelligence,
we have knowledge representation frameworks, and some of them contain a lot
of inference mechanisms, but the problem is that the operation of these
inference mechanisms is usually constrained by things like, well, how many
cycles are you willing to wait? And so they tried five inferences and if
they didn't come up with the answer by then, they failed. If you change
the representation only slightly five inferences is enough. In APL the
problem with eigenvalues is, of course, that depending upon the algorithm
you use and the number of bits in the machine you have, then you may get a
different answer because you have got to deal with numerical stability, and
there's just not a straightforward way to describe a correct procedure for
a lot of those numerical things, which is why I think a lot of numerical
analysts, at least, don't like including those complex things in languages.

What feelings do people have about this idea that we clearly separate as
language primitives those things that don't require this. In AI we call it
heuristic knowledge, but the analogy with numeric computations seems to be
pretty strong. We only put into languages as primitives those things that
we have very clear cut definitions for, and we force programmers to think
actively themselves about the other things, because otherwise, we lull them
into the belief that it's handled when, in fact, it isn't really. Anybody
who wants to answer that?

FALKOFF
To some extent, perhaps, I made an attempt to address that when I talked
about my concept of how systems are designed, and that as you progress with
the design and you identify those objects and functions on them which are
meaningful for the application, then you have, in effect, designed another
language. Now, if you have enough applications that come up with the same
kinds of functions, so that it becomes evident that these are universal
things, then perhaps they should be made primitive in some language. And
to a large extent, that is what has happened, particularly in APL. You

know, we have adopted all the mathematical functions. Most other languages have also.

BARSTOW
That's exactly what I would like to see people doing. There is a problem, though. As you get to higher levels of abstraction, there are often efficiency issues that effect the implementation of those abstract things, and that's why I'm interested in building automatic programming systems that can produce more efficient code.

CARBONELL
Well, one other way of thinking is that one can have a little bit of an a la carte language. The basic core of the language with all the primitives, like basic APL, or standard LISP, is what you get automatically and then given a nice enough environment, one can think of it in the simplest cases, just as in the library routines, in the more complex case, direct extensions to the language that actually actively change the syntax or change its selection of how some of the other primitives work to give them more power because routine assumptions can now be made about this problem class, because a programmer has explicitly chosen to build those assumptions in by loading in the particular 2 or 3 items from the additional extended language capability. If one views additional capabilities as a tool chest to draw from, it might be more productive than viewing them as either an on or off procedure from the basic language itself.

CARLTON-FOSS
I would like to respond to a part of that. I think this may be the time to emphasize the importance of documentation. I guess I would take the position that there should be no black boxes. I recognize that it may be a nuisance for users to have to know too much about the languages and tools they are using, but also it is crucial to be able to find critical information when it is needed. I learned this dramatically in the late 1960s when I was computing a Monte Carlo model of multiple scattering of charged particles through a sequence of lead bricks. The cross sections kept coming out the same on repeat runs with the same input parameters. Finally, I traced it back to IBM's random number generator, RANNO, which actually computed a non-random number that was hopefully "random enough" for most applications. But it wasn't for my applications, and I rewrote the routine to use the computer's clock instead. This gave numbers that were "random enough" for my purposes, although I have had conversations with others who have had to go further, measuring and digitizing the thermal noise in the computer chassis to develop numbers that were "random enough." All this says that it may be very important to have access to the insides of black boxes at one time or another. It can make the difference between a computer, its tools and its languages, being useful or a hindrance.

GOLDFINGER
With regard to APL, I think we can see what the problem is with regard to black boxes, and that is that there is really a spectrum of how black the box is. I just jotted down a few types of functions or operators, such as masking, which is something which you probably would want to do only one way. Then you can get into such things as sine and cosine, which we kind of regard as primitives, but really have numerical approximations involved in them; sort primitives, and certainly there are wide variations in the efficiencies of sort routines; and then we can go to something like the domino operator; and by the way, most APL manuals that I've seen don't tell

you what type of algorithm is being used for matrix inversion. So that we
have degrees of blackness, and I would agree that documentation is one way
to deal with this. The type of algorithm that's used should be published.

It might also be interesting to have some sort of system variables that
enable you to choose what particular implementation of a given function or
operator is used in your work space, so that if you don't like one type of
domino operator because of the type of matrix you're using, you change your
system variable, and you get a different implementation of it. Let me just
say that an idea came to me while we were talking about these primitives.
If you include enough primitives in your language, you can make it very
easy when you compare your language to others to make it look like it does
a good job of solving something just by making a problem you've already
solved primitive in your language, and I wonder to what extent when people
write these bench marks and show you how easy it is to write something in a
language they're really being fair. Maybe the way that we compare languages should not be on how well they do what they've been designed to do,
but how well they do what they've not been designed to do.

GUIER
I would like to second what several of you said about the modality or the
slowly building up hierarchies of programs. If you have a language that
has the proper primitives, and I think APL is one of the better examples
now, you can build up other programming languages; that is, other syntaxes
and other semantics and so on. It looks like a very different language,
and build it up on top of that original one, and accrue some, anytime you
want to, go back down and use the primitives in the language. You saw at
least one example in Bob Jernigan's APLLOG. He has another one, APLDOT,
that is a relational data base retriever. I think if we thought in terms
of sort of modalities or hierarchies of these kinds of things, if I need a
special kind of a language, turn that kind of a higher level component on,
and then maybe turn it back off if I don't need it again. Going one step
further, if one could find somewhere a really universal set of primitives,
it would be the ideal, where if you like a particular language and somebody
else likes a different language, there would exist an automatic translation
from one to another. I think we're getting very close to that in some of
the relational data base systems because that language is a very restric-
tive language in the sense of a relational calculus. So I would like to
end with reinserting the idea of some automatic translation from language
to language.

TEMIN
I would like to address my question to Dr. Carbonell, in particular, since
he brought up programming environments. Of the things you mentioned and of
many of the things Dr. Shneiderman mentioned earlier, these have been
implemented in various systems, not all of them at once. But, in particular, the INTERLISP system contains a lot of things you might like; even the
UNIX environment, including the source code control system, contains some
things you might like; and yet, take LISP, in particular, you see INTERLISP
come along which has been around for a few years, and Franz LISP which I
think came afterwards doesn't include a lot of the things that I think you
might like to see in INTERLISP. So here is a newer system which a lot of
people use which doesn't include some of the older stuff which we may
consider to be better and at least would like to have around. So I see
that there are a lot of things that we want, but there are so many of them
and they're so distributed, how do we ensure that the newer languages and

the newer environments contain all the old stuff plus the new stuff. We really can develop these by their large but very useful environments. Do we as researchers have a responsibility for this and is there anything we can do at least to encourage it if not to cause it to happen?

CARBONELL
Okay, let me first address a little bit the INTERLISP issue. INTERLISP is an example of a language that is larger, clumsier, slower, and has many other such detrimental attributes compared to something like MACLISP or COMMONLISP or Franz LISP, yet a lot of people...

BARSTOW
That is not a universally held opinion.

CARBONELL
I haven't got to the good part yet. Yet there are a lot of people, especially in the West Coast or who are transplanted from the West Coast who will swear by it. I used to be a medium M-LISP user, which is a precursor of COMMONLISP, and I believe that the reason why it is still widely used is because of the programming environment that it comes with; it is more than enough to overcome some of the other detriments, including especially efficiency. This efficiency is the main point in which it is worse than the others. It's also worse in terms of uniformity than something like COMMONLISP, which was designed rather than grown by a accretion. So, in spite of all the negative things that I had to say about INTERLISP as a language, it's still useful because of the environment. So this is the trade-off. Now, the other point that you made had to do with the promulgation of many of these tools - version control and so on - some of which are used on UNIX that you mentioned. I think that the only real responsibility is to teach the use of many of these environmental tools and make them far more friendly and directly documentable and directly self-documenting in order to make them more universally available. I think it's just a matter of lack of distribution of information there, why some people use it with very good effect and other people don't seem to have heard of it.

FALKOFF
I think for the sake of the record and those people who are here who are not familiar with APL systems that I should note that this list and Professor Carbonell's first list are all included in APL: interactiveness, the trace, the stop, and so forth, come standard.

JERNIGAN
We've had several comments on APL and other languages and how they influence the way we formulate problems or think about problems. I find in my experimentation with PROLOG that it greatly influences the way I think about relational data bases. There is a kind of modality of thought that has been eluding me for a year and I thought I would like to throw it out and see if you can offer some programming language suggestions as to how one might master what is maybe my lack of ability to grasp a particular kind of problem solving mode. And it is illustrated with the Tower of Hanoi problem, the recursive solution to the Tower of Hanoi. A very simple, little, almost one-line APL expression or PROLOG expression, whatever recursive language you would like to use, can give you a very quick solution to the Tower of Hanoi problem. But when I sit down and try to work on this problem, try to grasp what is going on here to the point where I feel like I have mastered this thing, to the point where I might be able

to master chess at some point, it's elusive, that is, the language form itself does not tend to shed any light on how the solution is formulated. In Jaime's presentation yesterday, he was suggesting certain kinds of solution forms, and I think some of those seem to lie outside the language and require a kind of modality of thought that is, maybe, not too directly expressible. Let me throw it up as a question: Are they directly express-ible? Can you offer a language form that would help those of us who are trying to deal with certain modalities of thought, like recursive solu-tions, so that we can really master those kinds of problems?

FALKOFF
Are you saying that having a recursive solution is not satisfying to you?

JERNIGAN
Well, no, I find it very satisfying, but after contemplating some recursive solutions I've seen and even some I've come up with, I don't know how many I've missed. That if I sit and contemplate a chess board, I can't envision that it may not be practicable, but I can understand how someone could, if they played 100 simultaneous games of blindfolded chess; that doesn't seem outrageous, there's something in that idea that I can grasp. But to follow to a depth of moving three disks blindfolded using a recursive solution, I find it almost an impossible task. There is a modality of thought in this recursive thinking that makes me get lost almost immediately. I can't work it through and I wonder is this a lack of the language or is there a way we can come up with a language that could address maybe more innovative and new modalities of thought?

CARLTON-FOSS
I can take a first cut at that maybe. This may relate to an axe that I have ground for some years now. Basically, the issue is a person's natural mode of thought, one's cognitive style, or the schema which one naturally uses or feels combortable with. As an example, I was talking during lunch with Maria McAllister, who is a mathematics professor at Moravian College, about people's mode of thought with respect to mathematical problem solv-ing. As a person who had a cognitive match with advanced physics but not with classical physics, I found myself one day during my junior year as one of only three remaining physicists in Warren Ambrose's notorious course on advanced analysis, able to prove certain theorems, but dissatisfied, unable to find crucial meaning in what I was doing. I resolved the problem by finding some books to create a bridge between Ambrose's very abstract differential forms and the differential forms that had a recognizable application in physics. I think that this abstract-concrete dimension is very important, and is a part of the differing orientations of different people who embrace one language rather than another, one discipline rather than another. My impression has been that those who made a commitment to mathematics, as Maria suggested earlier today, were those who derived satisfaction from the pure manipulation of the symbols, who did not require grounding in concrete reality and applications, who found adequate meaning in the manipulation per se. This, incidentally, can and has been quanti-fied using psychological testing. So, I might suggest a reformulation of the question from this perspective: How can we make some of the computer languages more concreate (as perceived by the users of a given type), or at least seem that way?

CARBONELL
I just want to mention very briefly that in none of the computer languages
or even the AI languages, is recursion ever something that you reason
about. It is something that you reason with. It is a tool rather than an
object which you can introspect about, that you can analyze, and that you
can use as a constructor, as a declarative constructor, to figure out how
it works. That can naturally be some ornament of languages to some degree,
but that was one of the things that I was trying to get across in the
previous talk. It's important to have a language that can not only carry
out the recursions, but carry out whatever control structure or data
representation it has. I can also introspect about it and reason about its
properties so that it can try to actively construct a particular recursive
solution, or perhaps search the space of possible recursive answers to
problems of this type.

GOLDFINGER
I think you're getting on a very real problem that exists with recursion,
and I think it's an emotional problem. I was trying to teach recursion in
LISP to some gifted high school students. They were very bright, and they
were very good at solving problems and picking them apart and analyzing
them. And I realized that the way that I had to teach it to them was to
say, let us start out by assuming you have already solved the problem.
Then what do you do with your solution? And you build a more complicated
case. Then I said okay, don't worry about solving the problem now, which
is what you do in recursion, you reduce it to a simpler case, which is
already solved. And there's an emotional block. What do you mean, don't
solve? I've told you what I did with the solution; how did I get the
solution, how did I construct the solution. And there comes the leap of
faith that the algorithm will do it. That's something which you have to
give up when you think recursively, and I think that's very hard for people
to do.

CARLTON-FOSS
Has anything that any of us have said touched on your sense of the
situation?

JERNIGAN
No.

CARLTON-FOSS
Not even the one about it being emotional?

GUIER
Would you be satisfied with it both being emotional and having mystified
mathematicians for most of the historical world? My thought on that subject and one which I have done and helps me, with recursion in particular a
lot, is to get at an interactive terminal with an interactive language and
simply play with it, not just run the recursion, but alter the recursion
formula or put in a very small perturbation into the recursion formula and
watch how the structure changes. That is one of the best examples of how
to really get insight using an interactive terminal that I can think of.

JERNIGAN
I was not trying to concentrate so much on the recursion itself, just using
that as an example of a modality of thought and wondering how many others
are out there, including that one, which do pose some difficulties.

PIETRZYKOWSKI
Sorry, but I would like to proceed with this issue. I don't think that
it's so mystical. I think that problems which are arising are that many
people do not know exactly the underlying mechanism. I am talking about
logic programming, particularly. It becomes much more clear when you look
at the search space in terms of graphs; then the possible recursions, which
are visible, become completely visible, just as different paths in the
graphs which are available. I would believe that some heuristic devices
which will symbolically execute in PROLOG try to do something, like they
draw little trees of what they are searching - they will help to de-mystify
the whole thing.

GOLDFINGER
I want to say one more thing. One thing that bothers me sometimes when I
see people talking about the nature of thought processes by giving an
example is that usually it's a problem that they have solved, and they show
you the solution and then they show you different ways of thinking it
through, which are more elegant. I recall when I was a graduate student, I
did my thesis in physics on something called the "Kondo problem", about a
magnetic impurity in metal, and this was a very important problem in
physics, in theoretical physics, because there had been two classic
problems in many-body physics that had been solved many years before, which
are superconductivity and liquid helium, and each of them was solved by
some unique method. And after they were solved, people developed tremendous formalisms, better ways of doing it, which were quite general, but
they didn't have a new problem to apply those formalisms to. And when the
Kondo problem came along, they applied all these formalisms to that problem
and none of them worked. That became a problem that was eventually solved
by its own unique method. And sometimes I see, especially when I read an
AI text, there will always be a chapter on the use of constraints in
problem solving. And usually it will be the work by Waltz on blocks world
vision, and it's not really a general method, it's an example of how
someone was clever. Usually, when you run into your own problem, it's only
after you've solved it by some sort of really clever method that you come
up with what is a general approach that would have worked if only you had
known it to begin with.

FALKOFF
I think my own experience is somewhat limited, but I did, in preparation
for this, fool around with a couple of simple recursive things, and the
observation that I have is that the exercise of actually writing the
recursive function, where you're forced to contemplate and decide on what
are the actual outputs, helps to understand the problem. And I think that
when you speak about it in general - for example, to merely say you have a
recursive solution to the Tower of Hanoi puzzle - unless you come out with
some string of numbers that in some representation tells you what to do,
you haven't really gotten any insight into the problem except that it could
be done recursively. When you do have that string of numbers, then you can
look at them and perhaps develop an iterative algorithm that would produce
the same string, so then you have got another solution.

E. RICH
I guess I want to comment on Bob's notion of recursion because I think it
is special, and I think that there is a straightforward explanation for
it. It's something that hasn't come up so far in the last couple of days,
which is, we're thinking about problem solving in languages and computers

and people. The fact is that people have different hardware than computers do. And it turns out that there have been experiments, people simply can't manipulate stacks deeper than about two. Someone did an experiment about twenty years ago on nested sentences in English and it turns out that you can give sentences like "the cat that ate the rat that buried the mouse" or whatever, how deep you like and everything's cool. If you nest them, though, you can't do it if you say "the cat that the rat that the horse that the cow ate bit chewed, chewed, whatever." Now you're going to come back and tell me that German is a counter-example, to which I will say only in written German do people do it more than about two levels deep. And you put in all kinds of clues. And of course, in written language, it's different because you get to go back over things. There's a whole lot of evidence that people simply can't, in terms of their short-term memory which can only hold maybe four things, manipulate stacks very well, and computers can manipulate them very well. We can understand how to do it. We just can't carry it through; but you can do it fine if I give you a piece of paper and a pencil. So, I guess what I'm trying to say is I think the answer to your question with recursion is it's not mystical and it's not emotional; it's purely got to do with human short-term memory, and computers have different memories, and they manipulate stacks real well. But that doesn't mean we can't understand how they work; we can understand how they do long division, we just can't do it very well, because we don't have the right kind of memory.

FALKOFF
I would like to suggest that people have different hardware. I don't think that you can learn to play 40 games of chess blindfolded simultaneously unless you have some special hardware, and I think the same thing probably applies to many other fields. This was touched on earlier and we have had some corridor discussion about it. It leads you to the idea that maybe not all programming languages are for all people, and when we talk about programming languages to help solve problems, maybe they have different classes of audiences and we should take that into account.

CARLTON-FOSS
Yes, I agree, although I think it is also an issue of "software" or "operating system" programming, that is, that many people can also adapt to a new language even though it may not initially match up with their natural ways of thinking. This observation comes from work with students, but also from my personal experience. When I was first looking at recursion, when learning LISP, I was mystified by the radical differences that resulted from using one operator versus another in the recursion formula. But it was simply a matter of playing with recusion, tracing through each step of what the computer did in executing the algorithm, and it quickly began to make sense both logically and intuitively. And it was not an issue of how many steps there were in the recurision, but of getting accustomed to the process, the inner logic, of the recursion.

McALLISTER
I understand that the name of this session is "Language Requirements for Problem Solving," and I'm very much interested because the types of problems that I generally spend time on and that I would like the computer to help me solve require different aspects. I have a pet peeve that I'm supposed to know which is the correct language for the correct environment. First of all, I hardly know what the correct environment is because I don't know much about the problem, and I want to find a solution for it, I want

to have a mathematical method or an approach or a systematic type of
analysis that I hope will eventually lead me to a solution or a variety of
solutions if there is no unique one. What my question really is concerning
now is, if we want to have nice languages, they should not be geared only
to logical manipulation or to number crunching. Classically, I was raised
to believe that FORTRAN, if you wanted to do very fast number crunching,
well that's it. LISP, I understand, if you want to do some kind of logical
manipulation. I want both.

GOLDFINGER
Use APL.

McALLISTER
I have my own opinion about that, and you do not want to hear it.
(laughter)

BARSTOW
One more comment and then I'll get the last word on several issues.

BLUM
My idea is inviting some comments from the panel. Elaine raised the question of the different kinds of hardware, the different kinds of capability.
Having worked in medical systems for a long time, we've been developing
medical systems that complement what a physician can do. So, for example,
an EKG system is really very good at averaging, finding the spikes and the
quantitative things and the physician is very good at recognizing patterns
and seeing all those things come out. In the case of a clinical information system, the system is very good at remembering all kinds of data and
passing it through sieves of if-then-else clauses and so on, and the
physician is very good at integrating that and taking the things that he
gets that are not in the computer, that he recognizes subliminally, and so
on. So I just really wanted to ask of the panel, in terms of problem
solving languages, what can you think of or what can you say about having
the two systems complement each other? In other words, the capabilities of
the human brain are to be able to recognize parts of the solution but not
be able to do other parts especially well, whereas the computer as an
analog shouldn't necessarily try and do what the human does well, but
should have all kinds of capability-we mentioned recursion and so on.
Where does that fit into our concepts of language design? Or if you don't
want to answer that, just do your final question, your summary.

BARSTOW
Does anybody want to give a short answer to that?

FALKOFF
Problems are solved by people, and once you can do it just by machine, it's
no longer a problem in my opinion.

GOLDFINGER
I would like to disagree with that. There is something which we might call
the "blivet factor". I am never really quite sure what a blivet is but I
know that when you push it in and try to constrain it, something pops out.
There was an article that appeared within the last couple of years in
"Creative Computing" about an attempt to run a Turing test on some sort of
limited problem domain. A Turing test follows the suggestion of Turing,
that the way to determine whether or not a machine can think is to attach a

couple of terminals, one to a machine in another room and one to a person
in another room, and see whether a person sitting at these two terminals
and allowed to do anything they want could tell which was the person and
which was the machine. And according to this story, which "Creative
Computing" maintained is true, they called in a number of people to run
this test. One of the people they called in was a businessman. They
explained the rules to him, and he said, "Okay", and he sat down at the two
terminals and sat there for about ten minutes and didn't do anything. So
they said, "I don't think that you understand the nature of the test.
Shall we explain the rules to you again? And he said "Yeah." They said,
"Well, you can do anything you want on these terminals, and you have to
tell which is attached to a man and which is connected to a machine." He
said, "Okay." He sat for another ten minutes and didn't do anything. So
they said, "What is going on here?" He said, "Well you said I could do
anything I want, and I choose to do nothing." And after about another 15
minutes, one of the terminals starts typing and says, "When does the test
begin? Anybody out there?" And he says, "That's the man." (laughter)
Now we have done some work on biomedical systems and we've always taken the
philosophy that certainly at present (and we believe for all time) biomedical systems and clinical systems will be aids to the clinician, aids to the
person, because the person will always have that blivet factor. No matter
what you do, there's going to be something which the person's going to
stand back and come up with, some sort of an "Aha" that breaks the rules.
My personal bias is that that will always be.

BARSTOW
I'm going to comment on five things, but they'll each be about two
sentences. (1) On this notion of cooperation between people and machines,
I think that one of the things that is important to stress is that in order
for there to be useful communication between people and machines, they have
to have at least some kind of shared knowledge, and so we can't expect the
machine to be a dumb machine and not have some of the same knowledge the
person has, or else they won't be able to communicate. (2) Jaime mentioned
INTERLISP-D and the programming environment, I won't comment on relative
efficiencies, but I will say that the reason that I am still an INTERLISP
user and expect to be for many more years is because of the programming
environment. (3) Recursion - I think when you think about it recursion is
basically just a kind of problem solving skill. When I try to teach it, I
just teach it as another kind of problem solving and I don't even try to
call it a name. You say, well, you've got a subproblem, and by George,
you've already got a routine written that will solve it. So you're done.
(4) With respect to German, when I was living there, it is true that in
spoken German, they'll stack them very deep, and one of the things that we
used to find ourselves doing was predicting what the unwinding of the stack
would be, and a comment on the redundancy of human language is that we were
often right. (5) Finally, one of the things that I realized during this
discussion is that there is a major feature of several programming languages that hasn't been mentioned at all, and I'm not sure why. So let me
mention it now, and that's the inheritance hierarchies that Smalltalk
provides. They are to me a superb organizational tool. They help me think
about the problem domain that I am trying to understand, so I'm a little
bit surprised that none of us, including me, mentioned it earlier. So let
me suggest that the notion of inheritance is in fact a very important
feature for languages to assist us in understanding domains and in problem
solving.

PANEL SESSION:
COMPARATIVE APPLICATION OF
COMPUTER LANGUAGES TO
PRACTICAL PROBLEMS

The Role of Language in Problem Solving I
R. Jernigan, B.W. Hamill, and D.M. Weintraub (Editors)
© Elsevier Science Publishers B.V. (North-Holland), 1985

PANEL SESSION:
COMPARATIVE APPLICATION OF COMPUTER LANGUAGES TO PRACTICAL PROBLEMS

Panel Chairman: Bruce Blum, Applied Physics Laboratory, JHU
Panel Members: Roy Rada, National Library of Medicine
                  Elaine A. Rich, University of Texas at Austin
                  Jean Sammet, IBM Corporation
                  Ben Shneiderman, University of Maryland

BLUM PRESENTATION

This is the panel on Comparative Application of Computer Languages to Practical Problems. It may to help analyze that title one phrase at a time.
"Comparative Application of" suggests that we are comparing different
things. This raises the question of criteria which, in turn, brings up the
question of measurement. Most are familiar with Lord Kelvin's statement,

> When you can measure what you are speaking about, and express it in
> numbers, you know something about it; but when you cannot measure it,
> when you cannot express it in numbers, your knowledge is of a meager
> and unsatisfactory kind: it may be the beginning of knowledge, but you
> have scarcely in your thoughts advanced to the stage of a science.

The psychologist George Miller, however, offers a different view:

> In truth, a good case could be made that if your knowledge is meager
> and unsatisfactory, the last thing in the world you should do is make
> measurements. The chance is negligible that you will measure the right
> things accidentally.

Clearly, this was the case when John von Neumann remarked,

> There's no sense being precise about something when you don't even know
> what you're talking about.

Do we know what we are talking about? When we compare applications are
they indeed comparable? I think not. We have no taxonomy for systems and
applications. Our software engineering databases combine prestructured
programming applications with structured systems, business programs with
embedded projects, assembly language with fourth generation languages.
There are too many variables, and hence there can be little confidence in
many of the findings. Naturally, there are some phenomena which are so
strong that they become evident in spite of our poor measuring tools. The
fact that individual productivity is inversely related to project size is
one such example. Nevertheless, there are major differences among applications, and the magnitude of these differences may obscure any significance
associated with a program language effect.

The next two words are "Computer Languages." Does that mean something
processible by a computer? In that case, it includes natural language as
well as the traditional high order languages (HOLs). Or do we mean
instructions for computer processing? Here we include the HOLs and all

other kinds of user inputs. For example, when one uses an automated bank
teller, is one inputting data or writing a program with a language having a
very limited syntax? What of using a spread sheet program, or performing a
computation on a programmable calculator, or coding and executing a calculation in BASIC? Is it an assumption of this conference that computer
languages are those languages for which the programmer is assumed to be the
user? Is our concern the role of language in problem solving with respect
to (a) the user with the application problem or (b) the programmer with the
program implementation problem? These may (or may not) be different.

When one speaks of programming, several biases come to mind. Is programming a process or a product? Programming is therapeutic; when I get
depressed I write a program because it is fun. In this case the concern is
with programming: the activity becomes the product. But programming ought
to be a process that leads to something else. Too often we focus on the
things that make it an enjoyable (or intellectually rewarding) product as
opposed to its role in the process of developing some useful product.
Another question about programming: is it a skill or is it a profession?
We have finally emerged from a three-year period where it was common to
equate computer literacy with learning BASIC. We now are getting to the
point where third graders are learning how to program. When only professional programmers had the skill, the issue of professionalism was clear.
Are we now moving to a period in which programming will become a universal
skill? If so, what will distinguish the casual user of programming
languages from the professional user: the class of problem, the degree of
abstraction of the language, the requisite application specific training?
Will programming languages become general purpose problem solving environments, or a family of highly symbolic vehicles for solving a narrow class
of problems, or both? Certainly, the kind of problem will define the
language characteristics. Finally, we note that not only is programming a
molder of thought, but it is also a molder of emotion. Let anybody who
disbelieves that try and say something nasty about APL to this audience, or
go to a MUMPS users' group and say something nasty about MUMPS, or a FORTH
users' group and say something nasty about FORTH, or a UNIX group and UNIX,
or an AI group (in the United States) and LISP. Such emotion limits
rational (or scientific) enquiry.

We are now ready to examine the final two words of the panel's title,
"Practical Problems." Consider the following comments by Belady and
Leavenworth about a different field

> ...**software engineering** is polarized around **two subculture**s: the
> speculators and the doers. The former invent but do not go beyond
> publishing novelty, hence never learn about the idea's usefullness - or
> the lack of it. The latter, not funded for experimentation but for
> efficient product development, must use proven, however antiquated,
> methods. Communication between them is sparse, and questions as "Does
> data abstraction improve modifiability?" can be answered by neither.

First we must note that it is obvious that computer languages are able to
solve many practical problems. Yet the two subcultures concept suggests
two viewpoints. Some concentrate on the problem solving environment, while
others are interested in the problem space. The complexities in each area
are enormous, and it is difficult to bridge the gap between the subcultures. Naturally, it is hoped that conferences such as this help to
identify the issues and foster communication.

Having spent my time telling you what the panel may discuss, let me close
with the following observation. Suppose that we had the best of all
possible languages, our problems would not disappear. There would be
issues of transportability. Even with our advances in networking, we still
have difficulty in communicating at the level of applications. Programs
which work with one operating system cannot be transported to another;
concepts demonstrated in one environment may be impractical in other
environments.

Secondly, there is our investment in obsolescence: program libraries,
training, databases. COBOL continues to be the most common mainframe
language in use. This is a measure of its success. The large backlogs for
new applications are a measure of its shortcomings. If a perfect substitute for COBOL were to exist, how long would it take to replace the COBOL
investment? This barrier to transfer is deeply rooted in the two subcul-
ture phenomena previously noted.

Finally, there is an unpredictability about the future. As our computer
technology advances, new applications emerge. We have recently seen this
with personal computers. The interest in artificial intelligence is
strongly linked to improvements in equipment which have made some AI applications economically feasible. One can expect such growth to continue.
The result will be new perceptions about practical problems and - perhaps -
new languages to solve these problems. Nevertheless, there are enough
challenges in using the technology we now have, and I look forward to
hearing what the panel has to say about the subject.

RICH PRESENTATION

We have spent the last two days discussing ways that the structure of a
programming language interacts with problem solving using the language. I
thought that what I would do is to confuse things further by saying that
what we need to do next is to broaden our notion of what a programming
language is. What I want to suggest is that in the early days we started
out thinking that languages were assembly languages. We moved up to
FORTRAN and various other languages that we use to communicate with
computers, which we now think of as programming languages. The thing is,
though, that just the languages that we usually think of as programming
languages are not the only languages that we use to communicate with
computers. Every time I build a program and I want you to use it, there
has to be a way for you to make input to my program and to tell my program
what to do. The way you do that is in some language. Now it may be a
trivial language; there may be five commands to my program and they may
each have single letters that you type. It may not be what you think of as
a linguistic language. You may draw pictures on the screen and point at
things, but it is, nevertheless a language that you use to comunicate with
a computer.

So for an example, every time I make a data base program there has to be a
way for you to make queries to my program. Figure 1 shows an example of
one such query language, INQUIRE. This particular query accesses a data
base that knows about badges and which badges are allowed in which areas of
a building. As the result of the query, a table showing which badges can
go where and how many badges can go to each of those places, and so forth
is printed out. If you look at the query, it looks a lot like a program,

so you are probably not too unwilling to believe that it is written in a
programming language. Some data base query languages look a lot less like
programming languages and perhaps a lot more like English, but what you'll
certainly agree to is that we need some way to communicate with these
systems or we might as well not have them.


```
      SCAN
      ,SEARCH IN CA7
          (CA EQ (' 31' '230' '231') SET COL OF AJ)
       AND COND IN LOCATOR
      ,DEFINE AJ TABLE(SET ,)
      ,COMPUTE MM FORMAT (I 4) ( N )
      ,BREAK BEFORE GROUP
                    TAB SKIP 2 'GROUP IS :     ' GROUP * *
                BEFORE IDNUM
                AFTER IDNUM
                    COMPUTE N FORMAT (I 3) (N + 1)
                    TAB BADGE 1 NAME 8 TOTAL OF AJ
                      43(H* 4) EMPSTAT 61
                AFTER GROUP
      ,TITLE
            BADGE 1 NAME 8
              (BLDG3 NIGHT   31) 40(INNER DAY 230) 47
              (INNER NIGHT 231) 53(EMP STAT) 60
        TOTAL TAB SKIP 'TOTAL PEOPLE:' 8 MM 22
                    TOTAL OF AJ 40(IB 4)
```


FIGURE 1
INQUIRE

Let me give you another example of a language for communicating with a
program. Aaron Temin told you yesterday a little bit about the help system
that we are building for the document formatting program SCRIBE. I want to
claim that SCRIBE itself, as a language, is a sort of funny language
because it allows for commands and then large hunks of stuff that don't
look like commands in any language. This is because you interleave
commands, which in SCRIBE start with "@" signs, with actual text, which is
what you wanted to format, and SCRIBE then produces some formatted object,
which may or may not be what you wanted. But in any case the input to
SCRIBE is in some language. Figure 2 shows an example of such an input.
SCRIBE's input language is actually very powerful. It is essentially
equivalent to a Turing machine. Since there are variables and you can
define functions and macros, so it is probably, even using a very strict
definition, a full-fledged programming language. The thing is, that if you
went up to many of the users of SCRIBE, including many of our secretaries,
and said, "You know that what you do everyday is to program, when you write
this sort of thing," they would say, "No, no, no I don't program! I don't
know how to." Well the fact is, they do know how to, and they are doing it
every day.

```
@make(report)
@use(bibliography "ai.bib")
@style(font "computermodernroman10")
@define(sentence=quotation,indent 0)
@center(A.I. and Languages)
@center(Elaine Rich
@value(date))
```
The purpose of this paper is to survey the recent work
in the area of artificial intelligence @cite(Rich)
and programing languages. To see
the problem, consider the following example:
@sentence(Sort the list A)

FIGURE 2
SCRIBE

Once we broaden our notion of what a programming language is, then the
question is, "What kind of programming languages ought we to have?" I
would like to say in answer to that, "I don't know." And in fact, what I
want to say is, "There isn't any one answer." Consider Figure 3. The
important point is this. What it means to understand language, either when
you do it or I do it or a computer does it, that there are statements in
some language and they are mapped by the dotted lines somehow, into some
set of target actions. For example, when you understand what I have said,
what it means is that you have taken what I said and translated it into
some action that you can perform in response to my statement. Your action
may be direct, or it may be an unobservable action like storing a fact in
memory for later use. It only makes sense to be able to say things that
can be mapped into an available action. If you have the situation, shown
in the top half of the figure, in which there are many more things you can
say in the language than there are actions that can be taken by the
receiver of the language, then you have overkill. You have put a lot of
extra burden on your language understander. So, for example, if you took a
traditional line oriented editor in which one can insert, delete, move, or
whatever ten things one can do, and you defined the interface language to
be English, because people already know it, then the problem is that there
are only ten things on the right (actions) and there are millions on the
left (English statements) and it doesn't make a great deal of sense. You
will spend all your time trying to understand which one of the ten actions
is desired. Also, as was pointed out yesterday, you have a problem if
there are objects in the input language that don't even follow the dotted
lines. In other words they don't map to any action in the target program.
It does not make any sense, for example, to tell your text editor that you
want to get a ticket to fly to New York because it doesn't know how to get
airline reservations. So no matter how sophisticated the language's understanding is, it cannot understand what you say.

Now let us consider the other side of this question (since you always have
to have symmetry in these things). Suppose we were smart enough to make a
target program that knew how to do lots of things. Suppose artificial
intelligence were able to build a program that could actually simulate all
of the things your secretary does for you. We are talking about the whole

business – folks showing up in your office who are determined to see you
and you don't want to talk to them and how your secretary convinces these
people to go away – the whole range of things that human secretaries might
do for you. Now the problem is you have a lot of actions. If you have a
simple language, then there isn't any way to map onto each of these actions
uniquely because there are fewer things in the language than there are
actions to be done. This is the situation shown in the bottom half of
Figure 3. When this happens you are wasting a good deal of the performance
that is available from your target program. So what we need is, as our
target programs get better and better, we need to have fancier and fancier
languages.

Language   Actions

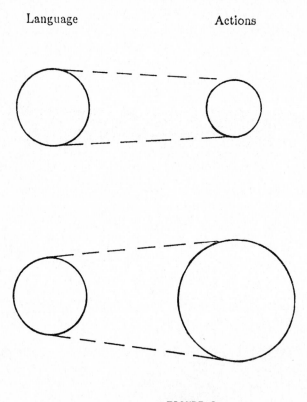

FIGURE 3
UNDERSTANDING

I think this is the answer to whether or not one ought to program in
English. My answer to that is that if your target program can do only a
small number of things (such as is the case for a text editor or the
underlying interpreter for most traditional programming languages), then,
no, you certainly should not program in English because it would be overkill. On the other hand, as soon as the number of things you want to say
gets very large, you are in a position of having to choose. Either you use
English or you invent some other language which allows you to say a similar
number of things. No one has yet succeeded in inventing an artificial

language that lets you say as many things as English does, but is as good as English at doing it. You can invent artificial languages when you only need to say a small number of things. For a very large number of things, though, English (or some other natural language) is probably the best choice.

So I would like to just conclude by saying the two points I wanted to make were that, when thinking about programming languages we ought to pull up several levels as we think about what they are, to consider that they are any way we have of communicating with a computer system. This system may be a very sophisticated performance program or it may be just a machine that can execute machine code. As we think about the language we ought to use to communicate with this system, we ought to make sure that we get a good match between the language and the target system. And that means that for most current programs we ought to design simple elegant languages that are relatively easy to learn and easy to interpret. But if we ever get to the point of being able to build very sophisticated programs we have got to confront the problem of selecting similarly sophisticated languages. The bottom line is that English is not as difficult as it is just to be perverse for folks who want to make programs to understand it. It actually is the way it is for a reason, namely that people need to say lots of things to each other, and if we ever get to this position with respect to people and computer systems then we may have to confront the same complexities there. I'm not claiming that we have reached this point yet, or that we are going to be there in the next few years.

SAMMET PRESENTATION

I want to talk to you very briefly about a class of languages that have not been addressed at all at this meeting in any way, shape or form. Yet they come closer than almost anything I've heard in the two days I've been here to try to really deal with the needs of the users and help them with their problem solving. This is a class of languages that I call "languages for specialized application areas," and are sometimes called "special purpose languages;" I will give you some examples in a minute. First of all, let me point out - for those of you who may think that they don't really exist - that there are literally hundreds of these languages and I could give you names, and citations. Thus, this is not some toy activity that is going on in only a university or research lab; these things really <u>are</u> important. Let me give you the names of just three which I chose because they were convenient; I want to see how many people have ever heard of COGO, APT, COURSEWRITER or any language you think might come into that same kind of category. Could I just get a show hands? Good, the results are better than I thought.

These languages use vocabulary and functions from an application area, and this now begins to touch on the point that Elaine Rich just made. You need to be very concerned with how much you can say and how much you can expect the interpreter or compiler or whatever is doing the translation to do for you. Now, many of you said you have heard of these languages, but how many of you have seen <u>examples</u> of any of these kinds of languages? Fewer but at least some.

Let me walk you through this example of COGO, (Figure 1) and then I'm going to show you another one, which I will not walk you through. COGO stands

for COordinate GeOmetry and is one of a whole class of languages for civil
engineering. I doubt if there are any civil engineers in the audience.
The problem to be solved is this: you are given the coordinates of a point
with coordinates 1,000 and 2,000, you are told that there is another point
which is a distance of 256.17 units, (whatever they may be) and at an angle
of 45°, 00 minutes and 00 seconds. Similarly you have another point with
its distance and angle. The problem is to find the coordinates of both
those points and the area of the triangle, and I guess if any of us were
smart enough to remember any of our high school geometry and trigonometry
perhaps this could be done. It turns out that there are civil engineers in
the real world who may or may not remember much about high school trigonometry but have a programming language which helps them. For example, there
is a command called STORE, which says essentially "record the coordinates
of point number 1 as being 1,000, 2,000." Now there is a command and
executable statement called <u>locate azimuth</u>, which says "locate point 7 by
going from point 1, a distance of 256.17 at the specified angle and locate
that point. You'll notice that the command does not say record the location, but it does say locate the point. Then there is another command to
do the same thing for point 95. The final command in this particular
problem says AREA and it gives the three point numbers. By now all the
necessary information has been calculated and stored so the area can be
computed. PAUSE is just the end of the problem.

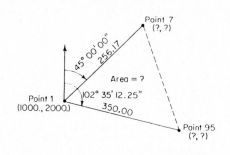

STORE	1		1000.	2000.		
LOCATE/AZIMUTH	7	1	256.17	45	00	00
LOCATE/AZIMUTH	95	1	350.0	102	35	12.25
AREA	1	7	95			
PAUSE						

FIGURE 1

SMALL COGO PROGRAM FOR FIGURE SHOWN. IN THE FIGURE ABOVE, GIVEN THE
COORDINATES OF POINT 1, THE LENGTH AND AZIMUTH (CLOCKWISE ANGLE FROM NORTH)
OF LINES 1-7 AND 1-95, 7 AND 95 AND THE AREA OF THE TRIANGLE. IN THE
PROGRAM, THE SECOND LINE READS: LOCATE POINT 7 BY GOING FROM POINT 1 A
DISTANCE OF 256.17 AT AN AZIMUTH OF 45 DEGREES 00 MINUTES 00 SECONDS.

Source: Fenves [FE66], p. 48, Reprinted by permission from <u>Proceedings of</u>
           <u>the IBM Scientific Computing Symposium on Man-Machine Communica-</u>
           <u>tion</u> © 1966 by International Business Machines Corporation.

A much more complicated and older language is called APT, for Automatically
Programmed Tools. I will show just a few things on this example (Figure 2)
and will not walk you through the whole problem. First of all I want to
point out that APT is one of the oldest languages in existence. It came
out around the same time that FORTRAN did, namely around 1956-1957. So if
any of you thinks this is a Johnny-come-lately, I guarantee it is not.
Secondly, APT is implemented on most computers. It is probably of no
practical interest what-so-ever to anybody in this audience, although as
computer scientists many of us are very concerned with such a system.
Again, we have vocabulary that is very relevant to the application which is
machine tools. Machine tool patterning involves a device that cuts out
some kind of path and you need to tell it what the path is; this is done by
programming paper tape or other input device which tells the tool how far
to go and in what direction, etc. This particular program starts out by
giving a whole lot of commands which are really declarations. For example,
it says the tolerance of the cut is to be 0.005 inch and the feed rate is
to be 80 inches per minute. Then for example, there are commands to turn
on the coolant and use a flood setting, whatever that may mean. The
important point is that the language means a great deal to the people who
are working in that particular application area. Then there are some
commands that have to do with cutting out this particular shape. I guess
one of my favorite ones says, "Go forward in the shape of an ellipse whose
center is PT1 and which has some specified major and minor axes and
relation to the bottom line". If you actually look at the diagram, this
command makes a little more sense. Again, these are statements of commands
which are useful and important to the people in the application area. They
are not very meaningful to people outside it and as far as I'm concerned
that doesn't matter.

There are literally hundreds of these languages that have existed over
many, many years. I haven't done any tracking recently, but for a number
of years in the late 60's and up to around 1977, some of you may know I did
quite a bit of tracking of what languages were in use in the United States,
and it turned out that that number was in the range of 160-170. What was
much more surprising to me, the first time I uncovered it, was that these
languages for specialized application areas - whether they be in graphics
or simulation or machine tool control, or civil engineering or whatever may
be considered as a specialized area - represented about half the complete
total of languages. So each year there were approximately 50% of the
languages I could identify in the United States that were aimed at those
very narrow application areas. I think the reason is that there are
certain needs to be met. We have been talking about languages for solving
problems, but I don't think anybody has crystalized what the users' needs
are in trying to do this. As far as I'm concerned there are really two
major needs; I have a whole paper that spells these out in a lot more
detail, [Sammet, An Overview of Languages for Specialized Application
Areas, Proc. AFIPS Spring Joint Computer Conference, 1972], but for the
purpose of this discussion, let me point out that the first of these major
needs is the functional capability, namely specific features. Now, you can
say - particularly in the example of COGO - that you could program that in
FORTRAN or PL/I or whatever your favorite language might be. I agree, but
it is awkward and you have to do it - it isn't done for you. So the first
thing that you need is the functional capability in which it is perfectly
legitimate to have a command that says "locate the azimuth", or "create an
ellipse". The second major need for a programming language and for solving
problems is a style that is suitable to the application area; vocabulary is

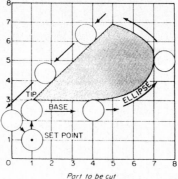

Part to be cut

Part Program	Explanation
CUTTER/1	Use a one inch diameter cutter.
TOLER/.005	Tolerance of cut is .005 inch.
FEDRAT/80	Use feedrate of 80 inches per minute.
HEAD/1	Use head number 1.
MODE/1	Operate tool in mode number 1.
SPINDL/2400	Turn on spindle. Set at 2400 rpm.
COOLNT/FLOOD	Turn on coolant. Use flood setting.
PTI=POINT/4,5	Define a reference point, PTI, as the point with coordinates (4,5).
FROM/(SETPT=POINT/1,1)	Start the tool from the point called SETPT, which is defined as the point with coordinates (1,1).
INDIRP/(TIP=POINT/1,3)	Aim the tool in the direction of the point called TIP, which is defined as the point with coordinates (1,3).
BASE=LINE/TIP, AT ANGL, O	Define the line called BASE as the line through the point TIP which makes an angle of 0 degrees with the horizontal.
GOTO/BASE	Go to the line BASE.
TL RGT, GO RGT/BASE	With the tool on the right, go right along the line BASE.
GO FWD/(ELLIPS/CENTER, PT1, 3,2,0)	Go forward along the ellipse with center at PT1, semi-major axis - 3, semi-minor axis - 2, and major axis making an angle of 0 degrees with the horizontal.
GO LFT/(LINE/2,4,1,3,), PAST, BASE	Go left along the line joining the points (2,4) and (1,3) past the line BASE.
GOTO/SETPT	Go to the point SETPT in a straight line.
COOLNT/OFF	Turn off coolant flow.
SPINDL/OFF	Turn off spindle.
END	This is the end of the machine control unit operation,
FINI	and the finish of the part program.

FIGURE 2
EXAMPLE OF APT PROGRAM FOR SPECIFIC PART TO BE CUT

Source: Hori, S., **Automatically Programmed Tools**, Armour Research
          Foundation of Illinois Institute of Technology, AZ-240, November
          1962. (Brochure.)

the best, but by no means the only, example of this. I have used these
examples of COGO and APT for many years in more general talks on programming languages and one time I was talking to a group that was primarily

engineering and scientific. Because the COGO example is such an easy one
to get the point across I used it, and was amazed at the end to have one of
the audience come up to me, and identify himself as a civil engineer (the
first one I've ever met). He proceeded to tell me that COGO was so useful
to them because it mimicked exactly the way in which they tended to write
and do things. And so there is a style associated with differing application areas in which you need to deal not only with the vocabulary, but also
with the format.

Let me end by pointing out again that this whole set of languages that are
very specialized (and don't look like the kind of languages that you see
because the users don't come to conferences like this), nevertheless turn
out to be one of the most useful devices that exist for solving actual
practical problems.

RADA PRESENTATION

My emphasis will be "What is relevant literature for trying to bring the
computer and the person closer together?"

First, a partial review of things discussed yesterday. A person who is
interacting with a computer, has some problem to solve, and we want to
transform the computer into one such that the computer is now more likely
to say to person X, "tell me your problem in a natural way." When Bruce
Blum gave his presentation yesterday afternoon, this person X was the
domain expert. In the case of building a medical information system, Bruce
would like the medical information specialist to be able to come to the
computer and give the problem to the computer in a natural way. Ben
Shneiderman's talk focused on an end-user that was a computer programmer.
For instance, Ben asked how can we document the program, so that the
programmer is more able to understand it. I would put Elaine Rich's talk
in roughly the same category as Ben's in terms of dealing with a computer-
type person. Elaine's system looks at a program and tells the user about
the program without the user having to look at the program directly.

As you might remember from my own talk, the word "natural" is key, and it
intrigues me as to what this word means in use by different people. Why do
we want the computer to be able to talk to the user in a more natural way?
Bruce Blum talked about shorter man-hours to develop an information system;
his system was more natural, if people took less time to develop the system
in it. Ben Shneiderman talked about errors as being an important measure
of success; the communication is more natural, if people make fewer errors.
That, in general, there is a faster, smoother motion to the goal might be
one way to generalize on these measures of "naturalness".

How do you make the system more natural? You might make models. If you
try to make models in the computer correspond to those in people more
closely, you might choose components that correspond to those in people.
Ben has a rough model with a syntactic and semantic component, where the
semantic component has a computer model and a domain model. What is a good
research strategy for getting models that are more like the models in
people? The general scientific method is to have a hypothesis that a
certain model is better than another. A good hypothesis is relevant to the
literature and testable. What does it mean to be relevant to the literature? From where do we draw to get good models? In this conference, there

has been talk about psychology but very little about biology or linguistics or philosophy. The fields of software engineering and artificial intelligence, which were strongly represented at this conference, are sources of models themselves.

One of the questions I would like to address to the audience and the other panel members is, "from where do we draw good models?" I would like to illustrate by giving you what I have done and asking for comment on my examples. In my approach to AI the hypothesis has been that the problem representation should show similar states to have similar values. My method is to fix a particular representation of the problem, and the fix is guided by principles. One principle is that I want extra knowledge about values and states because I hypothesize that similar states having similar values is an important characteristic of nature and people, and I want computers to be that way. One of the main insights that I have gained from this conference is that it might be useful to refer more to the literature about models that have been used. I got first excited about this in terms of talks here when Saul Amarel said there are certain models that we understand that seem to characterize cognition. As we do AI experiments we should refer to those few models that people seem to understand. As I have been working on my current project at the National Library of Medicine, which has to do with the retrieval of biomedical text, I've been thinking about models of retrieving biomedical text. In the library itself there is a kind of model of knowledge which is the Medical Subject-Headings graph structure. There are large manuals on how librarians classify text and should try to retrieve text. So I could use this as a model of how people behave and put that into the computer program, as I'm designing a computer system that improves retrieval of biomedical text. There are also physiological studies in the library science literature about how people retrieve literature. There are studies in memory, such as Walter Kintsch's on how text is organized in the mind. In my own strategy, I'm interested in some systematic way of starting with a system and making it progressively better. I look to the literature on how you make systems better and better. This is sort of meta-level from the retrieval problem itself. I look to learning about the literature. Jaime Carbonell's talk in this conference was about models of learning. So this is some of the literature I reference, and one of the things I'm struggling with now relates to the question I asked Ben Shneiderman after his talk yesterday, "are there models in the literature that can help us?" Should they be well-defined and well-accepted models? To what extent might we refer to the literature and get inspirations to make our hypothesis more precise, understandable, and testable.

SHNEIDERMAN PRESENTATION

One key issue in the practical application of programming languages is the reusability of software. It makes good sense to facilitate the process of reusing the work of our predecessors and colleagues when developing new programs, yet there is widespread belief that we can do much better in area. Programmers need to be more well motivated to reuse the work of others and to create programs which are more easily reused. The desire to encourage reuse emerges from the very practical basis that there are potentially vast savings in cost and time to produce new applications.

Reusability is not a new idea and I'd like to begin by reviewing the
diverse reusability strategies that were promoted by various programming
language developers. FORTRAN subroutine libraries were designed to allow
mathematical functions to be stored for multiple reuse. The IBM Scientific
Subroutine Package was an early and, arguably, successful effort. I had
the experience of using many subroutines, especially the Bessel function,
in my early work. I needed to make adaptations, but the code was designed
to facilitate such changes. The Scientific Subroutine Package and its
successors such as the products of commercial developers, e.g., IMSL
(International Mathematical and Subroutine Libraries, Inc.) or scientific
groups in chemistry, physics, etc., were valuable contributions. In the
PL/I and Pascal worlds there have been fewer subroutine libraries. The
FORTRAN successes came, I believe, because there was a well established
problem domain of scientific and numerical analysis problems.

In the COBOL world, the most important aspect of reusability is the INCLUDE
verb, an idea that has been mimicked in many other languages. This verb
allows a programmer to include a copy of programming language text which
has been stored away. If you have written a 600 line data declaration
section that is going to be used in three hundred programs written at one
site, then it is very important to be able to copy the declaration rapidly
and correctly. In this situation the semantics and the syntax are
precisely reused. A second form of reusability in COBOL is in the hundreds
of preprocessors that have been written for COBOL. These include local
variations to COBOL and substantial additions such as the CODASYL Data Base
Task Group extensions to support database management systems.

In APL and FORTH reuse occurs frequently in the form of program libraries
which contain hundreds of small functions relevant to a specific application domain. The designers of these libraries created a new environment
which raises the level of programming in banking applications, statistical
calculations, market research, or graphics displays. A knowledgeable APL
programmer may take several months to become fluent in the semantics and
syntax of these numerous small functions.

LISP programmers seem to build extensions to their local dialect, further
personalizing the code, and making it more conducive to the problems at
hand. In the world of BASIC, I perceive less circulation of subroutines.
Instead, entire programs are delivered and users often revise the code to
meet specific needs. Ada was designed to encourage reusability in many
ways, but we have yet to see how successful this will be. Ada packages are
conceptually appealing, but it is not yet clear whether programmers will
employ some of the elaborate possibilities.

The determinants of reusability are numerous, and many of them are human
factors issues which lend themselves to investigation and improvement. I
think a central issue is whether there is a clear specification of the
function provided and the input/output variables. The sine function is
often reused because its definition is precise and the input/output
variables are limited. I suspect not one in a hundred users of the sine
function know how it is computed, yet they have a high degree of trust in
its correctness. The modular boundary is clear and the "process-hiding" is
effective.

A second determinant of reusability is the benefit in reduced effort. If
the code to be reused will require substantial study and revision, then the

likelihood of reuse will decline. Therefore descriptions and documentation
of code for reuse should be exceptionally well done. The code itself
should be exemplary in its organization and clarity, so that if some
revision is necessary the effort will be low. A machine searchable
database of reusable code should speed identification of relevant code for
novel projects.

A third determinant of reusability is the trust in the source. If a
routine is created by a respected individual or organization, then a
potential user is more likely to invest the effort to learn about and reuse
a piece of code or a module. Further contributions to increase trust come
from the knowledge that other people have successfully reused this code.
The Certifications for the Collected Algorithms of the ACM are a successful
attempt to inspire greater confidence. A record might be kept to indicate
others who have reused a piece of code or a module. If someone is responsible for maintaining the code, then the user's confidence will also be
raised, because he/she knows that there is a source of consultation and
assistance.

In summary, the benefits of reusing code are substantial and increased
effort in this direction is warranted. Encouraging reusability will
require dedicated effort and planning, and a thorough understanding of the
technical plus the social issues.

QUESTION AND ANSWER PERIOD - GENERAL PANEL DISCUSSIONS

SAMMET
I want to make a couple of comments on what Ben Shneiderman said. First of
all, there is no doubt that the reusability is terribly important. I also
think it has very little to do with languages per se, in the sense of what
is being covered in this conference. Again, I do not want to denigrate the
importance of this particular concept. There are two important things that
Ben Shneiderman did not say; I don't know whether he would agree with them,
but I hope he will. When one talks about reusability I think it is very
important to make a distinction between planned and accidental reusability.
The case of scientific subroutine packages clearly is an example of planned
reusability. In any large project there are always routines that are
planned to be reused within that project, and perhaps beyond it, e.g.,
string handling, input/output. Those are easier to deal with because
people understand that they are going to be reused and therefore they are a
lot more conscientious about dealing with the things Ben Shneiderman said
one has to deal with. What is more exciting and remains to be seen, is
whether or not the unplanned reusability can occur, i.e., somebody writes a
routine, puts it in a library, describes it, and now maybe somebody else is
able to make use of it, and then indeed you have exactly all of those problems Ben Shneiderman delineated. Part of the problem is truly economic and
sociological because individuals and groups are not very highly motivated
to create accidentally reusable software. Thus, whether it is a graduate
student working on a thesis or a faculty member working on a grant, or an
industrial organization trying to create software that they need to use for
themselves or deliver, there is just not very much motivation for John Doe
to create software which has obvious reusability. It just might have if it
were created in a way where somebody else could come along and find that it
would have all the good paradigms and characteristics that one wants.

SHNEIDERMAN
Sure. I think it would be very nice if we could build the financial structures which could compensate people for making more frequent. If
programmers were rewarded by a penny per line of code that is reused by
somebody else, that might help, and that is the sort of market place
scenario which I think is worth pursuing. I would also encourage organizations to take a very constructive and aggressive attitude toward making
reusability possible. Reuse does not happen as well, as you pointed out,
by accident. If there is someone who is responsible for it, and their
professional career depends on the degree of reuse in the organization,
he/she may be able to find the policy and encourage the mechanisms to make
it happen. Yes, I think there is also the problem that it is not as much
fun to take somebody's else is code as it is to write your own. As the
field matures, people will come to respect more and more, those who can
take code from others rather than those who spend excess time having to
rewrite and reinvent.

BLUM
I would like to make a comment on reusability in terms of language
extension and the example that you gave in APL. That is, that you
essentially extend the language by defining a set of functions and having
people write their systems using those functions. The ADA package concepts
also represents similar language extensions in that one defines abstract
data types, for example, in a package and then writes programs using those
abstractions. I think that is both good and bad. On the one hand we have
language extension, which helps in a specific environment. (Jean Sammet
talked about languages for specific environments which are highly
tailored.) But in the process of doing that you also start to lose
generality. For example, the concept of the software factory is that you
have series of products each containing software. The software is written
and reused. It is reused at a level of semantic description, as patterns
where you make macro-like replacements, or as chunks of code per se. At a
recent conference, I was talking to one of the Japanese responsible for a
software factory. He kept talking about "cultures", meaning that the
products that use one software family had a different culture from products
that used another software family. Therefore, even though there was
reusability within a culture, there was not necessarily reusability across
cultures. One of the frightening things when you think of Ada, (which is
intended to provide uniformity of language for a large horizontal user
community), is the danger of producing a series of vertical cultures with
barriers for reusability across cultures. Will we find that this
"reusability" simply produces many cultures, or will we find that this is a
way of extending the language?

SAMMET
I would like to comment because while I agree with the spirit of what Bruce
Blum is saying (and I am also very worried about these cultures), I take
very strong exception to the use of the phrase "language extension". I do
not consider that adding subroutines to FORTRAN, or functions to APL, or
packages to Ada is changing the language. The language has a particular
syntax and semantics and every language that is worth its salt has some
kind of capability whereby one can create chunks of code that are identifi-
able and reusable; whether they are called subroutines or packages or
functions I really don't care, but I suggest that those are not, repeat,
not language extensions. I saw a paper once that said "new language for
banking" and when I looked at it, what it contained was merely a series of

subroutines to be added to FORTRAN. Now FORTRAN is a language and has its own syntax and semantics but adding subroutines to it does not add to, or extend, the language, it adds to the capability that is available to you when you use the language.

BLUM
I agree entirely, I probably should have said "environment extension."

E. RICH
I think maybe we actually agree on several things, because, what I noticed is that Jean Sammet, Ben Shneiderman and I were all saying much the same thing. If you assume that there is a close connection between the language and the set of things you can do, then in some sense if you add some procedures to a FORTRAN library what you have done is simply expand the system and it now knows how to compute Bessel functions and it did not before. But on the other hand, you also have to extend the language, whether you want to call it language or not, so that you can now communicate that this new thing is the thing you want to compute. Maybe the right analogy is with English. English gets extended all the time. Not that its basic structure changes although that occasionally does, albeit very slowly, but we invent new nouns and verbs all the time when we have new things to talk about, and we just fit them in to the structure that we have. I think the thing that we need to look at (pursuing the same ideas although a little farther) is to suppose that we are writing a program in a conventional programming language, and we want to reuse a module written in the language by someone else. If the only parameters to this module are two numbers that serve as inputs to the function that the module computes, then we get to keep the syntax of our programming language the same. We just call the module as a function of the two parameters, and we have extended in some sense our language, and certainly our capabilities. We make progress as soon as the module that I make allows you to tell it more things than simply the two numbers that you want. For example, suppose that Aaron Temin succeeds in building his SCRIBE help system. Now it's just a module. It is a very much bigger one, though, and the number of inputs that you might give to it is substantially more than the number you would give to a cosine routine. Then all of a sudden we actually have to think about extending the syntax of the language you are using because just thinking of my module as a function you can call with two arguments is not adequate to allow you to exploit it. What we all seem to be agreeing on is that we want to move away from programming machines at the levels that the architect builds them and towards higher and higher levels where more semantic objects have already been constructed and we use them. We also recognize that we have to think harder about how to use those objects. I do not think we are in disagreement on the fundamental issues, although I may find out that we are.

RADA
I had a comment, Elaine, which is germane to what you were just saying. You said there are many things that a person wants to tell the computer but only a few that the computer understands. The conclusion was that computer languages need to become more like English. I agree in one way, but I do not think it necessarily follows that as the computer becomes more powerful that its proper language becomes more like English. Necessarily, the proper language is more powerful. If one says that that computer's basic language should correspond to English (or some other native language), then one implies that the computer should work like people do. Yet, the two

work on different substrates. While the best language for the person to communicate to this complex machine inevitably includes part of native language, the ideal programming language remains to be characterized.

E. RICH
I guess what I was thinking was that the program evolves to mirror the problem domains that it has to solve, so in some sense it evolves to mirror the world that we are using programs to help us live in. Then it is not necessarily the case, but it strikes me that English has evolved to mirror that same set of problem domains and so you would have to convince me that we could do a lot better by inventing something else. I think the burden of proof would be on you since we know English and since English was designed, partially to mirror the psychological structures of people, also to mirror the real-world problem structures that humans talk about. As our programs become better at knowing about problem domains, they are going to have similar sorts of structures.

SAMMET
I was surprised that there has been relatively little discussion in this conference on the use of natural language for problem solving. Now that this has been brought out by other people, I would like to join in. I have long been a strong personal advocate of allowing people to communicate with a computer in the language which was natural to them. I say that to avoid quibbling about whether it is English or French or something similar. Furthermore I think it is very important to understand that if one is going to consider that concept as a methodology for solving problems, you need to allow the formal notation that is natural to the problem area, rather than the formal notation that we concoct as programming languages. So, mathematicians will write in mathematics with precedence operators (unlike some languages that don't have precedence operators), and chemists will draw whatever kind of molecular diagrams they draw, and those must be acceptable to the computer if the chemist is going to be able to get his or her work done. I think those people who are opposed to natural language - and I know there are experiments that show that this does not work very well - have never really had the full language domain available to them, because one has to postulate that the computer system, and the software sitting behind the terminal facade, have all of the world knowledge that is relevant to what you are doing, whether it be chemistry, civil engineering, or text editing. If you don't have that then you get no better results from the computer than you do by talking to an individual person who doesn't know anything about that field either.

SHNEIDERMAN
I would disagree, since I find that people are very different from computers. Just because people communicate effectively with each other through natural languages such as English, does not mean that they will operate computers most effectively with English. Precise, concise languages such as mathematical notations, musical notations or chemistry formulas should remind us that there are more effective approaches than English for many tasks.

SAMMET
I tried to say that very specifically.

SHNEIDERMAN
English, French, Russian, music?

SAMMET
No. No. No! I said that whatever notation is relevant so that it is
molecular diagrams for chemistry and special notation for music or whatever
is appropriate to the application area.

SHNEIDERMAN
OK, good, fine. I think that certainly precise, concise notations that are
familiar to people in the problem domain are very important areas in which
we should be designing interfaces. The Music Construction Set that Bill
Budge designed, allows the user to put notes on the screen by touching them
and they appear on the staff. Then it plays the music for you. I have
described a very natural mechanism for people to use computers under the
notion of direct manipulation.

I think that the suggestion that because people communicate through natural
language, English let us say, therefore people should communicate with
computers through English is a natural, appealing, and exciting idea, but
in the long run it won't work that way. That is a personal belief, that
tracks the historical imperative that Lewis Mumford outlined in what he
calls "the obstacle of animism". He describes how every technology goes
through this mimicry game of flying by flapping your wings or building
robots that have five fingers and arms whose dimensions match human arms,
but in the long run I don't think that the anthropomorphic image makes
sense. First tools and technology evolve to serve human needs and the
mimicry game fades after awhile. Designers need to find out what tasks
people need to accomplish and how best to serve that need, and I don't
think that natural language is the obvious way to serve those human needs.
The second aspect is that computers are very different from people. When
computers can display a thousand characters a second and people can only
type or speak at a much slower rate; when computers have remarkable capacities to compute rapidly, store potentially unlimited amounts of information, and display things on a screen in a graphic manner, then I think you
have to consider that the interface between computers and people has to be
designed rather differently than the way people relate to each other.

I do find the evidence of now maybe a half dozen experiments which show
that contemporary natural language systems are not as appealing as the
alternatives that were tested. I think that evidence can't be ignored.
One response is that these tests are on today's natural language systems,
and tomorrow's natural language systems will be better. Well tomorrow's
direct manipulation systems will also be better and tomorrow's notational
systems and visual displays will also be better, and I think the chance for
rapid evolution and effective use of computers lies much more in abandoning
what I consider the simpler notion of natural language interface and going
for much more elegant, rapid and powerful facilities.

BLUM
Can I just make one point and you can pick up after that. When we talk
about natural language, I'm not sure that you necessarily meant free text
natural language.

SAMMET
Oh yes I did. Let me say this now for the third time, natural language
includes whatever formalized notation happens to be natural to the problem
domain. I do include mathematics, I do include music notation, I do
include diagrams for chemists, and circuits for engineers, and anything

else that is similar. If the discipline or the problem domain has a formalized notation that they are comfortable with, then as far as I'm concerned, that's a subset of the natural language that they ought to be able to communicate in.

BLUM
Leaving aside natural text languages, let me say that I think that we have been limited in our expression by the prevailing technology. For example, we have restricted our concepts to think of superscript as being represented by 2 asterisks, and multiplication as always having to be explicit, and so on. Technology now allows us to have, inexpensivity, a much more natural representations of the way we think about some these highly specialized, nonanthropomorphic models. In that context there is not an issue of: is the machine really a person on the other end talking back to me. Rather it a matter of designing better user interfaces that are more natural, at least in terms of symbology, to the way people would like to express what they are saying.

SAMMET
All right first of all, my reply on the equipment. Equipment that enables you to type mathematics in normal notation has existed since the mid 1960's. You could type sigma signs, you make up integral signs, and you could put subscripts and superscripts where they belong, and not with these terrible double asterisks and so forth. So I agree with you.

BLUM
This is a matter of cost, and the impact of the costs which mold our perceptions of what is practical.

SAMMET
It is a matter of costs, it is a matter of psychology and a whole lot of other things. But the equipment is certainly not a problem. It wasn't a problem 20 years ago; the problem was psychological and economic even then. Secondly I would like to point out that, although Ben by implication says that I espouse natural language because of mimicry and anthropomorphism, those are not the reasons. The reason that I think one ought to be allowed to communicate with the computer in one's natural language is for all of those very good characteristics that Ben so elegantly delineated. Namely, the computer has very fast speed, a lot of memory and so on. And many things it does better than you or I do, in the problem domain. That is to say, I have somewhat more faith in the computer's ability to solve the problems of celestial navigation if that happens to be the area I'm in, or land description for civil engineering. I think that the computer has more capability of storing all of that material and responding properly to it than does another human being. In other words, the computer is better in the particular characteristics of speed and memory size, and it is for that reason that I want to use English, not because I want mimic a human.

E. RICH
I still think that I don't necessarily agree with Ben. He wants to talk about experiments that show that, for modern text editors and current data base retrieval systems, English isn't the right language, and I'll agree. But suppose that I knew how to build a program to do legal reasoning. Suppose I knew how, in a restricted domain of law, to simulate the behavior of a law clerk. I'm not saying I'm simulating that law clerk as a human being, but I learn about the legal rules of reasoning and I have all the

law cases. As Jean correctly points out, law clerks have difficulty
keeping law libraries in their memories, whereas computers could do that.
Now the question is that if I wanted to describe a case to the law clerk
and say, "go find me the precedents that might be relevant to this," I
would need a language. Lawyers do not have a formal language that is
substantially less restricted than English, and there is a reason why they
haven't got one. I don't know that there obviously is one, and I doubt
we'll come up with one. And as soon as we can build programs that can do
that sophisticated reasoning, that's when we need something as complex as
English in which to communicate with programs. So it's not the issue that
were talking about today's natural language systems that is the limit. We
are talking about today's problem solving systems, and most of them aren't
very sophisticated.

SHNEIDERMAN
Well I will simply suggest that precise, concise notations do sharpen
thinking in problem solving and let me not go further in the argument, but
suggest that the path for us all is to spend less time on panels arguing
over it and spend more building systems and doing the comparative studies.
There is a lot we need to do to understand the nature of interactive
systems. I offer again the notion of a benefits test to compare approaches
as a way of getting past this arguments. We really need to be thinking
about the speed of performance on a benchmark set of tasks, rate of errors,
subjective satisfaction, retention of commands over time, learnability of
the system, cost, and reliability. These are things we can measure about a
system. I think we are past George Miller's stage of confusion and reach
Lord Kelvin's stage of measurement and to create a science and get past
political arguments.

SAMMET
The one problem with what you are saying Ben - and I'm fully in favor, as
you know, of being able to measure these kinds of things - is that we are
still very far, in any of the domains that I know about, from having what I
would consider a natural language system to be able to measure. And the
moment that you have some subset thereof, and a person that has to learn a
subset of a natural language, that is not what I have in mind. The
measurement will always come out against the so-called natural language,
because we are an awfully long way from having these natural language
systems to communicate with in even a single domain.

SHNEIDERMAN
That is, such a frustrating hypothesis to put forward. That suggests that
we go on forever until natural language systems succeed. There is no way
that I can prove that it won't happen. I would turn around and ask someone
in the artificial intelligence community to pick a domain narrow enough and
useful enough that it can beat some other system. Let us see a commercially successful natural language system. INTELLECT is out there, it sold
copies to the researchers and the system programmers and it is being used
in a couple of hundred sites, but it hasn't taken off. CLOUT is now out
there and it seems to be not even as good as INTELLECT. By saying that AI
researchers have only had 25 years, give us another 50, you wear out my
patience. I want to see progress much sooner.

BLUM
We've driven enough people to their feet, so why don't we start taking some
questions from the audience.

FALKOFF
Well it seems obvious that the natural way to cause a vehicle to go forward is to say giddy-up. I would like to reinforce what Ben said. In general, referencing Elaine's comments about the lawyers, it seems to me you have to distinguish between the subject matter and the commands of the machine. You may indeed get to the point where you can analyze legal text, but I doubt if you will be happy at the time going to the machine and say, please if you don't mind would you look this up and analyze it? So there is a distinction there. The one point I would like to make, however, is that in general this all has been a kind of one way, going from people to machines. We have these machines, we have the ideas that they have brought forth. Maybe we have to learn from that and get some feedback.

BLUM
Comments from the panel? (None)

WEINTRAUB
Two comments. First of all, reusability in my opinion is a question of memory. What I mean by that is, if I had a function which is so easy to use that I don't have to go through and read the documentation every time I want to use it, I will continue to reuse that function, especially if I'm convinced that its not going to blow up. Which is why I see who wrote it. Also what I mean by that is that if you make things modular in a sense that, for example, that the direct expression shows in APL where you can immediately understand what anything does without having to go through the documentation, then it will be reused. However, reusability can only be carried so far and there is an analogy not due to me, from the Computer Magazine, which used to have a forum of some kind. Basically, if I'm an electrician and I'm building a circuit box and I bend it, I do not sit there and pound it out and try to make it straight again. I throw it away (unless I'm a very cheap electrician). (A nail is even a better example.) I throw it away, and I buy a new box and use it. There is only reusability when it starts getting into the maintenance level. In other words, maintenance starts to be a problem. I would throw in one other analogy, if I go into a hardware store and I say I need a fastener, then the clerk asks if I want a screw, nail, a rivet, etc. I say I want a screw, and he says OK now do you want flathead, roundhead, etc., and you can have a menu 16 levels long. The point is, that over the years the hardware industry has developed an awful lot of fasteners. For us to expect even one class of computer languages to solve all problems, when the hardware industry has demonstrated that not one fastener will solve all problems, I think is unrealistic. Now I have a direct question for Jean Sammet. Did you or did you not a few minutes ago show this language with the azimuths in it, and put it forward as a language? You showed us a civil engineering application language.

SAMMET
Yes.

WEINTRAUB
At the end you pointed out yourself, and I think someone out here would have pointed it out quickly otherwise, that such a language could easily be constructed as a set of routines as for example, in APL or in FORTRAN. APL being most straightforward because you don't need parentheses.

SAMMET
But you also don't have the vocabulary of the words "locate azimuth", which
seems to be useful to civil engineering.

WEINTRAUB
That's not my point. That is not my point at all. You could write a
function in APL called "locate azimuth" if you allow a slash or underscore.
And if you do that, you will have then have exactly the same language you
showed on the screen.

SAMMET
I can write the same thing in FORTRAN or PL/I with the language.

WEINTRAUB
Right, and therefore your comment about the banking system, where you said
that when the person wrote a set of routines he wasn't writing a new
language, is incorrect. Any language which can be expressed so straight
forwardly and turns to another language, by your conjecture a few minutes
ago, can't be called a new language.

SAMMET
No, because what I'm saying is that, in essence, what you need is notation
which invokes that function or that procedure. And that is very different.
There is a big difference to me between saying "**locate azimuth**" and saying
"CALL locate azimuth."

WEINTRAUB
That is a question of syntax, not a question of semantics.

SAMMET
I can not agree with you. I am very concerned with the syntax, and I find
a very big difference in notations; there is a simpler example than that.
Let us talk about matrix multiplication. If I am in any language in which
I have the ability to call functions or procedures, and I write CALL MAT
(A,B,C) where MAT is the function or procedure name, and then my parameters
are A,B,C, I really don't know from looking at that line what is supposed
to happen. Suppose the function is MATADD, so it is matrix add, and I have
parameters A,B,C. I do not know whether it is A = B + C, C = A + B, or B =
A + C. If on the other hand I change the syntax of the language, in whatever legal mechanism I'm allowed to do that, and I declare A, B and C to be
matrices, and then write A = B + C, or whatever I want, then I submit that
this is far, far, more readable and better than simply invoking the subroutines, which we all know we can do.

WEINTRAUB
Jean you've missed my point. You said that for one to simply write an
application which uses the specialized vocabulary or subset of the vocabulary of an application field is not inventing a new language. Yet, I claim
that using APL I can write your azimuth location system and do this very
quickly. It will be just as usable and in exactly the same form except
that you may have to use an underscore instead of a slash (because of
syntax). But by your comment about the banking system, you would say that
this is not a new language. Yet you present it as a new language. My
point is simply that the definition of a language involves a lot more then
the syntax and the application. It also develops the whole question of
what we've been discussing the last couple of days, which is what addi-

tional features or power do I get that I did not have before? And that is
why I have problem with language extensions. I was wandering a little bit
here, but let me finish the thought out. I can write an APL interpreter in
ADA. Have I invented a new language if I do that? Or have I rewritten a
preprocessor for ADA? The distinction is not clear.

BLUM
A brief comment. I think that there is a difference between a language and
it's implementations. I think that what Jean was talking about was a
language which is well thought out, a means of communication with a special
set of users. I think that what you were talking about was an implementation. Clearly that language could be implemented in APL. It could be (and
may be actually) implemented in FORTRAN or PL/I. Obviously in some deriva-
tive sense it's always implemented in assembly language. I think we have
to take a look at these as two different issues. In the process of doing
that we will see that there is a difference between the user interface,
which is a language, and the implementation of that interface, which also
requires a language. There may be some languages which are much better for
implementation purposes and some environments which are better for supporting implementations than others. I think that we have to keep the two
things straight in our minds. We also might address the philosophical
question, which I don't think we should try to answer, what is the difference between a user interface and a user-oriented language. Let us pass
that by.

TEMIN
Given that application languages are a good idea, why are there not many
many more of them? Why does not every field have its own program that
talks in its own language and does wonderful things? Things which we know
computers are capable of doing; I'm not asking for pie-in-the-sky stuff.
There are, in fact, things computers are capable of doing, for which there
don't exist application languages to allow people to do easily. Where is
the bottle neck?

SAMMET
I said there are hundreds. I can list them by name and give you citations.
There are a great, great, great many fields which have developed their own
languages.

TEMIN
And there are lots that haven't. Why not?

SAMMET
Why haven't you done many things? I will now add a list of 25 things to be
done.

TEMIN
So it is just a question of adding man-years then.

SAMMET
Question of what?

TEMIN
Is it then just a problem of man-years. There are no cognitive or unsolved
problems. We know how to solve all the problems, we just haven't taken the
time to do it?

SAMMET
I think (in some cases), yes. I think in some fields there have not been
persons who are highly motivated to do this kind of work. I don't know the
reasons since I have not gone to investigate all of these. However, I can
probably name you about 25 areas that are moderately specialized that do
have specialized languages. I mean languages, not subroutines added to
other languages. There are probably hundreds of other application areas
that do not, and if you are asking me why they haven't, I just don't
know. Make up your own answer.

BLUM
I think in part it's maybe a question of perceived needs. There may not be
a large enough audience to have a perceived need, and as a result, no one
will go ahead and implement it. Also, I think there's a tremendous NIH
factor, not-invented-here for you non-medical people. People like to do
their own thing, and I think that's one of the reasons why people prefer a
locally developed product to a superior, external standard product. I
believe that helps to explain the proliferation of UNIX editors and
formatters.

JERNIGAN
I would like to make a couple of comments - one about software reusability,
and the other about the discussion of about natural language. I think I
agree with Jean when she says that you do not extend a language when you
just put things into the subroutine or function library, I think that's
entirely correct. On the other hand, I think that this notion of software
usability is key to the extensibility of the languages that we use with
computers themselves. I will cite for that, the fact that several years
ago, and I think that this is something that Jean has also pointed out,
that at IBM some people wrote lots of routines for doing matrix manipulation. We can multiply vectors, add vectors, we could invert matrices, we
could transpose and rotate them, but instead of putting them into a subroutine library, they put a syntax and a symbol notation on top of that and
now we call that APL. I don't have to think of it in terms of calling it
anything under the subroutine library, but like you were saying, you
provide a kind of a language syntax, and suddenly I've got software
reusability. Now, I think if I had to think in terms of calling these
things out of a subroutine library, I probably just wouldn't even use them.

SHNEIDERMAN
You're saying that putting something into the language syntax, like matrix
multiply APL, gives authority to it, it legitimizes it; rather than the
subroutine library which has sort of a dangerous aspect to it, you don't
know where it came from.

JERNIGAN
No, I think that when it's in the language syntax, I think I can use it,
that is, I can think in the syntax.

SAMMET
Exactly.

JERNIGAN
And I don't know how to think in a syntax that involves a bunch of
subroutine columns.

SAMMET
Exactly.

SHNEIDERMAN
Okay, that is interesting. So language does shape thought, huh?

SAMMET
Which is exactly the reason that I said that I draw a very big distinction
between changing the language and its syntax, as contrasted with having
just the subroutine library. I think you and I are in complete agreement
on that point. Is that right?

JERNIGAN
That's right.

E. RICH
There's a funny thing, though, which is that, for example, in ADA, without
changing the syntax of ADA, you can do exactly what you want. Because you,
as a user, are allowed to overload the infix operators like times, and you
could make a package and define times for matrices to be the standard
matrix multiply operation. Now you can write it exactly the way you want
to. So one of the interesting things is that now we're recognizing that
whether it's new syntax or just an extension to the subroutine library...

JERNIGAN
You've got a self-extensible language

E. RICH
Some syntaxes allow for more new things to be added without having to
change the existing syntax, and essentially that is what ADA does.

JERNIGAN
Which is what you do when you build a function inside of APL. You can put
them into a function library, or you can incorporate them into a syntax
that you are using.

E. RICH
APL also allows the user to define infix functions, and that just makes it
seem that much more natural to use.

GUIER
I would like to make a comment on the natural language issue, and it's
certainly not new. I just thought I would bring it up again, primarily
because it's one of my pet prejudices. Namely, I think it's absolutely
right to think of a language as a natural notation of the professional
doing his thing in his environment. I think a human language like English
or a human discourse language is so imprecise in trying to talk to computers the way you want computers to do precise things that that analogy may
be going too far. I've noted that about the last ten years in the United
Nations, for example, diplomats use intentionally the ambiguity in languages to get around starting new wars or to get an agreement or a treaty
signed. Where there's intentionally left a little slop, the language is
explicitly allowed to do that and does it very well. I think that when you
are trying to communicate with something as literal as a computer, you may
need something that just has a little more precision in it.

SAMMET
Well, I disagree with that for a couple of reasons. First of all, I've
never thought that worrying about where semicolons went added to my
precision of thought. In fact, it deflects from the logic of what I am
trying to convey, because I am worried about where the parentheses are;
even if Ben's system will put them in for me, I'm still worried about
whether or not I need them at all! And I don't want to worry about where
the semicolons are and how I write and distinguish between comments and
action items, and a whole lot of other things of that kind. So I've never
thought that the programming language requirements added to the precision
with which I was trying to think about the problem. On the contrary, they
slow me down because I have to worry about all that stuff. As far as the
fact that one can deliberately talk ambiguously, I completely agree. It
happens time and time again, and I'm sure if anybody parsed the sentences
that we were exchanging this morning, probably 75% of them would be
construed as being ambiguous. Nevertheless, when people communicate, it
often turns out that the wording is very ambiguous but understood in spite
of that. If you look at directions on a shampoo bottle, it turns out that
they are quite ambiguous about what it is you are supposed to do. They
take for granted a certain domain of knowledge, so that if it says repeat
you know what you are supposed to repeat. And, you know, since they
obviously want to sell more shampoo, they tell you rinse the hair, wet the
hair, rub in the shampoo, rinse it out, and repeat. On the other hand, the
human being understands exactly what that is supposed to mean, and I submit
to you that in this natural language system nirvana that I postulate, the
computer will have that same kind of knowledge. It will understand what it
means when I say repeat. And if it doesn't, it will ask me the same way as
I may ask you to clarify your meaning if you are giving me instructions to
carry out a task.

GUIER
Yes, yes, I certainly agree with you about the semicolons and some of the
bad syntax that has been put into languages to try and make them precise
and exact. I still believe that the human discourse languages are so rich
in emotional content and color and mood and emotion as an intention to
display that emotion. This is simply an inappropriate thing to be telling
a computer. I don't think I need to tell a computer that I want a priority
of 1.99999 if 2 is the maximum priority and then say it in a highly
emotional loud way, "I think you know 2.0 is good enough."

SAMMET
But, let me pose a question to you. Surely, at one time or another, you
have given essentially technical instructions to somebody, whether it be
the carpenter who was to build something for you, or a secretary, or
another programmer, or even your boss. I am using technical in a very
broad sense, and presumably you try your best to dispose of the emotion
that is associated with that to convey very clearly to these other
individuals what it is you want them to perform for you and under what
circumstances. And I suggest that that is the way you ought to communicate
with this mythical computer.

GUIER
I would suggest that that's harder than in some other notation.

R. RICH

Let me make three very brief comments in order. You, Bruce, raised the question of whether programming was a profession or a skill, and from what you said about all these people learning BASIC, that might have some bearing on it. I think it doesn't. Probably most of the people here learned how to play "Claire de Lune" on the piano as youngsters. I don't know whether piano playing is a profession or not. But the fact that I learned how to play "Claire de Lune" two notes at a time has no bearing on that question at all. The second point has to do with reusability. At the Laboratory here, we get paid twice a month. There is a payroll program which is reused twice a month. That's one end of a spectrum. The other end of the spectrum is that if I have some little engineering problem to solve, I write a program from scratch in some language in which I'm fluent and I reuse the subroutine library, but not much else. In between that, there's a whole spectrum of things between taking an existing program and changing it, and using a program generator, which is an excellent approach when I'm in a special field of application. I'm with Jean that the more special fields of application we can well serve, the better off we are.

CARLTON-FOSS

I would like to place another chip on the pile of using psychological science to aid in the development of parsers and natural language. I think that the general issue of natural languages may be undoable right now because of the ambiguities, etc. in certain parts of people's expression. This is not entirely an issue of precision. I think that if, in addition to looking at evaluation issues of the success of natural languages, we also look at how to define restricted domains, then there can be some significant development. Maybe some people I don't know have done work in this area. But I'm soon to start a job in which we're going to be looking at a legal library and how do the people who are responsible for the library keep track of what's in it and where it is. It turns out that if you look at the general problem, you could create a huge data base and you wouldn't even really know how to file everything in it. There are issues of the history and the key parties and the cases, the politics of the cases, the function of the given case in the larger organization or within the legal department, what are the legal precedents, and what are the implications of a decision on a case. You can continue the list. It turns out that, at least from the information that I have at the present time, the legal argument for the cases, the key way that the lawyers actually organize this information in their own way of thinking (at least within this department) provides some way of determining how the people think about it. Then you can design your natural language system which does have to be at least somewhat English, but in the restricted domain. It can be a subset of English in which there's a lot of precision, perhaps, and or at least much more specificity and the natural language problem isn't as difficult. Next, picking up on what Ben said, you test it, and make sure that you got something good. I think the one area where we might disagree is that I agree very strongly with the laboratory approach. I think I might put a little more emphasis than you do on the, well I call it the ethnographic approach, to finding out some of the key information.

SHNEIDERMAN

Fine. I offer that the notion of a controlled experiment as an extreme, but certainly other steps in between are useful: observations, thinking aloud studies, or what are called action studies using more anthropological styles. To repeat, natural language only relieves users of some burden

with syntax and it does not help clarify the problem to them. I find
natural languages being used in cases where you don't know how to structure
the dialogue so the designer hopes natural language will take care of it.
And it just doesn't work out.

Even the promoters of INTELLECT (or the other natural language systems that
are sold commercially) do not advocate it for novices off the street. They
must come with sophisticated knowledge of the problem domain and the
semantics of query formulation. So the question remains: when someone
comes with sophisticated knowledge of the semantics of the application
domain, how much benefit is there of relieving them of the syntactic
details?

CARLTON-FOSS
Yes, it has to be tested. It can't be done with ESP or with an ad hoc
assumption.

SHNEIDERMAN
Right, or long sessions of hand waving.

FALKOFF
I would just like to add that, when I talked before about feedback from a
computer, the point that I really wanted to make is that the computer, in
our attempt to address it, imposes a discipline on us which is beneficial.
It tends to take some of the sloppiness out of our thought. I think we
ought to really take advantage of that. Another point is that while semicolons in programming languages may be a pain, I hope that Jean Sammet puts
commas and semicolons in the right place in her natural language. These
are important things.

SAMMET
I try I think I'm a better punctuater than many, but it turns out there are
ambiguities in that as well. People will dispute whether there should or
shouldn't be a comma in certain places.

FALKOFF
I agree with Ben in the sense that predicting that natural languages for
addressing computers will not take hold, because I think it's a bore. Even
in other contexts we find people inventing acronyms and shortcuts in order
to say as little as possible to convey their thought. And if you have to
go through natural language sentences where a few symbols will do, people
who use these things will get bored with the longer thing. One more point
and then I'll quit.

SAMMET
Could I just respond to that and then I'll let you make your point. I have
said over and over and over again, although I think you don't want to hear
it, that my definition of natural language in this context includes
whatever is appropriate notation for the domain. So if the domain has
abbreviations or its own particular jargon, that's perfectly appropriate
and to be included with the natural language that this mythical computer
should understand. You can't keep ignoring that point. I do not mean that
if I say "A+B" that I want to write p-l-u-s. I want to write a plus sign.

FALKOFF
That's fine. I couldn't agree with you more and my comment was a more
general one. It wasn't really addressing the point that you made. There
are people who, for example, in query systems, would like to use so-called
natural language constructions, and I think they are a bore. Now as far as
the last little point, that here we have a matter of fact that there is a
COGO in APL, which is an old product, and in fact, the first version of
that, which was reasonably complete, was written over a weekend. So that
what Weintraub said is borne out by experience.

E. RICH
Somehow I did not make my point clear about language. I don't argue that
we should use natural language for current query systems. I know all the
same experiments that everybody else does, that show that people write
their first query in English and by the twentieth they've got shortcuts and
so forth because they don't want to write so much. If English is such a
poor vehicle for communicating things, why is it that people keep using it
to talk to each other? The reason is they have lots and lots and lots of
things to say. When we have programs for which there are only five things
to say, there are better languages and no one would sensibly, I think,
argue that we ought to use English. I am only arguing that we are to
consider English as a possibility when we have programs that are so sophisticated that no shortcut language allows you the set of options that you
need. And only in that circumstance are we considering English; we are not
considering it for the set of programs that we have now, at least I'm not.

BLUM
Unless there are any further comments from the panel, that seems to be an
excellent statement to close this session.

SUMMARY REMARKS

The Role of Language in Problem Solving   I
R. Jernigan, B.W. Hamill, and D.M. Weintraub (Editors)
© Elsevier Science Publishers B.V. (North-Holland), 1985

SUMMARY ASSESSMENT OF THE SYMPOSIUM

J. C. Boudreaux

Center for Manufacturing Engineering
National Bureau of Standards
Room A-351 Metrology
Washington, D. C. 20234

The conveners of this symposium thought that it would be useful to conclude
our work with a summary assessment of some of the more interesting points
that we have been discussing for the past few days. Rather than presenting
yet another technical session, it will probably be more profitable to use
this opportunity to state for future reference those issues that we believe
could use more careful exploration and discussion. In keeping with the
spirit of this session, I will now offer a few general remarks about the
issues that I found most interesting and I will soon encourage you to do
the same during the discussion period to follow.

An issue which surfaced many times and in many different guises during this
symposium was the need for new paradigms to explain the variety of cognitive roles that can be meaningfully assigned to computers and computer-
based systems. It seems to me that there are three main proposals, which
overlap and interdepend in many complicated ways.

(1)   Computers as Assistants
      This paradigm is the one that is most likely to be invoked by
      those who believe that computers, if properly programmed, can be
      made to behave in a way that is similar to human agents. This
      belief has long been associated with some workers in the field of
      Artificial Intelligence who accept the general point of Turing's
      Test as a more or less conclusive vindication of the thesis that
      there are no convincing differences between the intellectual
      abilities of humans and machines. Though it is impossible to
      prove the connection on purely formal grounds, those who accept
      this paradigm also tend to prefer what is usually called a
      "natural language" interface to the computer.

(2)   Computers as Environments
      Though related to the first paradigm, this one is more circumspect
      in that it does not attribute human-like personality traits to the
      computer. Several examples come to mind, probably the most obvious
      of which is the role of the computer in text editing. In this
      case, the computer presents the writer with the image of a text
      file, and the writer's task is to use the text editing commands to
      modify this image into some more satisfying form. Another very
      clear example of this paradigm is the video game. As a frequent
      player, I am always astonished by the fact that this computer-
      driven system allows me to slip into a parallel universe in which
      I am completely surrounded by an artificial environment of the
      programmer's own devising.

(3) Computers as Tool Kits

This paradigm may seem the most pedestrian of all, but it is one
that I believe will be the most fruitful in actual practice. The
basic insight is that computers are destined to be cheap and
plentiful and will be as easy to carry about as present day calculators. Thus, we should come to view the computer as a more or
less well integrated tool box that allows us to get on with our
business. Obvious instances of this paradigm are any one of a
number of recently built operating systems. But I think that a
better analogy is provided by the ubiquitous automobile. Though
at some deep emotional level it may be hard for some of us to
accept, cars only have instrumental value, that is, they are one
method of getting from one location to another, and the brakes,
accelerators, and other tools that they provide drivers are
subordinated to this primary function. Success or failure is not
a matter of using the tools more or less skillfully, but of
avoiding obstacles, including other vehicles, and steering a
rational course to our destination.

The point of this inventory is really to suggest that paradigms, even if
very roughly sketched out, are enormously important to the design of
programming languages. What I would very much like to see at the next conference is a more thorough analysis of alternate paradigms, deepening and
widening those that I have already mentioned, and perhaps defining others.

The dispute this morning posed another issue that has very far-reaching
ramifications. Recall that the disputed point was the utility of adopting
the "natural language" style as a design goal for future programming
languages. In my opinion, the discussion of this issue should be divided
into two closely related issues.

First, we need an accurate description about what we mean when we talk
about natural language. This may seem obvious but I assure you that it is
not. It will raise issues that are more difficult and subtle than those
who have not studied the issues can readily imagine. One indirect measure
of these difficulties is the host of problems that beset the creators of
Common English in the 1930's. Though they did severely restrict the degree
of syntactic complexity of English, still the end result contained far
richer structure than we could easily account for. In order to make the
case that computers ought to be addressed in natural language, the
proponents should clearly specify a "natural language" framework with the
same attention to syntactic and semantic details as is typically displayed
by designers of more conventional languages.

Second, a great deal of work is needed on the question of how best to
represent the technical vocabulary which binds together the experts in all
application domains. Once again this may seem to be a simple matter but I
can assure you that it is not. At the present time I am doing a project of
this kind at the National Bureau of Standards. In this case, it is the
lexicon of the world of automated manufacturing, that is, the world of
machinists, process planners, industrial engineers, to name but a few of
the technical domains. Similar work has become the focus of a thriving
industry under the unfortunate rubric "Knowledge Representation." In any
event, the difficult problem that has yet to be solved is the proper method
of eliciting useful information from those that are expert in the given
domain.

Another issue that has made more than one appearance here is the notion of program schemata. In general, program schemata are cognitive structures which mediate what a programmer knows about the programming language -- that is, the general template of an assignment statement, or an iterative loop, or variable and type declarations, and so on -- and the programmer's understanding of the application domain. Though it is commonly accepted that skilled programmers are acquainted with many schemata and in fact that it is this which separates them from their less experienced colleagues, very little is known about how these schemata are formed and about the best method to elicit and record them.

Finally, I cannot resist the urge to conclude with an issue that I believe is enormously exciting, namely, the attempt to distinguish between language as text and language as graphic sign. Let's see if this distinction can be stated in sharper form. A text may be conceived to be a linearly ordered sequence of words, that is, if one points to any arbitrary word in a text, then that word has at most one predecessor and one successor. Our ability to understand and manipulate texts is very finely honed. After all, the acoustic blast that I am presently aiming at you is obviously a linearly ordered text file, and I submit that you are all quite easily able to do some fairly complicated syntactic and semantic processing with it "on the fly". As natural as this model seems to be, several speakers have given us reason to believe that matters need not be this way. In particular, we could build a model of programming languages that is based not on the one dimensional aural model, but upon a model that is directly based on our visual apparatus, which is inherently two dimensional. In a sense, the use of a graphic notational scheme is very ancient, certainly going back to the hieroglyphic notation of ancient Egypt.

In conclusion, I should emphasize that the time has surely come for computer scientists to break free of the narrow professed limits of their field and turn their attention to the four thousand years of recorded history that preceded the development of that most human-like of machines. With that slightly insulting remark, let me now ask you to describe the issues that you would like future speakers to address.

WEIFFENBACH
I would like to make a suggestion and an insulting comment. I would like to suggest to everybody in this audience that there is in fact a community that exists and moves outside of the world of computers. I would like to support what Jean Sammet was suggesting and point out that ultimately you had better address the people who are going to be your users in terms that they can understand. You cannot avoid that. The question was raised as to why computers aren't being used in many areas where they are not being used. Well, let me tell you, you keep trying to talk to those of us outside of your little circle in a language we do not understand. I'll cite an example from my own experience which isn't quite identical to what I'm saying, but it illustrates a point. I had to argue with people who were sending me computer printout and who worked for me that I did not want an ephemeris that was printed in Julian seconds. That is a point, and I think you have made a very serious error in not understanding. I think that is what Jean Sammet was trying to say. Ultimately you have got to talk to the user at the end of this whole chain. Use whatever language you want and that you might choose when you design and implement your programs, but there is a different language which you must couch your output in for those of us who sit outside of your little circle. You cannot avoid that.

I would suggest when you look at this in the future, don't try to denigrate
natural language. You've got to make that translation.

BOUDREAUX
Thank you.

KEESE
Partial overlap, but in a slightly different direction. I would like to
suggest that you add to your list of paradigms the computer as a member or
as a part of society, in that there are real uses, such as giving out telephone numbers or explaining to people why their bills are so big, in which
your user is a customer, not a user in the traditional sense; and this
requires a very different area of language, particularly, the real commercial use for natural language input.

BOUDREAUX
Excuse me, just to understand that last point. The natural language input
would be used for the benefit of the computer? Did I misunderstand you?

KEESE
No. The natural language input would be, when the computer answers the
telephone, which has an irate customer at the other end, or when the computer gives direction, there will be people interacting with the computers
who are not computer programmers, who do not consider themselves computer
users, who do not care that it is a computer they are speaking to, who are
going to be angry whether it is a computer or not.

BOUDREAUX
Thank you.

LYON
I come here not to praise users, and, in fact, I have a criticism of this
conference on that point. Users tend to be often very parochial, as we
have, over and over, heard about APL at this conference. They often don't
understand very well what they are doing, and for a specialist in our area
they can unfocus you. They keep dragging you into quantum electro dynamics
and every other sort of thing that you could imagine. Let me give you an
example; it has been kicked around. In programming languages, the idea of
dynamic and static binding is quite important, and so is context. We heard
the virtues of APL as a language extolled, and yet, I believe Barstow
brought up a point about inheritance in SMALLTALK, which is static, so that
in APL you have dynamic contexts, and in SMALLTALK, he gave a quick-think
phrase to what is basically a static method of establishing context, which
for error recovery can be very important. That is a dimension of programming languages, that level of talking about it I think is far more important than talking about how APL helped me to write ovals, or circles, or
squares. That is what happens when you focus too much on the users,
because you forget about a dimension like that, and there are any number of
others that apply equally as well. I would like to see the conference
focus on some things that have to do with a higher level of talking about
programming languages and the environments.

BOUDREAUX
But you don't mean to say, do you, that the users are wholly irrelevant?

LYON
No. They are quite important, but we have met the users and we cannot become them. You must constantly maintain that distinction.

McALLISTER
I have a comment which may be interpreted as a question, or certainly as a suggestion. I am a teacher of mathematics. I am doing research in applied mathematics. I use the computer when I teach my courses and I use the computer as an assistant, as a tool kit, and as a user environment precisely as you have identified very aptly today. However, even though I am in a very definite area, I believe that a Nirvana, and I borrow your term, Jean Sammet, as a Nirvana, I am unable to envision the likelihood of having any two mathematicians agree what is the correct term, what is the feasible term, and what is the purpose of the "natural language" to be used in mathematics. I want to be a little more specific. When I teach and I use the computer, I call into the course, and the purpose of the course I am using the computer for, different languages, some because of the efficiency, time-wise, storage-wise or whatever; also because of the purpose of the course perhaps in future employment, it is beneficial for the students to do and use and become very knowledgeable in a certain area. So I don't think that I would like even the consideration of the creation, even it were possible, of a language that is called universal in mathematics. I find that when I hear a suggestion, do it. I like when my student has absolutely no background to hack the subject that I'm going to teach in a special seminar and wants to sign up for that seminar.

BOUDREAUX
Thank you very much.

DOTY
What I would like to see at a future conference like this is more discussion of the conceptual part of solving a problem. We've heard a lot of discussion on the implementation aspect and arguments over which languages are better. To me, the hard part is the conceptual part. Once I know what form my solution must take I can write it in any language. The problem comes when my problem is so big that I myself can't get a single conceptual answer. For example, writing the avionics for the B1 bomber. Big problem. No one person can hold it all in mind. How can we organize a group of people or a group of people and machines to solve a big problem and not worry about whether we do it in APL, or whether we do it in ADA, or anything else?

BOUDREAUX
I think that was the point of the conference. I think the only slight problem is that we have a very hard time addressing that fundamental question. Maybe next time you should submit a paper.

KING
The next conference, I think one thing I would like to see is a harder line driven between two different interfaces, what I find to be fairly clear interfaces; number one, the interface between the computer and the programmer; and number two, the interface between the programmer and his applications world, and what we popularly like to call the user. I saw a great deal of discussion about environments and about natural languages, and no one really seemed to pin-point where those things would be good. An environment is ideal for a programmer and, of course, for a user, but they

are two different animals. And a natural language, although it might be
very well used for a user application, for example, an airline reservation
system or some other such thing, it would find a very hard place to be in a
programmer's environment. So, we could sort of concentrate for the next
session on the two different aspects which I feel are very important and
are fairly easily delineated.

BOUDREAUX
Thank you.

O'CONNELL
I pride myself to think that I am one of that 1% of professional programmers who can produce 1,000 lines of error-free code. (laughter) I have so
many comments, it has been very frustrating the whole time. When I originally heard the title of the conference, the image that it brought to mind
for me was that when I write a computer system, and usually it takes a year
or two, I write cascades of programs and data files, and to me each one of
those data files essentially represents a language syntax and each one of
the programs represents a transformation of that language into some further
language which accomplishes a certain end, generally, in the application.
I think that there is a point I think Lieberman (something to that effect)
once suggested that part of the reason we find it so difficult to understand the nature of language and how it relates to human thought is basically that there is an isomorphism between the two. The nature of human
thought is the nature of language.

The whole concept of transformation of deep structure, you know, comparable
again to this cascade, in certain circumstances, of repeatability and of
recursion, seems to me are all essentials to the functions of language and
to functions of problem solving. I've been involved lately in a program
which has as one of its input files a syntax description of a markup language. So that it gets rather recursive in that as I approached the problem
of parsing the language and was developing the tools to use in parsing, I
recognized that I might as well parse the syntax itself and create a tree
to direct the subsequent operation of the whole system. This whole
approach of essentially creating a parsing tree in which certain nodes are
repeatable, certain nodes are optional, and certain nodes may occur
recursively, I think is a thread which has appeared throughout a lot of
discussion here today. I'm wrestling with these currently, and like you
said I should probably write a paper. A significant point in this aspect
also is the whole question of natural language and formal symbolisms. I
think that Russell in Principia starts out actually justifying why it is
that he is using the formal notation to do what it is he is doing, and he
essentially says that it is more appropriate to the task at hand, that to
attempt to elucidate the nature of the number two in natural language would
take forever. Similarly, however, he says that to attempt to make the
statement "A whale is big" in a formal notation, would take forever, and to
me that was an implication that it is the very ambiguity of language, of
natural language, which grants it this power, and that it is the context in
which a specific ambiguous expression occurs which allows us to sufficiently narrow down the ambiguity to use it as in Wittgenstein's sense of
language games, perhaps a more ecological approach to language than a
logical one, and so on and so forth.

SHNEIDERMAN
I'd like to add to your list of metaphors and also to respond to earlier speakers here with their frightening suggestion that computers be included as members of society. Well, if you would include a telephone answering machine computer or telephone calling computer, then you would include tape recorders, and various other devices as members of society.

BOUDREAUX
Yes, Ben, but getting back to....

SHNEIDERMAN
The point is that computers to me are very, very different from people. When I think of designing computer systems, I design them to serve people's needs, and that is the underlying principle. One of those needs is communication with other people, and one of the metaphors that I'm very sympathetic to in the design of computers is how a computer acts as a medium of communication across time and space, not only in allowing me to use electronic mail in a very natural way, but also in creating a program which somebody else uses, especially obvious in a CAI computer system structure environment where I'm using the computer as a medium of instruction, just as a book is a medium of instruction. That metaphor I find to be a useful one that I encourage you to add to your list.

BOUDREAUX
Thank you.

CARLTON-FOSS
I'm going to add another metaphor with a look at Chinese, which started out as a pictographic language and then evolved into the characters over several thousand years. I'm similarly fascinated with the various metaphors of what computers might be, and we could probably create a huge list of such metaphors. A colleague recently wrote something up about computers as members of an organization. I think that as modification of languages becomes increasingly accessible to at least programmers, not the top 0.1% of programmers, but perhaps the top 20 or 30% of programmers and maybe even some users, we are going to see a natural evolution of the symbology used, and similarly that type of thing happens in people's relationships with computers within organizations. So that is another metaphor that may be irrelevant, but if it is included I might push this colleague to talk next year.

BOUDREAUX
Thank you.

O'CONNELL
The concept that a computer as member of society actually calls up in my mind, one of the things that really impresses me about programmer or programs and executing programs, is the fact that once you get it to work, it will do the same thing every time. Just about every time (every once in a while there is a blown fuse somewhere), there is a reliability on which it is possible to build, and the whole transition from machine language to assembly language to operating systems to compilers is essentially, to me, analogous to the accretion of a cultural knowledge base which has so much daily effect on our lives.

JERNIGAN
A distinction of AI is the tacit acceptance that a machine is potentially
capable of human thought. With this acceptance comes not only the vocabulary of psychology but its entire metaphysical orientation. Traditional
computer engineering represents a different metaphysical standpoint. The
computer is a machine doing only what it is programmed to do, and that with
the vocabulary and the thinking of the engineers. Several years ago when I
first developed a modeling system, the major problem was the lack of a
vocabulary for describing the modeling system and what it did. My own
tendency in dealing with this was to draw on my background in philosophy.
But it was quite strange, if you can imagine, to try to discuss economic
models in terms of phenomenological reduction. However, it does seem to be
acceptable to use more mundane words when we speak of machine thinking and
knowing, such as "machine knows" and "machine thinks". Perhaps the science
of scientific or mechanical problem solving is so new that it hasn't established its own domain, its own vocabulary. Hence it is forced to look
outside for understanding. I don't accept that. When we do a phenomenological reduction, or mechanical problem solving, we do find that the
structures, the mechanics, the local phenomena themselves dictate the
linguistic notions appropriate to the problem. This is evident if you look
at the work of people like Euclid, Babbage, Newton, McCarthy, Iverson, and
Codd.

I'd like to end up by inviting further comments but I'll only accept them
only in the form of written papers, about 12 pages in length, which we can
discuss at the next symposium. I thank you for coming, the conference is
adjourned.

CONTRIBUTED PAPERS

The Role of Language in Problem Solving I
R. Jernigan, B.W. Hamill, and D.M. Weintraub (Editors)
© Elsevier Science Publishers B.V. (North-Holland), 1985

A LANGUAGE FOR THE CONCEPTUAL SCHEMA OF A DATA BASE

Sudhir K. Arora, Surya R. Dumpala, Sei-Jong Chung and Bob Weems

Department of Computer Science
Northern Illinois University
DeKalb, Illinois 60115

Several data base machine architectures have been
introduced in the literature. Of these only two address
the problem of multiple model support - GDBMS and WCRC.
Of these two only WCRC supports a full language at the
conceptual level - WCRL. In this paper we describe the
complete WCRL which is based on a data structure called
Elementary Well Connected Relation.

I. INTRODUCTION

Several data base machine architectures have been studied in the literature
- CASSM (Copeland, 1973), RAP (Ozkarahan, 1975), DIRECT (DeWitt, 1978),
RARES (Lin, 1976), DBC (Banerjee, 1979), RELACS (Oliver, 1979), SEARCH
PROCESSOR (Leilich, 1978), XDMS (Canaday, 1974), IFAM (DeFiore, 1973),
etc. Some of these machines address only specific operations on a data
base; some implement only one of the major data models - Relational,
Network or Hierarchical; some implement all three data models but not
simultaneously on the same physical data.

There is a need for a data base machine architecture which can support
different data models simultaneously on the same physical data. In such a
machine users should be able to view the data according to a relational,
network or hierarchical data model. This is called "logical data independence." Further, changes to the physical data should have minimal effect
on the user's view of the system - "physical data independence." Such an
architecture would conform to the ANSI/X3/SPARC (Ansi, 1975) proposals or
the coexistence model (Nijssen, 1976) which envisage three levels for a
data base system - External, Conceptual and Internal. At the conceptual
level a stable common view of data and its semantics must reside. The
physical data is stored at the internal level and may be altered to take
advantage of evolving technologies.

To the best of our knowledge, there are only two projects in the world
addressing this problem - Generalized Data Base Management System (GDBMS)
(Dogac, 1980) and Well Connected Relation Computer (WCRC) (Arora, 1981).
WCRC is one hardware version of the ANSI/X3/SPARC architecture or the
Coexistence Model. It has an external level, a conceptual level and an
internal level. It can handle queries from several users simultaneously on
the same data at the internal level. It may be used as a back end to a
host computer or as an independent data base computer to do non-numeric
processing.

In this paper we introduce the complete conceptual level language, WCRL. This is a language that supports set processing and is equally well applicable to the three major data models, Network, Relational and Hierarchical.

II. CHOICE OF THE LANGUAGE

WCRC supports several user languages at the external level. This makes it necessary to have a common data model independent language at the conceptual level. Some of the data model independent languages proposed in the literature are LSL (Tsichritzis, 76), FQL (Buneman, 79), QUEST (Housel, 79), and WCRL (Arora, 80). We use WCRL as the language for conceptual level of WCRC for the following reasons.

1. The language is based on a data structure called elementary well-connected relation (EWCR), which can be easily mapped into the storage structure at the internal level, a binary pseudo-canonical partition (explained later in detail).

2. It allows complex conditional expressions involving set comparators, aggregate functions, and recursion, some of which are not available in the other languages.

3. It is algebraic in nature and also allows navigation through the model.

4. It applies equally well to the three major data models - Network, Relational and Hierarchical.

The language as described in (Arora, 80) provides only data retrieval facilities. In this report, we extend the language to accommodate data definition, data manipulation (update) and storage definition. In the sections which follow, these aspects are discussed in detail.

III. WCRL - DATA RETRIEVAL LANGUAGE

WCRL is based on the theory of well-connected relations (Arora, 79). Those definitions and the language commands which are relevant for the discussion, are presented below.

Definition III.1:
A Well-Connected Relation (WCR) is a binary relation W, on two sets A and B such that

$$(\forall)\quad (a \epsilon A)\quad (\forall b)\quad \{b \epsilon B\}\quad (aWb)$$

The sets A and B are called the first and the second constituents of the WCR. Two WCR's are compatible if their constituents are based on the same domains.

Definition III.2:
An Elementary Well-Connected Relation (EWCR) is a WCR in which the first constituent has a single element. The second constituent is then called the 'Image set' of the first constituent.

Definition III.3:
A Trivial Well-Connected Relation (TWCR) is a WCR in which both the constituents have a single element.

Definition III.4:
A relation R[A, B] can be expressed as

$$R[A,B] = \sum_{i=1}^{n} R_i[A_i,B_i]$$

$$= R_1[A_1,B_1] \cup R_2[A_2,B_2] \cup \ldots \cup R_n[A_n,B_n]$$

$$= \pi(R), \text{ a } \underline{\text{partition}} \text{ of } R$$

where $R_i[A_i,B_i] \cap R_j A_j,B_j = \phi$

for $i \neq j$, $i \geqslant 1$, $j \leqslant n$ and $A = \bigcup_{i=1}^{n} A_i$ and $B = \bigcup_{i=1}^{n} B_i$

Definition III.5:
A partition of a binary relation R[A,B] is a $\underline{\text{canonical partition}}$ (CP) if,

$$R[A,B] = \sum_{i=1}^{n} w_i[A_i;B_i]$$

where a. $w_i[A_i;B_i]$ is a WCR for $1 \leqslant i \leqslant n$,
       b. A_i is a set with a single element for $1 \leqslant i \leqslant n$,
       c. $A_i \neq A_j$ for $i \neq j$ and $1 \leqslant i \leqslant j \leqslant n$

A graphical illustration of these definitions is shown in Figure 1.

WCRL provides two levels of operations based on set processing. This means, any condition on the first constituent of an EWCR is evaluated only once for all the tuples in the EWCR. The two levels in WCRL are the low level and the high level. The former uses only EWCR's as operands while the latter uses canonical partition. The operations fall into three classes, namely, set reconstitution, set join and pseudo set operations. These operations are briefly discussed below.

1.   Set Reconstitution (σ_{sr}):
      The algebraic representation of this operation is given by,
      σ_{sr} W[A ; B] (First Expression (x); Second Expression (y)) = C[X ; Y] where expression may be,

      a.   A list of attributes belonging to (A U B)
      b.   Null

354   *S.K. Arora et al.*

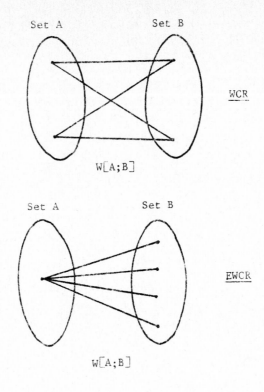

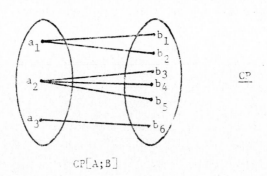

FIGURE 1
ILLUSTRATION OF A WCR, AN EWCR AND A CP

c.   A list of attributes where some attributes are restricted or
     selected according to a relationship.

A qualification may involve arithmetic (+, -, *, ÷) operators and aggregate
functions. For example, X may be of the form,

EX: $X = A_1$, $(B_1 = 1000)$, $(B_2 = AVG(B_3) \div 100)$

The result of this operation is a CP, $C(X;Y)$.

2.   Set Join (σ_{sj}):
It is of the form,

$\sigma_{sj}\ w_1[A_1;B_1]\ w_2[A_2;B_2],\ldots,w_n[A_n;B_n]$
(conditional expression) $= C[X;Y]$

where $X = A_1,A_2,\ldots,A_n$

$\qquad\qquad$ Codd's Join (σ_{cj})

and $Y = B_1,B_2,\ldots,B_n$

or $X = A_1,B_1,A_2,B_2,\ldots,A_n$

$\qquad\qquad$ Brooming Join (σ_{bj})

and $Y = B_n$

The conditional expression is a set of conditions connected by boolean operators $(\wedge,\vee,\neg)$. Each condition may involve relational operators $(=,>,\geqslant,$ etc.) set comparison operators (current $A_1 > $ All B_2, All $A_1 \leqslant$ some B_1, etc.) and set containment operators $(A_1 \subseteq B_1, A_2 \not\supset B_2,$ etc.) on any two attributes from 1 A's and B's.

EX: A conditional expression may be of the form,

$(A_1 = 50) \wedge (A_2 \subseteq A_3) \wedge (B_2 > 100) \vee (B_2 \cdot \geqslant \cdot \cdot B_3)$

where '$\cdot \geqslant \cdot \cdot$' stands for some B_2 is greater than or equal to all B_3. (NOTE: 'one dot' corresponds to 'some', 'two dots' to 'all' and 'no dot' to 'current').

3.   Pseudo Set Operations
These operations are 'selective union', 'selective intersection' and 'selective difference'. The arguments of these operations should all be compatible EWCR's.

a.   Selective Union $\left(\sigma_{su}\right)$

$\sigma_{su} w_1 [A_1;B_1], \ w_2 [A_w;B_2], \ldots, w_n [A_n;B_n]$
(conditional expression) = w[A ; B]

b.   Selective Intersection $\left(\sigma_{si}\right)$

$\sigma_{si} w_1 [A_1;B_1], \ w_2 [A_w; \ B_2], \ldots, w_n [A_n; \ B_n]$
(conditional expression) = w[A ; B]

c.   Selective Difference $\left(\sigma_{sd}\right)$

$\sigma_{sd} w_1 [A_1, \ B_1], \ w_w [A_w, \ B_2], \ldots, w_n [A_n, \ B_n]$
(conditional expression) = w[A ; B]

In selective union, the tuples from the EWCR's that satisfy the conditional expression are unionized into one EWCR. Selective intersection collects the common tuples from the EWCR's that satisfy the expression and forms a new EWCR. Selective difference removes the common tuples from w_1 and retains only those that satisfy the expression in w_1.

The high level operations are similar to the low level ones except that the arguments are canonical partitions instead of EWCR's. When CP's are involved, the low level operation is applied to each EWCR/set of EWCR's in the CP.

In order to demonstrate the suitability of WCRL as a conceptual language, the language must be further developed. The operations for data definition, data manipulation (for updates) and storage definition are developed further.

IV. DATA DEFINITION LANGUAGE (DDL) FOR WCRL

The DDL of WCRL presented here, is intended for defining the schema and the constraints of the entity-relationship model. The various statements are given below.

1.   Entity Definition Statement
     This statement takes the form,

$$\sigma_{de} \ a_1, a_2, \ldots, a_n \ (<E\_SPECIFICATION>) = e\left(E_1\right)$$

where a. $a_1, a_2, \ldots, a_n$ represent the attributes of the entity,
       b. E_i is the name of the entity,
and   c.   <E_SPECIFICATION> is of the form (t_e, K, R) where t_e indicates the type of the entity (weak, normal), K denotes the key and R specifies the supporting relationship in the case of a weak entity.

       EX: σ_{de} DNO, Budget, CE (Normal, DNO, Φ) = e(Dept.)

2. Relationship Definition Statement:
    The algebraic representation of this statement is given by,

$$\sigma_{dr}\ a_1, a_2, \ldots, a_n\ (\text{<R\_SPECIFICATIONS>}) = r(R_1)$$

where   a. $a_1, a_2, \ldots, a_n$ are the attributes of the relationship,

       b. R_1 is the name of the relationship,

and   c.   <R_SPECIFICATION> is of the form (d,<spec.list>)
Here d indicates the degree of the relationship (i.e. binary, k-ary), and the <spec.list> constitutes $(s_1, s_2, \ldots, s_i, \ldots, s_k)$ where each s_i is of the form (e_1, e_2, t_r); e_1 and e_2 denote the entities involved in a relationship while t_r indicates the type (i.e., 1:N; 1:1 or N:M). Further, the existence of some entity instances that do not participate in the relationship may be indicated by a '~' sign on top of the corresponding entry in t_r. Such a relationship is called partial relationship. They are made total by inclusion of null values.

EX: To define N:M relationship 'Work' between entities 'Employee' and 'Project' where every employee is not required to work on at least one project, the following statement may be used.

$$\sigma_{dr}\text{No. of Hours (binary, (Employee, Project, } \widehat{N}\text{:M)}) = r(\text{work})$$

3. Attribute Definition Statement:
    This statement takes the form

$$\sigma_{da}\ v\ (\text{<V\_SPECIFICATON>}) = a(A_i)$$

where   a.   v denotes the value set that forms the domain of the attribute,

       b. A_i is the name of the attribute,

and   c.   <V_Specification> is of the form (t_d, u, p) where t_d represents the data type of the value set (real, integer, etc.), u specifies the units (dollars, years, etc.), and p denotes the range of allowable values or uniqueness

EX: The attribute 'Salary' may be defined as follows:

σ_{da} Amount (Integer, Dollars, 10,000-60,000) = a(salary)

4.   Constraint Specification Statements:

The constraints on E-R model may be specified using the following security and integrity statements.

a. $\sigma_{SC}\ a_1, a_2, \ldots, a_n (SE_1, SE_2, \ldots, SE_n)$

where   (1) a_i is an attribute or an entity

and   (2) SE_i is a security constraint expression on a_i. In general, SE_i is a list of expressions of the form (U#,(<A LIST>)) where U# indicates the user identification number and A LIST consists of the allowable accesses (Read(R), Write(W), Modify(M), Read statistical values (S) such as 'Average', etc.). Thus, several constraints can be specified in one statement.

b. $\sigma_{IC} O_1, O_2, \ldots, O_n (IE_1, IE_2, \ldots, IE_n)$

where   (1) O_i is an object name, namely, an entity, a relationship, an attribute or a value set name.

and   (2) IE_i is an integrity constraint on O_i of the form, <Agr_fn> <object name> <opr> <opd> where <Agr_fn> : = AVG|SUM|COUNT|MIN|MAX|Φ(optional), <object name> : = entity name|relationship name| value set name|attribute name, <opr> : = set containment operator|comparison operator, <opd> : = constant|<Agr fn> (<object name>).

EX: 1.   The user access to an attribute 'Salary' can be specified as σ_{SC} Salary ((U_1,(R,W)), (U_2,(M,S))). This means that user 'U_1' is allowed to read and write while U_2 is permitted only to modify and read the statistical results. Of course, modify implies read and write as well but not vice versa.

2. The integrity constraint, "Manager's salary should be greater than average salary of the employees" can be specified as,

$$S_{IC} \text{ Mgr. Sal (Mgr. Sal > AVG(EMP. Sal)}$$

3. The constraint "The managers should be a subset of employees" may be specified as,

$$S_{IC} \text{Mgr (Mgr   Emp)}$$

4. Schema Change Specification Statement:
A change in the objects of the schema may be effected by the following statement,

$$S_{cs} O_1; O_2; \ldots, O_n [\text{<object type list>}] (\text{<condition list>})$$

$$= 0(0_1; 0_2; \ldots; 0_n)$$

where a. o_i is an object name list or an object name
        b. <object type> is either an entity, a relationship, an attribute or a value set; the ith object type corresponds to o_i
        c. <condition> is of the form
            <object name> <opr> <constant> and
            <opr>: = →|=|>|_|°|<
            <constant>: = Integer|Real|character string,

and   d. 0_i is a list containing the new name(s) of the object 0_i. If 0_i contains two names, the instances satisfying the condition are given the first name. The rest are given the second name.
           NOTE: all the schema change statements of the DBA Language split, merge, shift, rename, and delete, can be expressed by S_{cs} statement. The other statement 'add' is taken care of by the data definition statements.

EX: 1. Change the entity 'Persons" into two entities called 'Male persons' and 'Female persons'; change the attributes S_{cs} <Person; (home phone, office phone)}

[Entity ; attribute] (Person . Sex = 'Male';Φ)
           = O{(Male person, Female person) : (phone)}

2.   Delete attribute 'Age' of the entity person σ_{cs} person . Age
     [Attribute] (Φ) = O(Φ)

V. DATA MANIPULATION LANGUAGE (DML) FOR WCRL

The DML fo WCRL for updates, based on the entity-relationship model,
comprises the following statements.

1.   Insertion:
     Insertion of a tuple into entity or a relationship is accomplished
     by the following statement,

$$\sigma_I A_1, A_2, \ldots, A_n \ [\text{<object name>}] \ (t_1; t_2; \ldots; t_m) = O(O_i),$$

     or $\sigma_I \ A_1, A_2, \ldots, A_n \ [\text{<object name>}] \ (\text{<Expression list>}) = O(O_i),$

where a. A_i is a list of some or all attributes of the object type
            (unspecified values are taken as null values)
       b.   <object name> is either an entity or a relationship name
       c. t_i denotes a tuple which is of the form $(a_1, a_2, \ldots, a_n)$ is a
            value
       d.   <Expression list> gives a tuple or a set of tuples, as
            follows,
            <Expression list>: = <expression>; <expression list>|
                              <expression>
            <expression>: = <iexpr>, <iexpr list>|<iexpr>
            <iexpr>: = <aggr func> (<atname>)|
                      <aggr func> <atname> <op> <constant>
      where   <atname> is an attribute name
              <aggr func> is one of (MIN,MAX,SUM,AVG,CAR)
      and   <op> is one of (+,-,x,÷).
       e. O_i is the 'rename' of the object type (Renaming is optional).

2.   Deletion:
     The following statements can be used for deleting the tuples from
     an entity or a relationship

$$\sigma_D\ A_1, A_2, \ldots, A_n\ [\text{<object name>}]\ (t_1; t_2; \ldots; t_m) = O(O_i)$$

or

$$\sigma_D\ A_1, A_2, \ldots, A_n [\text{<object name>}]\ (\text{<rexpression list>}) = O(O_i),$$

where   <rexpression list> is as follows
        <rexpression list>: = <rexpression> <rexpression list>|
                    <rexpression>
        <rexpression list>: = <rexpr>, <rexpr list>| <rexpr>
                    <aggr func>
        <rexpr>: = <aggr func> <atname> <op> <constant>
                |<atname> <cop> <aggr func> <atname>,

where   <atname> is an attribute name,
        <op> is one of $(+, -, x, \div)$
        <cop> is one of $(=, \ne, >, <, \geqslant, \leqslant)$

and   <aggr func> is one of (SUM,AVG,MAX,MIN,CAR).
        The syntax is similar to that of insertion statement.

3.   Modification:
        The statements to modify tuples in entities or relationships take
        the form,

$$\sigma_M\ A_1, A_2, \ldots, A_n\ [\text{<object name>}]\ ((a_1, a_2, \ldots, a_n),$$
$$(b_1, b_2, \ldots, b_n)) = O(O_i)$$

or

$$\sigma_M\ A_1, A_2, \ldots, A_n\ [\text{<object name>}]\ (\text{<mexpression list>})$$
$$(b_1, b_2, \ldots, b_n) = O(O_i)$$

where a. a_i's are the values of A_i's before update
          b. b_i's are the values of A_i's before update
and   c.   The value for <mexpression list> is a tuple or a set of
             tuples. The syntax of <mexpression list> is same as
             <rexpression list>. NOTE: all the three update statements
             provide an option to rename the object after update.
        NOTE: WCRL also allows for recursion which can be used to define
        new operations based on the other retrieval operations.

VI. STORAGE DEFINITION LANGUAGE (SDL) FOR WCRL

The SDL part specifies the mapping between the logical and physical organization of the data (i.e. the conceptual and the internal levels). It also provides facilities for formatting, creating, destroying and renaming of the internal level storage structures. Before developing these language statements, a discussion on the mapping between conceptual and internal levels is presented here.

It may be recalled that the logical objects at the conceptual levels are entities, relationships, attributes and value sets while the storage structures at the internal level are canonical partitions (CP's). There exists a simple mapping between the objects of E-R model and the CP's, as follows.

1. The attributes are functional mappings from entity sets and relationship sets into value sets. Therefore, a binary relation consisting of the key of an entity or a relationship set and a value set would conveniently form a CP representing an attribute.

2. All binary relationships which are 1:1 or 1:N type directly correspond to the respective CP's.

3. Any N:M type binary relationship may be transformed into a set of two CP's by associating a system defined key with it. For example, a relationship, $R[A,B]$ would become CP's, $[K;A]$ and $C_2[K;B]$ where K is the system defined key. On similar grounds, a k-ary relationship may also be transformed into a set of CP's.

These mapping rules would form the basis for the following SDL statements.

a. Mapping Statements
    (1) $\sigma_E \; c_1, c_2, c_3, \ldots, c_n \; (\text{<M\_Expression>}) = e(E_i)$

    (2) $\sigma_R \; c_1, c_2, c_3, \ldots, c_n \; (\text{<M\_Expression>}) = e(R_i)$
    These statements define the entity|relationship sets respectively, in terms of the CP's. <M_Expression> stands for a Codd's join statement. The c_i's denote the canonical partitions and E_i/R_i the name of entity|relationship set. The language also allows for "logical CP's" to be defined in terms of the CP's at the internal level, by the following statement. A logical CP may have more than two attributes.

    (3) $\sigma_{CP} \; c_1, c_2, \ldots, c_n \; (\text{<M\_expression>}) = cp(C_i)$
    where c_i's denote the CP's or other logical CP's already defined.
    C_i denotes the logical CP

and   <M_Expression> stands for a set join or a pseudo set operation.

For example consider the CP's, C_1[ENAME;SAL] and C_2 [ENAME;SAL] where C_1 corresponds to all the employees and C_2 corresponds to only managers. Suppose we need only those employees who are not managers but earn more than \$30,000. Then alogical CP, C_3 can be defined as

$$\sigma_{CP}\ C_1, C_2\ (\sigma_{sd}\ C_1[\text{ENAME;SAL}],\ C_2[\text{ENAME;SAL}]\ (C_1.\text{SAL} > 30,000))$$

$$= cp\ (C_3[\text{ENAME;SAL}])$$

NOTE: These statements may involve some data retrieval statements such as set join in their <M_expression> part. The same statements may also be used for renaming by letting M_expression to be null.

b.   Storage Definition Statements

    (1)   Format:

           The following statement is used to specify the <u>format</u> of the canonical partitions,

$$\sigma_{FM}\ c_1, c_2, \ldots, c_n\ (<F_1,\ F_2, \ldots, F_n>)$$

where c_i's are the CP's and F_i is of the form $<f_1, f_2>$. f_1 represents the first constituent and f_2, the second constituent. Each f_1 is a 3-tuple of the form $<n, t_d, n_b>$ where n is the name of attribute, t_d the data type and n_b the number of bytes.

<u>EX</u>: σ_{FM} SAL,BUD (((ENO, Integer, 4), (Salary, Integer, 5)),
                                 ((DNO, Integer, 3), (Budget, Real, 6))

This statement defines the format of two CP's, SAL(salary) and BUD(Budget).

    (2)   Create:

           This statement takes the form,

$$\sigma_{CR}\ cp\ (\ \text{DATA}\ )$$

It creates the CP with the data specified; the data must follow the format specified earlier by a format statement.

(3) Destroy:

This statement is of the following syntax,

$\sigma_{DS}\ c_1, c_2 \ldots, c_n.$
All the CP's specified would be physically deleted from
the data base.

VII.   CONCLUDING REMARKS

In this paper we have introduced the complete Well Connected Relation
Language (WCRL). It consists of the following:

1. Data Retrieval Language
2. Data Definition Language
3. Data Manipulation Language
4. Storage Definition Language

WCRL is a data model independent language which is well suited for use at
the conceptual level of a multi-level ANSI/X3/SPARC data base. We use it
in our data base computer architecture, namely, the Well Connected Relation
Computer (WCRC).

VIII.   REFERENCES

1. ANSI/X3/SPARC Interim Report on Data Base Management Systems, FDT 7(2),
   Vol. 7, No. 2, 1975.

2. Arora, S. K. and Smith, K. C., "A Theory of Well-Connected Relations,"
   J. of Information Sciences, 19, 1979, pp. 97-134.

3. Arora, S. K. and Smith, K. C., "WCRL: A Data Model Independent Language
   for Data Base Systems," Int. J. of Comp. and Inf. Sciences, Vol. 9, No.
   4, Aug. 1980, pp. 287-305.

4. Arora, S. K., Dumpala, S. R. and Smith, K. C., "WCRC: An ANSI SPARC
   Machine Architecture for Data Base Management," Proc. Eight Int. Symp.
   on Comp. Arch., Minneapolis, May 1981, pp. 373-388.

5. Banerjee, J., Hsiao, D. K. and Kannan, K., "DBC - A Database Computer
   for Very Large Databases," IEEE Trans. on Computers, Vol. C-28, No. 6,
   June 1979, pp. 414-429.

6. Buneman, P. and Frankel, R. E., "FQL - A Functional Query Language,"
   ACM SIGMOD, Boston (1979), pp. 52-58.

7. Canaday, R. H., et al., "A Back-End Computer for Data Base Management,"
   CACM 17, 10, Oct. 1974, pp. 575-582.

8. Copeland, G. P., Lipovski, G. J. and Su, S. Y. W., "The Architecture of
   CASSM: A Cellular System for Non-numeric Processing," First Annual
   Symposium on Computer Architecture, 1973.

9. DeFiore, C. R. and Berra, P. B., "A Data Management System Utilizing an
   Associative Memory,"Proc.ACM National Computer Conf.,1973, pp. 181-185.

10. DeWitt, D. J., "DIRECT - A Multiprocessor Organization for Supporting Relational Data Base Management Systems," Proc. Fifth Annual Symp. on Comp. Architecture, 1978, pp. 182-189.

11. Dogac, A. and Ozkarahan, E. A., "A Generalized DBMS Implementation on a Database Machine," ACM SIGMOD, Santa Monica, CA, May 1980, pp. 133-143.

12. Housel, B. C., "QUEST"" A High-Level Data Manipulation Language for Network, Hierarchical and Relational Databases, "IBM Res. Rep. RJ2588(33488), 7/25/79 (1979).

13. Leilich, H. O., Stiege, G. and Zeidler, H. Ch., "A Search Processor for Data Base Management Systems," Proc. Fourth VLDB, West Berlin, Sept. 1978, pp. 280-287.

14. Lin, C. S., Smith, D. C. P. and Smith, J. M., "The Design of a Rotating Associative Memory for Relational Data Base Applications," ACM TODS, Vol. 1, No.1, March 1976, pp. 53-65.

15. Nijssen, G. M., "A Gross Architecture for the Next Generation Database Management Systems," Modelling in Data Base Management Systems, North-Holland, 1976, pp. 1-24.

16. Oliver, E. J., "RELACS, An Associative Computer Architecture to Support a Relational Data Model," Ph.D. Thesis, Syracuse University, 1979.

17. Ozkarahan, E. A., Schuster, S. A. and Smith, K. C., "RAP - An Associative Processor for Data Base Management," National Computer Conf., 1975, pp. 379-387.

18. Tsichritzis, D., "LSL: A Link and Selector Language," ACM-SIGMOD, Washington, D. C., June 1976, pp. 123-134.

The Role of Language in Problem Solving I
R. Jernigan, B.W. Hamill, and D.M. Weintraub (Editors)
© Elsevier Science Publishers B.V. (North-Holland), 1985

SILICON APL - THE TRANSLATION OF APL LANGUAGE
CONCEPTS INTO PROCESSOR ARCHITECTURES

Andrew Goldfinger

The Johns Hopkins University
Applied Physics Laboratory
Laurel, Maryland 20707

It is presently possible to fabricate VLSI chips with hundreds of thousands
of individual components. How are such complex devices to be designed?
Noticing the analogies between hardware chip design and computer
programming, many people have begun work on "silicon compilers", higher
level languages in which chip architectures can be specified and then
automatically translated ("compiled") into detailed designs of the masks
necessary for fabricating a chip.

We have found that the APL language has great promise as a basis for
silicon compilers, and we have begun to develop an extension of it
("Silicon APL") that can be used to specify chip design. In the process we
have found that

(1) mere consideration of the nature of the APL language has led
     to ideas of processor architectures that might not have
     otherwise occurred.

(2) modes of thinking about data and operations that were learned
     through APL programming are readily and productively
     transferred to hardware chip design.

THE SILICON APL CONCEPT

In considering the transfer of language concepts to hardware, it is
dangerous to make categorical predictions of the future, but it may be safe
to make the following observations concerning the APL-silicon
compatibility.

1.   Some functions, such as ∧, ∨, ~, ∧/ etc. have already been
     implemented on chips. (These are merely standard logic
     gates.)

2.   Other functions, in particular ⍳, will never be
     implemented. (silicon would have to manufactured during
     execution!)

3.   Therefore, at most a subset of APL will be useful.

Below, we will present a small subset of APL that could be compiled into
silicon. While this subset is perhaps too small to be practically useful,
it should form a basis for future development.

DATA TYPES

In APL one does not distinguish data types in the same way as in PL/1 or
FORTRAN (FIXED versus FLOATING for example). In silicon APL such
distinctions do appear necessary to enable the programmer to express
control over data flow. In silicon, data are carried by one or more
conducting paths which have trains of bits propagating along them, and data
type declarations seem to be necessary to allow for proper decoding of data
and interconnection of functions. We define data to mean the following:

    1.   A number of conducting paths on the chip along which

    2.   a train of bits propagates in

    3.   synchronization with a system clock.

Data are described by their spatial and temporal qualities as indicated in
Figures 1 and 2.

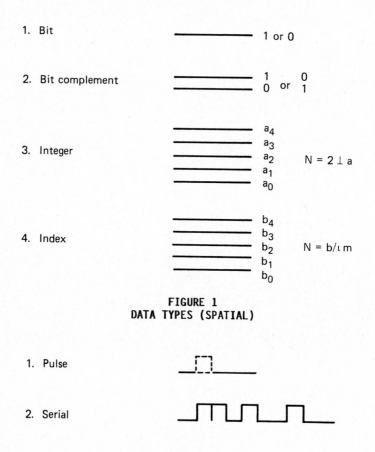

FIGURE 1
DATA TYPES (SPATIAL)

FIGURE 2
DATA TYPES (TEMPORAL)

Spatially, a single conductor represents a single bit. Data can be single
bit, or "bit complement", i.e., a doubled line carrying the data bit and
its inverse. It can also be multiple bit forming a binary integer, and the
bits of this integer can also be complemented (not shown in the figure).
Another type of data is "index", the utility of which will be apparent
below. Index data consists of a number of parallel bits all but one of
which are zero, the non-zero bit indicating the value of the index as
shown. Index data can also occur in complemented form.

Temporally, data can be "pulse", i.e., single bits (although perhaps
spatially parallel) or serial (a number of bits in sequence). Just as data
shapes are controlled by the function ρ data types are controlled by the
new (overstruck) function $\natural$. It has monadic and dyadic forms, defined in
analogy to the reshape operation ρ:

$$A \leftarrow B \; \natural \; V$$

indicates that V should be formatted into data type B, and

$$B \leftarrow \natural \; V$$

assigns to B the data type of V according to the following format:

$$\natural V \leftrightarrow N_1, N_2, N_3$$

N_1 = number of serial bits (N_1=1 for pulse data)

N_2 = number of spatial bits (N_2=1 for bit data),
positive for integer data, negative for index data.

N_3 = 1 if data are complemented, 0 otherwise.

DATA SHAPES

Shapes of scalar, row and column vector, and two-dimensional matrix are
allowed, as shown in Figure 3. Unlike standard APL, row and column vector
are distinct objects. Arrays of rank higher than 2 are not supported. The
busses in this figure represent a number of parallel lines as determined by
the data type, and the data "packets" indicated by the cross-hatching are
either pulse or serial. For example, the scalar data could be uncomple-
mented integer serial (2 8 0 = $\natural$ X) in which case the packet would consist
of eight parallel lines carrying two serial bits each. (This could be a
representation of a complex number with eight-bit real and imaginary
parts.) Row vectors are represented by serial packets, while column
vectors are represented by parallel busses. Arrays are parallel busses
with serial packets.

The standard ρ operator defines and modifies data shapes. The following
conventions are used.

(ρx) = null vector	→ scaler	
= positive integer	→ column vector	
= negative integer	→ row vector	
= 2 integers	→ matrix	

1. Scalar

2. Row vector

3. Column vector

4. Array

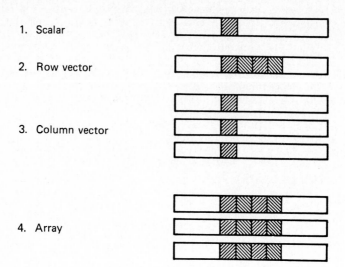

FIGURE 3
DATA SHAPES

We note that a number of parallel conductors carrying a train of bits could
be interpreted as data in many ways. It is the ρ and $\mathbb{n}$ operators that give
meaning to the data.

DATA CONVERSIONS

Since data have various types, data conversions are necessary. These are
accomplished by the dyadic $\mathbb{n}$ function. This function can also be used
reflexively to establish (in PL/1 language: declare) the data type. Thus,
for example, the reflexive expression

Y ← 1 3 0 $\mathbb{n}$ Y

establishes Y as a three-bit non-complemented pulse integer, while

Z ← 1 ‾8 0 $\mathbb{n}$ Y

converts it to an eight-bit non-complemented pulse index and assigns this
value to Z. A brute force way of implementing this conversion is shown in
Figure 4; more efficient ways are no-doubt possible. We note that a non-
complemented to complemented data conversion occurs in the left-most part
of the figure (Y' ← 1 3 1 $\mathbb{n}$ Y).

Another important type of data conversion is parallel to serial. For
example:

W ← 3 1 0 $\mathbb{n}$ Y

Such a conversion is accomplished by the latch/shift register combination
shown in Figure 5.

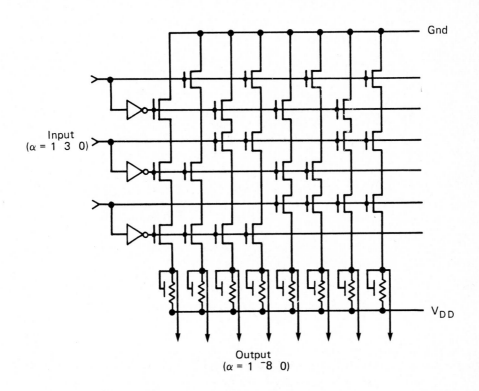

Input
(α = 1 3 0)

Gnd

V_{DD}

Output
(α = 1 $^-$8 0)

FIGURE 4
INTEGER TO INDEX CONVERSION

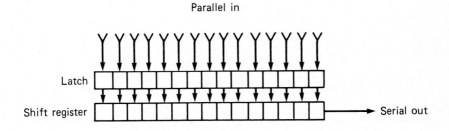

Parallel in

Latch

Shift register

Serial out

FIGURE 5
PARALLEL TO SERIAL CONVERSION

INDEXING

The motivation behind the "index" data type is shown in Figure 6. This
figure shows an implementation of the index operator [ ] applied to
parallel data (a column vector). Through the use of latches and shift
registers, indexing can also be applied to row vectors. In silicon APL,
the index operator becomes more than just a means of selecting data. By
using the indexing structure to control entry to various operators, such as
shown in Figure 7, it is possible to define an index operator that applies
to functions. Thus

    Y ← X[I]

would be the orthodox use of indexing to select data, while

    Y ← A[I]

would be the new use of indexing to select operators as shown in Figure 7.

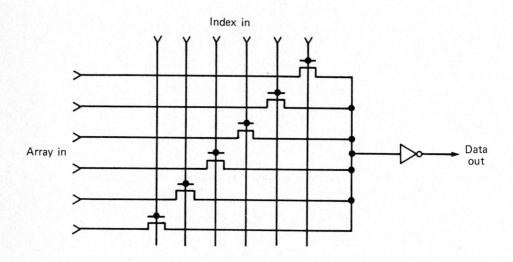

FIGURE 6
INDEX OPERATOR (PARALLEL ARRAY)

FUNCTIONS

Functions come in monadic, dyadic and even multiadic forms. Some are pre-
defined, such as ∧, ∨, ~, etc. as shown in Figure 8, but it is also
possible for the user to define his own functions. This can be done in two
ways.

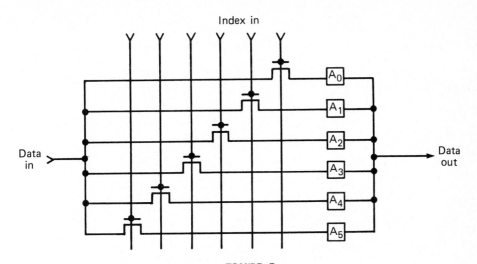

FIGURE 7
INDEX OPERATOR (APPLIED TO FUNCTION)

1. The user can define APL functions in the standard ∇ FUNCTION
    manner which serve as macros on the chip.

2. The user can define new primitive macros directly through
    layout. Often, such macros will not be expressible as APL
    functions, but once defined will be combinable into APL
    architectures as shown in the next section. This type of
    definition is analogous to writing assembly language
    subroutines to be called by APL functions.

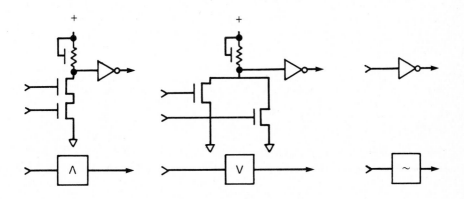

FIGURE 8
PRE-DEFINED FUNCTIONS

ARCHITECTURES (OPERATORS)

While the APL or user-defined functions in monadic or dyadic form are the
building blocks of the chip, the APL operators serve to define the chip
architecture, that is, to combine the functions into useful structures.

Examples are given in Figures 9-12. Figure 9 shows an implementation of
the reduction operator, in which "a" is an arbitrary dyadic function. The
"left-to-right" execution rule is observed. If a is an associative
function, this does not matter. The scan operator is shown in Figure 10.
As shown in Figure 11, it is relatively easy to extend these ideas to a
full inner product.

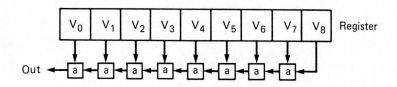

FIGURE 9
REDUCTION a/v

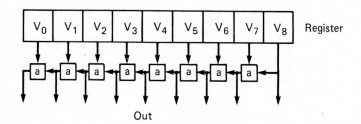

FIGURE 10
SCAN a\v

The outer product is a bit more complicated, since the temporal dimension
comes into play. An example is shown in Figure 12. The vectors X and Y
are latched into registers, but the X is also transferred into a shift
register. The y vector is kept static as the X vector is shifted and
dyadically combined with the Y's through the "a" functions. A spatial-
temporal matrix emerges from the right.

USE OF THE CONCEPT IN ACTUAL DESIGN

Armed with the above language/hardware concepts, we set out to design a
fully parallel multiplier for unsigned N-bit integers. We had already
found that designs for such multipliers abounded, but that generally they
occupied chip areas that increased as N^2, and we wanted to find a spatially
more efficient architecture.

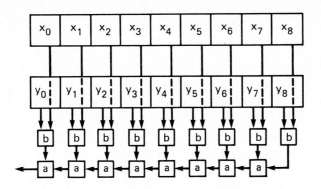

FIGURE 11
INNER PRODUCT x a.b y

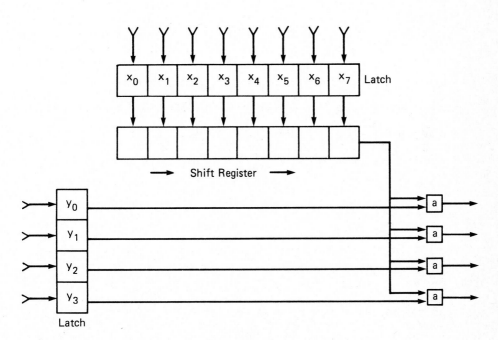

FIGURE 12
OUTER PRODUCT x °.a y

In attempting to represent existing multipliers by an APL expression for potential implementation by silicon APL, another candidate architecture was discovered. This architecture is described below.

Before presenting the concept, we wish to make the following points:

1.   The thought process of expressing the multiplier in APL
     inspired the architecture.

2.   In particular, the concepts of outer product and the use of
     time as a dimension for rank two matrices were germinal to
     the design.

3.   While the architecture may or may not be a new development,
     an entirely new mode of thinking was used in arriving at it.

MULTIPLICATION IN APL

Let A and B be vectors representing the binary digits of two numbers a and
b, so that

$$a = a_0 + 2a_1 + 2^2 a_2 + \ldots 2^N a_N \tag{1}$$

and similarly for b. The unsigned product of a and b is then

$$v = ab = a_0 b_0 + 2(a_0 b_1 + a_1 b_0) + 2^2(a_0 b_2 + a_1 b_1 + a_2 b_0)$$
$$+ \ldots + 2^{2N} a_N b_N \tag{2}$$

Notice that all possible cross-products between the elements of a and B
occur, therefore, when expressing this in APL we use the outer product.
Since multiplication of single bits is required, we use the logical and
function to form A∘.∧B   The summation in eq. (2) can be represented as a
series of shifts and adds. To accomplish this, we first define a
vector C←⍳N+1 (with ⎕IO ← 0) to represent the various shifts. We then pad
the outer product with left zeroes to allow room for the shifting, to get

$$M2 \leftarrow C \phi ((\rho M) \rho 0), \; M \leftarrow A\circ.\wedge B \tag{3}$$

and finally, we sum the rows, treating them as binary integers:

$$V \leftarrow +/2\perp \lozenge M2 \tag{4}$$

For a 6-bit by 6-bit multiplication, the final expression is

$$V \leftarrow +/2\perp \lozenge C \phi (6 \; 6\rho 0), \; A\circ.\wedge B \tag{5}$$

ARCHITECTURE

As shown in Figure 13, we divide this expression into three parts. The
outer product is implemented as in Figure 12, with A latched into a single-
fixed register and B entered into a shift register. The shift left operation is realized by a barrel shifter, with padding by zeros accomplished by
initialization of the register. The shift control vector C is also stored
in a shift register, placed as shown for a compact layout. Finally, the
binary conversion and addition is accomplished by an accumulating adder.

The area of this layout increases as N, and the execution time as N^2.

CONCLUSIONS

Our principal conclusions are the following:

1. APL is a useful language for silicon chip design.

2. APL language concepts and ideas can be inspirational in existing silicon chip design techniques.

3. Cross fertilization occurs: hardware and software people should work together so that this conceptual transfer can occur.

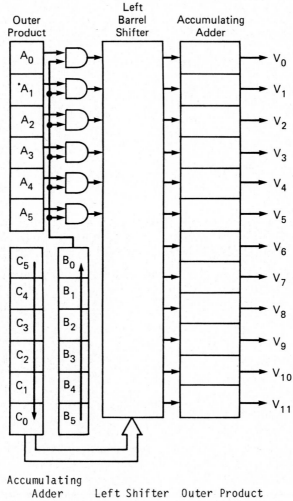

FIGURE 13
MULTIPLIER ARCHITECTURE

BANQUET ADDRESS

The Role of Language in Problem Solving I
R. Jernigan, B.W. Hamill, and D.M. Weintraub (Editors)
© Elsevier Science Publishers B.V. (North-Holland), 1985

THE FUTURE OF PROGRAMMING LANGUAGES

John W. Carr, III

Department of Computer Science
Moore School of Electrical Engineering
University of Pennsylvania
Philadelphia, Pennsylvania 19104

I met Al Perlis, as I remember, at MIT as a graduate student. He was
assigned as a teaching assistant in Beginning Calculus (or something like
that). Some student spoke up in the class and said, "Mr. Perlis, did
anybody ever tell you that you look like Harpo Marx?", which I thought was
a pretty fair statement. I kept that to myself, so Professor now Doctor,
now Chairman of the Department at Yale, Alan Perlis said, "You know, you
look just like a guy I flunked last semester." Perlis would, therefore,
have served you much better tonight, but we can at least try accordingly.

This projector display is going to be a problem of analog computing - fuzzy
in the details. (If you can't see from the back, I will read whatever I
can and try to stand out of the way). The printing on (Figure 1) says "The
Future of Programming Languages." I put on this slide, first, a quiz. Out
at the side, having become enamored by all you people with Prolog, we've
placed questions and colons. I don't know what is a question and what is
an answer in many of these cases. Therefore, there may be an extra colon or
one question too many in some of these. Please interpret what you see in
whatever way is most suitable.

THE FUTURE OF PROGRAMMING LANGUAGES

FIRST, A QUIZ:

:- Q. WHAT IS A P.L.?

:-

PUSH DOWN THIS QUESTION

:- A TRIBUTE TO J. CARSON

FIGURE 1

"What is a programming language?" (Figure 1) That is a salient question
when it comes to futures. I said to myself: it's too early for that,
let's push down this question and go on to the next. This slide pays
tribute to Johnny Carson; I think that Carson seems to embody the nature of
Prolog, which is a different type of programming language. Carson public-
ized the "Ali Kazam" routine, in which you bring out the sealed envelope
and it gives the answer, and then Carson comes up with the question. This
in a way is what one may call "Super Prolog". We'll see whether computers
can use it in terms of this frame of reference of programming languages.

Please picture, if you will, me not as what I am but as Johnny Carson
before he got divorced. I'll start off with an answer to a question which
my introducer didn't credit to me. Back about 1958, I organized with some
other computer people in Europe, Friedrich Bauer, who was the German representative, and Phillipe Dreyfuss, the French representative, a committee to
develop an international language, which you may or may not have heard of
since then - or at least any more. As a result of this, the following
(Figure 2) question was an answer. It reads "A. What is ALGOL and who is
John Carr?" And it, of course, was a result of a question: "Q. Dear Mr.
Watson, Jr. Why is IBM not supporting the international effort to standard-
ize on ALGOL? (signed) John Carr, President ACM, 1958." Mr. Watson has
gone on to greater and even higher things, but I think that was one of the
more puzzling of the questions which was posed to him. Some "fink"*,
inside of IBM called me up and told me that is what happened, so I swear
its true for whatever it's worth. That says something about my bona fides,
whatever they are, in programming languages.

          :- A. WHAT IS ALGOL AND WHO IS JOHN CARR?

          ?:- Q. DEAR MR. WATSON JR:

          WHY IS IBM NOT SUPPORTING THE INTERNATIONAL EFFORT TO
          STANDARDIZE ON ALGOL?

                              JOHN CARR
                              PRESIDENT
                              ACM
                              1958

                    FIGURE 2

We'll go on to the next envelope (Figure 3) which really has nothing to do
with questions except to say that you can question an answer with a
question, which is, I think, super-super Prolog, but maybe not. Many of
you are too old and do not know that Alice B. Toklas may have the first
credits to have put pot in brownies. She is also known for other things,
but she was close friend, confidante, all that sort of thing, to Gertrude
Stein and lived on Rue de Fleurus in Paris the same time I lived there in
1951. I passed by her house every day where she lived with Gertrude Stein
(I never saw either of the two ladies, but they were there, apparently).
Alice on her, either on her death bed or Gertrude's death bed, and you can
correct me, said, "Gertrude, is this the answer?" That's a question,
obviously. She was asking, "Is this the answer," so maybe it was also an
answer. Now that is the Prolog problem I'm giving you; and Gertrude then
answered, "Yes, but what is the question?" Well, so much for Gertrude
Stein; I think she would have been good on Carson's show.

---

* Fink: Dutch, German, or Yiddish origin for finch, a chirping bird.

```
?:- Q.  ALICE B. TOKLAS (ON HER DEATH-BED):

        GERTRUDE, IS THIS THE ANSWER?

 :-  A.  GERTRUDE STEIN:

        YES, BUT WHAT IS THE QUESTION?
```

FIGURE 3

The next one (Figure 4) of course, has the "colon minus sign" up there, which indicates that I've extended the notation a little bit: we're putting in ordered pairs, which are X and Y. And if I don't get it straight, I'm sure that most of you will recognize this. The envelope I have here in my hand; and the answer is "A. (X/Y pairs): APL/Iverson (he has been mentioned a number of times); Cobol/Hopper (she really hasn't gotten a good press; I think it's a sexist group); PASCAL/Wirth (he received a prize recently from the ACM and certainly deservedly); and three dots vertically in both directions." And when we open the envelope, it says "Q. There is no language but X, and Y is his (her) prophet." That's what Ali Kazam said.

```
 :-  A.  (X, Y)

        (APL     , IVERSEN)
        (COBOL   , HOPPER)
        (PASCAL  , WIRTH)
            ●           ●
            ●           ●
            ●           ●

?:- Q.  THERE IS NO LANGUAGE BUT X AND Y IS HIS PROPHET.
```

FIGURE 4

All right, now we're really beginning to stretch things because this perversion of Prolog has been going on a long time, and I'm still not sure what is going to happen. This next slide (Figure 5) came on because we have representatives of the Bureau of Standards here. I felt that we should at least recognize them. Of course, you probably haven't heard of "A. ANSI II"; but perhaps you have, and when we open the envelope it says, "Q. Banned use of Cobol on Sunday morning."

```
 :-  A.  ANSI II.

?:- Q.  BANNED USE OF COBOL ON SUNDAY MORNING.
```

FIGURE 5

And, of course, this one (Figure 6) we talked about has ":-A. ADA!" (The
A is for answer.) You know this is "show biz" sort of like "Oliver!", or
"Mame!", or what else? ADA! Now for those of you who really haven't ever
seen the play about Byron in which he is half in love with his half sister
and half in love with his pregnant wife: he wanders around slightly
crippled, and the child is waiting to be born. The child, of course, is
Ada, the daughter of Byron and the mathematician mother who has been
tutored in Cambridge without actually going to class. She decides that
it's either the half sister or her, and they go through a lot of convolu-
tions and you wonder if the child ever got her name on anything after all
this pre-natal influence and so forth. So does the language follow the
namesake? But here we have the envelope, and when we remove the question
it says "Q. Whom the gods would destroy, they first make mad." Now that,
of course, is not about the daughter, the Countess of Lovelace, but has
something to do with the idea of programming languages. Perhaps the people
with ADA just had a "window in time" in which the young woman was to rush
in and receive her due credits, but maybe the window didn't close or open
at the appropriate time.

                    :- A. ADA!

                    ?:- Q. WHOM THE GODS WOULD DESTROY, THEY FIRST MAKE MAD.

                                    FIGURE 6

Now, I remember the first time that I saw FORTRAN (Figure 7). I read about
those powerful instructions and so forth, and I never thought I would be
publicizing it many years thereafter. Of course, the answer is ":-A.
FORTRAN!" and when we open up the envelope, it says, "?:-Q. The work for
which John Backus was given an IBM fellowship." Now I like that but
perhaps some of the rest of you do not. I think that John Backus should
have been recognized in any case, and whether the language really has that
much future is not ... the question.

                    :- A. FORTRAN!

                    ?:- Q. THE WORK FOR WHICH JOHN BACKUS WAS GIVEN AN
                            IBM FELLOWSHIP.

                                    FIGURE 7

I hope this is the last one (Figure 8). The answer is, (you see it has an
exclamation point, so it is not a question, it's an answer), ":-A. What is
past is Prolog!" and the envelope says, "?:-Q. What Shakespeare said about
Japan's fifth generation computers." A very prescient man, the bard of
Avon, if he said it.

```
:-  A.  WHAT'S PAST IS PROLOG!

?:- Q.  WHAT SHAKESPEARE SAID ABOUT JAPAN'S FIFTH
        GENERATION COMPUTERS?
```

FIGURE 8

Having said that, we're through with our quiz. This (Figure 9) is a
picture, whether you know it or not, of a very famous event in Europe and
America in the 1820's. Edgar Allan Poe wrote a short article (published I
think in the Baltimore Sun at the time) describing the very famous
Maelzel's chess player which was being carted around Europe. Perhaps some
of you read about it? In the midst of this there is a drawing of an Arab
with a turban, obviously a pre-computer automaton. The hand of this
automaton would come out at each move and pick up a chess man and move it
to another part of the board. Because chess is now played by computers,
and because Maelzel's chess player was really a fraud, it turns out, of
course, that it worked only because there was a "little person" in the
box. A dwarf crawled into the mechanism and pulled handles, and the arm
came out and moved the chess pieces. The dwarf was not very good at chess;
he couldn't beat the Russian grand masters. Therefore the critics figured
out that the machine was really a fraud rather than a genuine machine, as
you can imagine. Maelzel's chess automaton, the beturbanned Arab, sits on
a box. Inside, there is a "little person".

> TAKE A PHOTOGRAPH OF ANY PERSONAL COMPUTER WITH AN
> ARAB SITTING ON IT LOOKING AT A CHESS BOARD, AND A
> SMALL HUMAN BEING ("LITTLE PERSON") PEEKING OUT FROM
> INSIDE THE P.C.

FIGURE 9

The reason I put the little person in the box, here, is that this is one of
the key differences between programs today and programs five years ago.
(Maybe you'll say three years ago or maybe you'll say ten.) This, then, is
the new programming environment for programming languages. I say to my
students, "Look, you are to transport the essence of yourself, as far as
this problem or system is concerned, into this box, which is a Rainbow, PC,
Apple, Macintosh, or whatever you're dealing with. You are to be the
'little person' there in essence, in spirit, although you are far away in
the Caribbean, or wherever you are. You are to talk and emulate your true
behavior to the people that are interacting with you through the computer.
Now you no longer have merely a program; you have, in a certain sense, the
essence of this little person inside the PC coming out in relationship to
you.

"The program is the message. The user who sits outside is dealing with
you. The programs inside the computer are the extrapolation of one or more
human beings into these procedures."

You might say this is what some people call "user friendly", but in truth
that computer has a little person in it, even though its chess playing is
still not very good after all these years. When you put a human being into
the box, and that human being is interacting with the outside world, this
changes the nature of what a programming language is.

What is the interface, then, between the user and the little person in the
box? It is entirely different because the bits that used to pass across
that interface have changed in bandwidth character. In particular, if all
of the intellectuality of Dave Weintraub and his freshman cohorts at
Carnegie-Mellon could be put in the box, when I sat down in front of it, I
could let them do all of the work. As far as passing the bit stream from
the programming language into that system, there would be very little of it
for me to do. The end result is that technology actually drives
programming languages. What the programming language of today will be is
different from that of five years ago and is different from that of five
years hence. I can't predict it, but I can tell you a few things which
you've recognized. I speak not for the elitists of NBS, or APL/JHU.
(Someone said, "you're going to Johns Hopkins?" And I said "Johns". She
said, "you don't mean John Hopkins?" I said "Johns!" This lady is a
feminist; she said, "I want my daughter to go to Marys Washington.")

A cheap graphics terminal like the Macintosh, which is "a computer for the
rest of us," has changed the shape of computer interaction. I think the
word "cheap" is the first point. You may or may not like the "mouse" we
are talking about, but if you don't like rodents, there will be something
else. There will be interfaces between human and machine which are taking
bits in a different fashion entirely from that sequence of keys that once
were being punched on IBM's 80-column card key-punches. I have a program a
graduate student named Harvey wrote. The only reason I mentioned it here
is because the program is called Harvey. It's a good program, but it's
called Harvey, it's not called FORTRAN; it's got the name of the writer, so
it's the little person in the box. Students use it and talk about what
Harvey is doing. I think I may have got the key idea out of that. Harvey
has a set of graphics menus which are added to our Univac, which has nothing at all. You interact with a mouse equivalent. You have these languages
which are interfaced through no keystroke. All you do is point to things
as you do on the Hewlett-Packard touch-screen. Bits going in here and
there, nothing said or written. You could argue even now that most programming (by volume) at the present time (and I haven't done the statistics, so
I could be completely crazy) is in menu languages. You can add an 8087 to
every 8088, and they change the nature of the programming; attach a VLSI
search chip to a cheap Winchester disk for a thousand dollars, put the
search chip right at the disk head, and you have a different type of data
base system from what you have ever had before.

How does this change the nature of a programming language? Are what we are
doing here and the problems that we are solving really germane? Or are we
going to be undercut by some violent change in technology which we can only
guess at? Someone today mentioned that the Japanese are dealing with 40
layers on a VLSI chip. That is a good number, it's larger than one would
expect. I've been to conferences on VLSI design in which there were three
and four hours of discussion about the best way to get a routing wire solution for one or two layers of metal. Will these solutions be important if
you have 40 layers all together? The technology, then, is one of the

elements which I ask you to think about in the future of programming languages.

We're almost back to the quiz. We're going to try to see what a programming language is. At IFIP in Paris in 1983, Carver Meade said, "it is a serious question whether there will ever be a massively concurrent architecture which is general-purpose in the same sense the sequential engine is." He doesn't believe it can be done. Does this also mean "No programming languages?" Meade has had at least one excellent idea, the idea of the "silicon foundry". (It may never work, but it probably will.) It's the idea that you have a publishing house, in which you go to the publisher and you hand in your manuscript and sometime later the book comes back to you. You don't have to know about "printers' devils" and pis and ems and printing presses and ink or anything like that. It just comes back. Similarly the same for VLSI chips! VLSI-oriented architecture and parallel-processing control mechanisms, not Prolog, mind you, are the main research topics for the fifth generation computer technology in Japan. What sort of programming languages are they going to be having above and beyond Prolog?

The final set of arguments that I want to give you, an approximate truism which I'd like you to try to remember and challenge, is "The Logical Equivalence of Hardware and Software." What does this mean? It means that the method of thinking, the method of writing, the method of talking about hardware and software, is the same, logically, to a first approximation. Now one can say, "I know the differences between them, and those differences are a higher order of approximation." We add the word "firmware", which, of course, you recognize as microprograms, and "humanware", which, of course, I hope you don't recognize because you can say you first heard it here tonight. This is the activity that the operator on the PDP-1 did when he went down the line and threw those switches one after another. Then somebody else in DEC walked in and said, "gee, we can copy down what he does and put it in a program and it will run the PDP-1 the same way that he is doing it. And it will do it a lot better." So the guy worked himself out of a job, but fortunately had another one. "Humanware", of course, is that set of processes, procedures, whatever you want to call them, algorithms inside the human brain that you're trying to copy over into the computer. Finally you have the one which is probably the most important long-term, which is "natureware". All these, in a certain sense, are the same up to these somewhat sophisticated approximations, which you can point out, where the non-logical equivalences don't really hold.

If one says that almost every subroutine can be put on a chip, if one puts subroutines on a chip, how does this affect your programming language? There are people around who say you can take LISP and you can run it through a compiler, and you output a silicon bus. On this bus you attach a LISP subroutine for "append" or something like this, a new procedure. You run the procedure through the compiler; this new subroutine comes out as a new area on the chip. Now you keep feeding the LISP routines that you write in through the compiler. They come out on the next version of the chip. They all work perfectly because all the code is debugged, of course, all the time, and the programming language is no longer a programming language? The word is engraved in silicon.

To answer the question about programming languages pretty quickly, I just collected some contrarian "programming languages" to show you my opinion of

them. This first (Figure 10) example of a programming language is a
procedure for designing and verifying a hardware system with the Motorola
6800 microprocessor, which I had on a slide. It is a flow diagram describing a process which interacts with various artifacts, some of which are
computers, some of which are devices for burning ROMS, some of which are
typewriters which punch tapes, etc. Every once in a while the human enters
the procedure. To the extent that one is willing to define programming
languages in this way, this is a programming language example. If you are
interested in computer aided manufacture, and you have to have a programming language to do that, then something like this is going to end up as an
example of such a procedure.

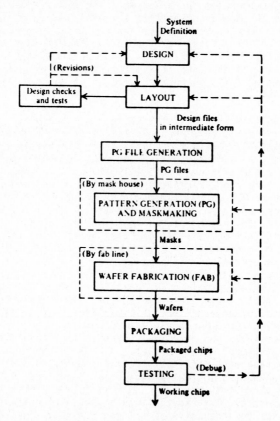

A. THIS IS A PROGRAMMING LANGUAGE EXAMPLE AND AN EXPERT SYSTEM?

FIGURE 10

We can choose the topological "sticks diagram" of something called a "stack
data engine" by Rick Mosteller at Cal Tech. This is a programming language
example. Or the photograph of a chip; it has about 20,000 rectangles on
it. Each of those rectangles represents a certain artifact. Inside a
computer there is an isomorphism of this chip into a language which
describes this particular chip. Is such a language a programming language

example? And is it important whether it is or not in terms of this conference?

Take the diagram of the process for designing wafers using computer aided fabrication and computer aided design. Is this a programming language example? It is a computer aided manufacturing structure which at some stage will be completely automated using computers, so if it's going to be automated there will have to be some sort of description of the result of the automation process. It may not be a program, but is this a programming language example?

Finally, I want to give you one which is more like a program. Inside a computer is a description in a language called CIF (Cal Tech Intermediate Form). Outside the machine, sent through an appropriate fabrication device, it becomes a chip, but this program is generated by another program, of course, which is called "Make Finite State Machine."

If MKFSM makes a finite state machine, the output must be a finite state machine in some abstract form. A finite state machine represents a chip, and we've said that there is a logical equivalence of hardware and software and therefore that chip must "be" a program. So here we have a program generator which generates a program which is a chip abstractly and whose description is a finite state machine. The system, which is due to Landman at Xerox PARC, asks you a sequence of questions: "what should I use for the subsymbol that I'm creating; how many inputs; how many outputs; how many terms; and the Boolean expressions, what is lambda, which is to be the size of the rectangle in this case; do you want to program this PLA, (program logic array, which is an and/or sequence interface); do you want the input of this to be clocked; do you want the outputs to be clocked; do you want to make this PLA a finite state machine?..." And after you put in the last "yes" or "no", the machine goes and creates the "program" and sends it out in a form which can be shipped out to be fabricated.

Now, there are lots of slips between cup and lip, and you can put the wrong equations in here; that certainly is a semantic failure. Syntactically, apparently, you don't get any errors in this process. The process of creating this program is one which prohibits syntactic errors, which I would consider to be those of the design constraints, design rules on the rectangles themselves: they can only be so close to each other, they cannot be too narrow. These are the fabrication design rules which are necessary, and I think in fairness one could say that they are analogous to syntax. Certainly the semantics of this is that if we give the wrong behavior to this finite state machine, then semantically it is incorrect. We really have to take it and verify it and see whether it does what we want, but here is a program which is the most simple of abstract structures, the finite state machine, which does the job of a portion of a chip. Is that a program? Is the finite state machine description, in whatever form, in circles with arrows, or columns of states, is that a programming language?

I hope that you will tell me, if not tonight, sometime in the future, exactly what you think a programming language is. In the coming of the automatic design and fabrication of the concurrent systems which make up a micro chip, which is permeating this sort of foundry derived from publishing, you have an example of a change in programming tactic. You're dealing with concurrent programming, not on just several Cybers connected together, or three, or five thousand, but on an asynchronous system which is going to be fabricated and then connected into the machine, is hard and no longer soft in the sense of the wire itself, and which creates an entirely different frame of reference. When you make a chip, it either goes or it doesn't; you can't go back and correct it easily like a program. You can't put "fixes" in, you can't call somebody up on the telephone and say "what do I do now?" You either throw it away, or use it in a crippled fashion, or start over again and spend a lot of money.

So the investigation of the programming tactics for producing "programs" that are chips, can, I think, lead many of us to a better understanding of the constraints that one has on programming. And in particular, when you program a chip, you end up with constraints that are sometimes like software, but sometimes not. You have problems on area, which, of course, are analogous to, but more complicated in two dimensions than the constraints on the number of bytes, the number of bits that you store your program in. You have constraints on power dissipation which certainly are not around inside a computer. That constraint has been handled by the manufacturer. He says "you are going to have so much power dissipated;" we agree that that is that; "don't worry about it anymore." But in the chip, you have to worry about it beforehand. You also have the problem of timing, how fast does it go? You also have the problem of timing in some of the real-time programs that you have described. There are analogies, but again, the approach to hardware is a much more engineering type of approach than has been in software.

There may be a large number of people -- predicted in the tens to hundreds of thousands -- who will be using small, cheap computers and sort of cottage-like industries, apparently, to produce small circuits or programs or a mixture which will be able to be fabricated, presumably very cheaply, with some sort of foundry process like the Mosis process in California. These people will be programming a different type of program with a different set of constraints, with programming languages which may very well be the same - or different - programming languages of today. A student brought his chip design in for a divider and said "here is the simulated chip in PASCAL." I couldn't tell whether it worked. I said, "have you run it?" He said, "yes." I said, "how do you verify it?" He said, "well nobody ever taught me how to verify a PASCAL program."

We've had Johnny Carson and Gertrude Stein. I guess we'll go back and have Dave Weintraub.

QUESTION AND ANSWER PERIOD

WEINTRAUB
I've got a question for you. I have maintained for a very long time that there is no such thing as hardware in the traditional sense anymore in computers. Except for the cables and the pins and things like that, as soon as we went to CAD/CAM, or just CAD even, people were no longer tracing things through in the same sense that hardware had that firm (no pun intended) feeling about it, that you could check every circuit. When things became complicated enough that we needed computers to help design them, we could no longer call them hardware in the traditional sense of

something which could be tested with every path well defined and well known, but hardware only in a sense that theoretically it can be done. But from the point of view of software, you can do that too, if you know the state of the machine at one instant, precisely, and you say there will be no more I/O, you can predict what will happen from there.

CARR

I think these hardware people are more serious than the software people are about making sure that they verify what they have. They have giant meetings on testability in which they describe all the technology and all the theory that they've had. You can go to IBM at Poughkeepsie and Yorktown and talk to people like Paul Roth, who has a complete theory of verification procedures which are being used in the gate array processes there. I think that there is a distinct cleavage (there shouldn't be but there is) between hardware and software on this verification business. It's something that software people should investigate and query themselves about: is there basically a difference between the two? There is a difference in the non-logical equivalence of hardware and software, but does that difference mean that the product that goes out the door has to be violently different? (Of course TI doesn't always check their chips, but many people always don't check their programs, as well, so it can only be a little different.)

GUIER

I would like to suggest a thought on the definition or distinction of the two. I think the hardware people are the coders. We programmers are the ones who will have fun with it.

CARR

A lot of people have fun, I think that that is a good self-image and I accept it. If I have fun I'm in the top-down crowd.

AUTHOR INDEX

LANGUAGE/SYSTEM INDEX

SUBJECT INDEX